Technological Choices for Sustainability

Springer

*Berlin
Heidelberg
New York
Hong Kong
London
Milan
Paris
Tokyo*

Subhas K. Sikdar · Peter Glavič · Ravi Jain
(Editors)

Technological Choices for Sustainability

With 83 Figures

Springer

Editors

DR. SUBHAS K. SIKDAR
US Environmental Protection
Agency
26 W. Martin Luther King Drive
Cincinnati, OH 45268
USA

PROF. DR. RAVI JAIN
University of the Pacific
School of Engineering and
Computer Science
3601 Pacific Avenue
Stockton, CA 95211-0197
USA

PROF. DR. PETER GLAVIČ
University of Maribor
Faculty of Chemistry and
Chemical Engineering
Smetanova 17
2000 Maribor
Slovenia

ISBN 3-540-21131-4   Springer-Verlag Berlin Heidelberg New York

Library of Congress Cataloging-in-Publication Data Applied For

A catalog record for this book is available from the Library of Congress.
Bibliographic information published by Die Deutsche Bibliothek
Die Deutsche Bibliothek lists this publication in die Deutsche Nationalbibliographie; detailed bibliographic data is available in the Internet at <http://dnb.ddb.de>.

This work is subject to copyright. All rights are reserved, whether the whole or part of the material is concerned, specifically the rights of translation, reprinting, reuse of illustrations, recitations, broadcasting, reproduction on microfilm or in any other way, and storage in data banks. Duplication of this publication or parts thereof is permitted only under the provisions of the German Copyright Law of September 9, 1965, in its current version, and permission for use must always be obtained from Springer-Verlag. Violations are liable for prosecution under the German Copyright Law.

Springer-Verlag Berlin Heidelberg New York
Springer-Verlag is a part of Springer Science+Business Media
springeronline.com

© Springer-Verlag Berlin Heidelberg 2004
Printed in Germany

The use of general descriptive names, registered names, trademarks, etc. in this publication does not imply, even in the absence of a specific statement, that such names are exempt from the relevant protective laws and regulations and therefore free for general use.

Cover Design: Erich Kirchner, Heidelberg

Typesetting: Camera-ready by Damjan Krajnc

Printed on acid free paper   30/3141/LT – 5 4 3 2 1 0

Preface

This book was made possible by the exceptional support provided by NATO Scientific and Environmental Division, University of Maribor (Slovenia), Government of the Republic of Slovenia, British and the United States Embassies (Ljubljana, the Republic of Slovenia).

The authors, as listed in this book, took the time to prepare excellent manuscripts focusing on various issues related to technological choices for sustainability. These manuscripts were rigorously reviewed and refereed by scientists and engineers before inclusion in this book. An introductory chapter was prepared to provide an overview and to integrate technical issues covered in the book. A summary chapter is included at the end that provides a synthesis of panel discussions related to the three main sections of the book.

The editors are most grateful to the contributors, sponsor organizations, and many colleagues who were kind enough to assist us in making this book possible. We are particularly grateful to Damjan Krajnc of the University of Maribor for compiling all the manuscripts in the correct format, creating the index, and assuring that all the contents are faithfully presented in this volume. Background information about the editors and principal authors and contributors to this book follows.

Subhas K. Sikdar, US EPA
Peter Glavič, University of Maribor, Slovenia
Ravi Jain, University of the Pacific, California

Editors

Dr. **Subhas K. Sikdar**, as the Director of the Sustainable Technology Division of the US EPA, is the primary spokesman of EPA's R&D on clean technologies and pollution prevention. He directs research, both intramural and extramural on tools and methods for pollution prevention, cleaner process technologies, and demonstration and verification of cleaner technologies. Before joining EPA in 1990, Dr. Sikdar held managerial positions at the National Institute of Standards and Technology in Boulder, Colorado, and General Electric Corporate Research & Development Center in Schenectady, New York. He began his professional career as a Senior Research Engineer with Occidental Research Corporation in Irvine, California in 1975. Dr. Sikdar earned his B.S. in chemistry, a B.Tech in chemical engineering, and an M.Tech in polymer science from Calcutta University in India. He received his M.S. and Ph.D. in chemical engineering from the University of Arizona. Dr. Sikdar is a Fellow of the American Association for the Advancement of Science (AAAS), Fellow of the American Institute of Chemical Engineers, Honorary Fellow of the Indian Institute of Chemical Engineers, winner of three EPA bronze medals, an R&D 100 award (1990), AIChE's Larry Cecil Award for Environmental Chemical Engineering (2002), and University of Arizona's Professional Achievement in Engineering Alumnus Award (2003). In the past he was a member of the Vision 2020 Steering Committee for chemical industry, an action network leader of the Council for Chemical Research. He is a member of the Board of Governors of the Council for Chemical Research (CCR) and of the Green Chemistry Institute, member of AIChE's Research and New Technology Committee and the Chair of the Sustainable Engineering Forum. For some years he has been championing the concepts and methods for clean products and processes through a NATO Pilot Project, two NATO workshops, and an Engineering Foundation Conference. He is a current member of the Industrial Advisory Board of the University of Arizona's College of Engineering and of the Department of Chemical and Environmental Engineering, and of the Department of Chemical and Environmental Engineering of the Illinois Institute of Technology. Dr. Sikdar is the leader of the technical expert group for a Center of Excellence on Environmental Engineering and Hazardous Wastes composed of several universities in Thailand. He is the founder and the co-Editor-in-Chief of the international journal, Clean Technologies and Environmental Policy, published quarterly by Springer Verlag of Germany. Dr. Sikdar has published more than 60 technical papers in reputed journals, has 22 U.S. patents, and has edited 13 books.

Prof. Dr. **Peter Glavič** is a professor at the Faculty of Chemistry and Chemical Engineering, University of Maribor, Slovenia. He earned his B.S. in Chemical Technology from University in Ljubljana and B.S. in Business and Administration from the university of Maribor. He received his M.S. and Ph.D. in Chemistry from University in Ljubljana. He has worked for more than 8 years in industry, in paper manufacturing, ferroalloys, abrasives and fertilizer production. He has authored more than 100 scientific papers, coordinated 200 scientific and professional projects. He has published nearly 100 professionals papers and many textbooks. He is actively involved in the Working Party on Computer Applications in Process Engineering (WP CAPE) of the European Federation of Chemical Engineering, EFCE. He is a member of the Core Group of PREPARE (Preventive Environmental Approaches in Europe), a member of the Commission on physicochemical symbols, terminology, and units at IUPAC (International Union of Pure and Applied Chemistry) and a national representative at the NATO CCMS Pilot Plant Study on Clean Products and Processes. He has chaired the technology foresight group of the Slovenian chemical and process industries. Prof. Glavič is a member of editorial boards of *Resources, Conservation and Recycling, Chemical* and *Biochemical Engineering Quarterly* and was a guest Editor of *Computers & Chemical Engineering*. He has been a vice rector at the University of Maribor and vice dean at the Department of Chemistry and Chemical Engineering in Maribor. From 1990 to 1997 he was a member of the Slovenian Parliament. He is a member of several international scientific committees: ESCAPE (European Symposium on Computer Aided Process Engineering), PSE (International Symposium on Process System Engineering) and World Congress of Chemical Engineering. He is chairing the Association of Economists in Maribor and the national Section of Chemical and Process Engineering.

Ravi Jain, Dean and Professor, School of Engineering and Computer Science, University of the Pacific, Stockton, California. He received his B.S. and M.S. degrees in Civil Engineering from California State University, Sacramento and a Ph.D. in Civil Engineering from Texas Tech. He studied public administration and public policy at Harvard, earning an M.P.A. degree.

Prior to this appointment, Dr. Jain has been an Associate Dean for Research and International Engineering and Executive Director of Interdisciplinary Research Centers, professor of civil and environmental engineering, and director of the environmental engineering management graduate program at the University of Cincinnati, College of Engineering.

He has held research and faculty appointments at the University of Illinois (Urbana-Champaign) and Massachusetts Institute of Technology. He has been a Littauer Fellow Harvard University, and a Fellow, Churchill College, Cambridge University.

He has directed a staff of over 200 engineers and scientists, conducting interdisciplinary research for the U.S. Army, and was the Founding Director of the U.S. Army Environmental Policy Institute (AEPI). He has worked for the California State Department of Water Resources and industry and has been a consultant to federal agencies, international organizations and private industry.

Dr. Jain has served on numerous National Task Forces and Advisory Councils for the U.S. Department of Defense, NSF, Navy, Army, EPA and NAS. He is a fellow ASCE and Diplomate American Academy of Environmental Engineers.

He has published **twelve** books and over 100 journal articles, book chapters and technical reports. His three most recent books include *Environmental Assessment* (McGraw Hill, 2001), *Better Environmental Policy Studies – How to Design and Conduct More Effective Analyses* (Island Press, 2001) and *Management of Research and Development Organizations – Managing the Unmanageable* (Wiley, 1997).

List of Contributing Authors

Aivars Spalvins
Environment Modelling Centre, Riga Technical University, 1 Meza Street, Riga.LV-1048, Latvia

Andrzej Wasiak
Bialystok Technical University, Wiejska 45 A, Bialystok, Poland and Supreme School of Economics, Choroszczańska 31, Bialystok, Poland

Anna Christianova
Czech Cleaner Production Centre, Dittrichova 6, 120 00 Prague, Czech Republic

Annik Magerholm Fet
The Norwegian University of Science and Technology (NTNU), Department of Industrial Economics and Technology Management

Audrey L. Mayer
U.S. Environmental Protection Agency, National Risk Management Research Laboratory, Sustainable Technology Division, Sustainable Environments Branch, 26 West Martin Luther King Drive, Cincinnati, OH 45268, USA

Bernard A. Engel
Center for Advanced Applications of GIS (CAAGIS) Purdue University, West Lafayette, IN 47907 USA, 765-494-4600

Bojan Radej
Institute of Macroeconomic Analysis and Development of the Republic of Slovenia – IMAD; University of Ljubljana

Carlos A. V. Costa
Laboratory of Processes, Environment and Energy Engineering, Faculty of Engineering, University of Porto, Porto, Portugal

Chris J. Johannsen
Laboratory for Applications of Remote Sensing (LARS), Purdue University, West Lafayette, IN 47907 USA, 765-494-4600

Christopher W. Pawlowski1
U.S. Environmental Protection Agency, National Risk Management Research Laboratory, Sustainable Technology Division, Sustainable Environments Branch, 26 West Martin Luther King Drive, Cincinnati, OH 45268, USA

Damjan Krajnc
University of Maribor, Faculty of Chemistry and Chemical Engineering, Smetanova 17, SI- 2000 Maribor, Slovenia

David A. Landgrebe
Purdue University, West Lafayette, IN 47907 USA, 765-494-4600

David R. Shonnard
Department of Chemical Engineering, Michigan Technological University, 1400 Townsend Drive, Houghton, MI 49931

David T. Allen
Department of Chemical Engineering, Campus Mail Code: C0400, University of Texas, Austin

David W. Keith
Engineering and Public Policy, Carnegie Mellon University

Douglas M. Young
National Risk Management Research Laboratory, Office of Research and Development, U.S. Environmental Protection Agency, 26 W. Martin Luther King Drive, Cincinnati, OH 45268 USA

Elena García-Sandá,
Department of Chemical Engineering.
School of Engineering. University of Santiago de Compostela.
Campus Sur s/n. 15782 Santiago de Compostela, Galicia, Spain.

Felicita Briški
University of Zagreb, Faculty of Chemical Engineering and Technology
Marulicev trg 19, 10000 Zagreb, Croatia

Francisco Omil
Department of Chemical Engineering.
School of Engineering. University of Santiago de Compostela.
Campus Sur s/n. 15782 Santiago de Compostela, Galicia, Spain.

Gilbert L. Rochon
Department of Earth & Atmospheric Sciences, and Department of Agronomy, Purdue University, Young Graduate House, Room 420, 151 South Street, W. Lafayette, IN, USA 47906-3560

H. J. van der Kooi
Laboratory of Applied Thermodynamics and Phase Equilibria, Delft University of Technology, Julianalaan 136,
2628 BL Delft, The Netherlands

Heriberto Cabezas
U.S. Environmental Protection Agency, National Risk Management Research Laboratory, Sustainable Technology Division, Sustainable Environments Branch, 26 West Martin Luther King Drive, Cincinnati, OH 45268, USA

I. Lace
Environment Modelling Centre, Riga Technical University, 1 Meza Street, Riga.LV-1048, Latvia

Ivanka Zakotnik
Institute of Macroeconomic Analysis and Development of the Republic of Slovenia – IMAD

J. de Swaan Arons
Laboratory of Applied Thermodynamics and Phase Equilibria, Delft University of Technology, Julianalaan 136,
2628 BL Delft, The Netherlands

J. Slangens
Environment Modelling Centre, Riga Technical University, 1 Meza Street, Riga.LV-1048, Latvia

Janko Seljak
University of Ljubljana, Faculty of Administration, Gosarjeva ulica 5, SI-1000 Ljubljana, Slovenia

Janne Hukkinen
Helsinki University of Technology, Laboratory of Environmental Protection, PO Box 2300 (Otakaari 8), 02015 HUT (Espoo), Finland

Jochen Gassner
Graz University of Technology, Institute for Chemical Engineering Fundamentals and Process Engineering, Resource Efficient and Sustainable Systems Working Group, Inffeldgasse 25, 8010 Graz, Austria

Jonathan M. Harbor
Purdue University, West Lafayette, IN 47907 USA, 765-494-4600

Jordan Pop-Jordanov
Research Center for Energy, Informatics and Materials (ICEIM), Macedonian Academy of Sciences and Arts
(MANU), P.O. Box 428, Skopje, Macedonia

Juan M. Lema
Department of Chemical Engineering, School of Engineering. University of Santiago de Compostela, Campus Sur s/n. 15782 Santiago de Compostela, Galicia, Spain

Jurgis Staniskis
Institute of Environmental Engineering, Kaunas University of Technology, K. Donelaicio str. 20, LT-3000, Kaunas, Lithuania

Kristan Cockerill
Secretary, International Society for Industrial Ecology, P.O. Box 7731, Albuquerque, NM 87194.

Larry L. Biehl
Department of Agronomy, Purdue University, West Lafayette, IN 47907 USA, 765-494-4600

Marija Vuković
University of Zagreb, Faculty of Chemical Engineering and Technology
Marulicev trg 19, 10000 Zagreb, Croatia

Minh Ha-Duong
Engineering and Public Policy department, Carnegie Mellon University, 5000 Forbes
Avenue, Pittsburgh, PA 15213 USA

N. Theresa Hoagland
U.S. Environmental Protection Agency, National Risk Management Research Laboratory, Sustainable Technology Division, Sustainable Environments Branch, 26 West Martin Luther King Drive, Cincinnati, OH 45268, USA

Neil Winterton
Leverhulme Centre for Innovative Catalysis, Department of Chemistry, University of Liverpool, Liverpool, L69 7ZD

Nhan Nguyen
U.S. Environmental Protection Agency, 1200 Pennsylvania Ave., Mail Stop 7406M, Washington, DC

Niels Jensen
CAPEC, Department of Chemical Engineering, Technical University of Denmark, DK-2800 Lyngby, Denmark

Nina Horgas
University of Zagreb, Faculty of Chemical Engineering and Technology
Marulicev trg 19, 10000 Zagreb, Croatia

Nuria Coll
CAPEC, Department of Chemical Engineering, Technical University of Denmark, DK-2800 Lyngby, Denmark

Peter Glavič
University of Maribor, Faculty of Chemistry and Chemical Engineering, Smetanova 17, SI- 2000 Maribor, Slovenia

R. Janbickis
Environment Modelling Centre, Riga Technical University, 1 Meza Street, Riga.LV-1048, Latvia

Rafiqul Gani
CAPEC, Department of Chemical Engineering, Technical University of Denmark, DK-2800 Lyngby, Denmark

Ramón Méndez
Department of Chemical Engineering.
School of Engineering. University of Santiago de Compostela.
Campus Sur s/n. 15782 Santiago de Compostela, Galicia, Spain.

Raymond L. Smith
National Risk Management Research Laboratory, Office of Research and Development, U.S. Environmental Protection Agency, 26 W. Martin Luther King Drive, Cincinnati, OH 45268 USA

Roland Clift
Centre for Environmental Strategy, University of Surrey, GUILDFORD, Surrey GU2 7XH, Tel.: + 44 1483 689271, Fax: +44 1483 686671

Sander Lems
Laboratory of Applied Thermodynamics and Phase Equilibria, Delft University of Technology, Julianalaan 136,
2628 BL Delft, The Netherlands

Sarada Majumder
Information Technology at Purdue (ITaP), Purdue University, West Lafayette, IN 47907 USA, 765-494-4600

Sharon Austin
U.S. Environmental Protection Agency, 1200 Pennsylvania Ave., Mail Stop 7406M, Washington, DC 20460, Tel: 202-564-8523, Fax: 202-564-8528

Teresa M. Mata
Laboratory of Processes, Environment and Energy Engineering, Faculty of Engineering, University of Porto, Porto, Portugal

Valdas Arbaciauskas
Institute of Environmental Engineering, Kaunas University of Technology, K. Donelaicio str. 20, LT-3000, Kaunas, Lithuania

Vasil Simeonov, Chair of Analytical Chemistry
Faculty of Chemistry, University of Sofia "St. Kl. Okhridski", 1164 Sofia, J. Bourchier Blvd. 1, Bulgaria

Victor Teodor Petcu Nitica
Pollution Prevention Center, Theodor Sperantia Str. 98, Bl. S28, Sc. 1, Ap. 10, Bucharest 3, RO-74316, Romania

Vladimir Gheorghievici
Pollution Prevention Center, Theodor Sperantia Str. 98, Bl. S28, Sc. 1, Ap. 10, Bucharest 3, RO-74316, Romania

Zoran Gomzi
University of Zagreb, Faculty of Chemical Engineering and Technology
Marulicev trg 19, 10000 Zagreb, Croatia

Table of Contents

Preface v

Editors vii

List of Contributing Authors xi

Table of Contents xvii

Introduction 1

FRAMING THE ISSUE OF SUSTAINABILITY 5

Science and Sustainability: who knows best?
   Neil Winterton 7

Sustainability: Ecological, Social, Economic, Technological, and Systems Perspectives
   H. Cabezas, C. W. Pawlowski, A. L. Mayer and N. T. Hoagland 37

US EPA/Academia Collaboration for a Green Engineering Textbook For Chemical Engineering
   D. R. Shonnard, D. T. Allen, S. Austin, N. Nguyen 65

Innovative Industrial Ecology Education Can Guide Us to Sustainable Paths
   Kristan Cockerill 77

The sustainable industrial development: reality and vision
   Jurgis Staniskis, Valdas Arbaciauskas 91

SUSTAINABLE PATHWAYS 101

Clean technologies for wastewater management in seafood canning industries
   F. Omil, E. García-Sandá, R. Méndez, J. M. Lema 103

Kinetic Analysis of Aerobic Composting of Tobacco Industry Solid Waste
   F. Briški, N. Horgas, M. Vuković, Z. Gomzi 127

Preventative Measures in Production from the Point of Sustainability
   Anna Christianova 139

Environmetric strategies to classify, interpret and model risk assessment and quality of environmental systems
Vasil Simeonov 147

Carbon storage: the economic efficiency of storing CO_2 in leaky reservoirs
Minh Ha-Duong, David W. Keith 165

An Integrated Computer Aided System for Generation and Evaluation of Sustainable Process Alternatives
Niels Jensen, Nuria Coll, Rafiqul Gani 183

Pollution prevention and environmental management systems – tools to obtain a sustainable development
Victor Teodor Petcu Nitica, Vladimir Gheorghievici 215

SUSTAINABILITY METRICS 227

Technology sensitive indicators of sustainability
Andrzej Wasiak 229

Metrics for supply chain sustainability
Roland Clift 239

Quantifying technological aspects of process sustainability: a thermodynamic approach
S. Lems, H. J. van der Kooi, J. de Swaan Arons 255

Defining and Measuring Macroeconomic Sustainability – The Sustainable Economy Indices
Jochen Gassner 267

Environment as a factor of national competitiveness in manufacturing
Bojan Radej, Ivanka Zakotnik 283

Indicators for sustainable energy development from a negentropic perspective
Jordan Pop-Jordanov 305

Sustainability indicators for anticipating the fickleness of human-environmental interaction
Janne Hukkinen 317

Measuring Sustainability – Index of Balanced Sustainable Development
Janko Seljak, Damjan Krajnc, Peter Glavič 335

Evaluating the Environmental Friendliness, Economics and Energy Efficiency of Chemical Processes: Heat Integration
    T. M. Mata, R. L. Smith, D. M. Young, C. A. V. Costa 355

Eco-efficiency reporting exemplified by case studies
    Annik Magerholm Fet 371

Interpolation for creating hydrogeological models
    A. Spalvins, J. Slangens, R. Janbickis, I. Lace 387

Indicators of Sustainable Production
    Damjan Krajnc, Peter Glavič 395

Remote Sensing as a Tool for Achieving and Monitoring Progress Toward Sustainability
    G. L. Rochon, C. J. Johannsen, D. A. Landgrebe, B. Engel, J. M. Harbor, S. Majumder, L. L. Biehl 415

SUMMARY OF PANEL DISCUSSIONS **429**

Subject Index **437**

Introduction

The Brundtlund Commission report, *Our Common Future*, started a global discourse on sustainable development more than a decade ago. The report focused on the adverse environmental impacts of rapid economic development, especially in the most advanced economies. Our Common Future issued warning that the developing countries face grave environmental dangers if they emulated the current economic model of the developed world. The report reached a consensus: the current economic growth, which has been consuming natural resources at an ever-increasing rate, is not sustainable, and that the growth was causing alarming levels of pollution throughout the world threatening human health and the environment, and it created a huge disparity of living standards between the developed and developing nations.

The Brundtland report called out for united actions by all nations to create conditions under which the three pillars of sustainability – development, environmental protection, and socio-economic equity can be balanced for all mankind. Sustainable development was defined by the Brundtland Commission as "ensur[ing] that [development] meets the needs of the present without compromising the ability of future generations to meet their own needs" (*Our Common Future, 1987*). Our common future is, therefore, critically dependent on our collective ability to preserve our common right to the earth's natural resources, energy sources, and clean environment – for this generation and all future generations. Sustainable development focuses on *obligation to future generations, industrial practices, exploitations of resources,* and the role of *science and technology,* among others. Though this call to action is geopolitical, tools and instruments that could be used to begin a common journey to achieve sustainability must be scientifically, technologically, and economically sound.

The Brundtland Commission, and subsequent international meeting such as the Rio conference, while raising the awareness of the unsustainable nature of development and stressing the need for sustained growth, has not provided, save political measures, much guidance for determining possible paths for approaching sustainability. In fact the sociopolitical debates have overwhelmed the scientific discourse. In a recent report of the US National Research Council titled, *Our Common Journey: a transition toward sustainability* (National Academy Press, Washington, DC, 1999), the status of this journey towards sustainability is described in the following manner:

The political impetus that carried the idea of sustainable development so far and so quickly in public forums has also increasingly distanced it from its scientific and technological base. As a result, even when the political will necessary for sustainable development has been present, the knowledge and know-how to make some headway often has not.

A gap therefore is evident between rhetoric and knowledge needed for implementation. It is true that knowledge alone is insufficient, but knowledge must be at the disposal of the policymakers if we want to see progress toward the goal of attaining sustainability.

Many chapters of this book resulted from papers initially presented and discussed among the participants in a NATO sponsored Advanced Research Workshop held in Slovenia in 2002. The main purpose of this NATO workshop, first of its kind, was to bring together experts who are able to contribute implementable ideas that are scientifically credible, technologically possible, and economically justifiable. The workshop focused on knowledge – on the possibilities in the context of policy debates – but not on policy matters. The workshop provided guidance on what key questions we should try to answer in the future, what research must be undertaken, and how the ideas of sustainability can be integrated into decision making. This book encompasses three main thrusts: **Framing the Issue of Sustainability, Sustainable Pathways** and **Sustainable Metrics.** The final section of the book is a **Summary of Panel Discussions**.

Framing the Issue of Sustainability

First and foremost is the need for exploring the scientific basis of the idea of sustainability. What does sustainability mean from scientific viewpoint? Is it a utopian notion? If sustainability refers to a defined system, such as a product, a manufacturing process, a manufacturing site, a community such as a city, can the attainment of sustainability for one system be in conflict with another, when the former includes the latter? What can science tell us about the possibilities of attaining sustainability by balancing technological growth, environmental protection, and resource conservation (or intergenerational equity)? These related questions are examined in this section.

Chapters under this section attempt to cover **Relationship of Science and Sustainability**; **ecological**, **social**, **economic**, **technological**, and **systems perspectives related to sustainability**. In framing the issue of sustainability, does education have an important role to play? A chapter that describes a **green engineering textbook for chemical engineers** and another chapter that focuses on **innovative industrial ecology education related to sustainable paths** provide further insights into framing the issue of sustainability. Another chapter discusses **sustainable industrial development**, a **reality and vision**, with specific examples of progress that has been made in implementing sustainable industrial development in one of the eastern European countries.

Sustainable Pathways

We need to have a clear perspective of what each scientific discipline brings to this debate on technological choices. In other words, a discussion on the scope of contribution each discipline can make to this debate is helpful. This section provides perspectives from experts in selected disciplines, such as ecology, chemistry, engineering, and economics. Discussed here are advances in basic sciences and engineering such as chemistry, material science, energy and power, and biotechnology that offer dramatic improvements in environmental performance of products or processes that are more efficient yet economically viable. For instance, socioeconomic methods for achieving sustainability, such as market-based methods that achieve reduction in pollution, resource depletion or loss of biodiversity could stimulate certain technological choices for sustainability.

Chapters in this Section that provide guidance for sustainable pathways include discussion of **clean technologies** for **wastewater management** in **sea food canning industries, kinetic analysis** of **aerobic composting** of **tobacco industry solid waste** and **preventive measures** in **production from the point of sustainability**. Detailed technical discussions provide information and ideas useful for making progress towards sustainable pathways. Chapters in this Section also provide information about strategies one can use to pursue credible sustainable pathways. For example, chapters discuss **environmetric strategies** to **classify, interpret** and **model risk assessment** and **quality** of **environmental analysis**; **carbon storage** and its **economic efficiency**; **integrated computer aided system** for generation and **evaluation** of **sustainable process alternatives** and sustainable development using **pollution prevention** and **environmental management systems**. Chapters in this section attempt to synthesize and integrate information that should be helpful in exploring sustainable pathways.

Sustainable Metrics

The ability to make an assertion on progress towards sustainability will depend on methods used to measure progress towards sustainability. Relevant questions in this context are: can we define sustainability in terms of measurable entities (metrics)? How can we ascertain that we are moving in the right direction? How can we determine that we are near there, or that we have attained sustainability? Discussion on measuring progress toward sustainability focuses on resource energy use, environmental impacts of activities in manufacturing, mining, agriculture, and service sectors of the economy. The environmental impact process necessarily uses the cradle-to-grave life cycle assessment – the method that has been in increasing use in product design. Chapters in this section attempt to synthesize and integrate information that should be helpful in explaining sustainable pathways.

The **Sustainability Metrics** Section of this book is extensive and includes 13 different chapters. This Section includes chapters that describe **technology sensitive indicators** of **sustainability, metrics for supply chain sustainability**, **quantifying technological aspects of process sustainability**, and **defining and measuring macroeconomic sustainability and related economic indices**. Other chapters focus on issues related to **industrial competitiveness, energy** and **human environment interaction**. There is always an issue of measuring sustainability. One chapter focuses on **measuring sustainability** as an **index of balanced sustainable development**.

At times there is a need to evaluate the **environmental friendliness** and **economic** and **energy efficiency** of various industrial and chemical processes. One chapter specifically focuses on this issue. Another chapter describes **eco-efficiency reporting** using various **case studies**. Three last chapters in this Section cover **hydrogeological models, indicators** of **sustainable production**, and use of **remote sensing as a tool for monitoring progress towards sustainability**. In this last chapter presented are sustainability indices based upon case studies that can be useful for rural and urban ecosystems. This concept is based upon using historical data collected utilizing remote sensing techniques. Information and concepts presented in the Section provide ideas and ways to focus on this very essential and difficult task of identifying and defining sustainability metrics.

Summary of Panel Discussions

This last section of the book provides a summary of panel discussions. Important questions raised during the conference related to various subject areas presented were discussed by a group of panelists. Key ideas and issues are summarized. In addition, comments, recommendations and suggestions that might be useful for future research are identified.

Sustainability is a global concept, and hence will have global appeal. Both the Western and the partnership countries can benefit from the materials discussed in this workshop. This is particularly true of the partnership countries, where new economic development will take place, and many of the sustainability tools, methods, and technologies could prove to be good candidates for testing and evaluating. The book is timely, and the topic areas covered are unique. Considerable scientific discourse took place on sustainability in a multidisciplinary gathering which generated much of the information presented here.

This book should serve as a resource for science-based knowledge for future researchers and policymakers and as a catalyst for collaboration among interested experts from various disciplines. In this sense, the effect of this book may indeed be significant.

Subhas K. Sikdar, Ph.D., US EPA
Peter Glavič, Ph.D., University of Maribor
Ravi Jain, Ph. D., University of the Pacific, California, USA

References

Sikdar S, Glavic P (2002) *Technological choices for sustainability*. Abstracts, NATO Advanced Research Workshop, Maribor, Slovenia
World Commission on Environment and Development (1987) *Our common future*. Oxford University Press, Oxford, UK

Framing the Issue of Sustainability

Science and Sustainability: who knows best?

Neil Winterton

Liverpool Centre for Materials and Catalysis, Department of Chemistry, University of Liverpool, Liverpool, L69 7ZD, UK

1 Introduction

Practicing chemists and chemical technologists, such as myself, will not necessarily be experts on sustainability, though the questions we ask, in attempting to understand its significance, scope and implications, are ones that all scientists and technologists, and indeed all citizens, should also address. However, whilst the title poses an important question for scientific inquiry related to sustainability, it is one than cannot readily be answered. Instead, I raise a number of issues to ponder.

My own experience grows out of 25 years in product and process research and development, primarily in the chlorine industry. During this time the chlorine industry, despite its major contributions to human health and well-being, was subject to increasingly critical scrutiny (Thornton 2000), some of it justified, but much of it prejudiced and myth-making. Because of what I know in relation to chlorine and the environment (Winterton 2000), it was apparent that there was, and still is, a mismatch between public perception and reality, between belief and knowledge and not just in the public mind. More disturbingly, this was evident also among many scientists who I would have expected to be more sceptical. I was concerned about how this arose, the consequences, if not countered, and the potential implications of some of the responses.

Furthermore, it was apparent (Winterton 2001) that these concerns were not limited to chlorine, but extended to much of technology and indeed to questioning of the role of science itself, just at the time when, it seemed to me, science and technology were needed to help steer human society towards greater sustainability. As I looked further into these matters, I realised that associated controversies had been raging about me for some time and that I was woefully ill-informed about matters beyond my own immediate scientific interests and pre-occupations.

One of the basic questions, therefore, is framed in the title of this paper: 'Science and Sustainability: who knows best?'

2 Further Questions

Many associated questions arise:

- What do we mean by 'knowing' or knowledge? (and, indeed, who are 'we'?)
- What do we mean by scientific knowledge?
- What is the relationship between scientific knowledge and other forms of knowledge?
- Is scientific knowledge special and should it be privileged?
- Does the lay public accept this?
- How do members of the lay public arrive at their views on scientific matters?
- Is science too closely identified and linked with its technological application?
- How should critical scientific and technological controversies be resolved?
- To what extent should decision-making on such matters be based on a democratic process or on rational judgement by experts?
- How can a balance between these be achieved and be made acceptable?
- What happens when scientists become activists?
- How can science and democracy be reconciled?

This paper will attempt to address, if not fully answer, some of these questions, particularly in relation to sustainability and sustainable development.

3 What is Science?

The standard dictionary definition of 'science' taken from the Oxford Concise English Dictionary is *'systematic and formulated knowledge'*. However, as Ziman has so clearly and comprehensively explained in his recent book (Ziman 2001), things are much more complicated than this. He concludes that science is not simply a body of knowledge: it is also a social institution, without a written constitution, though with its own ethos, rules and methods, usually associated with the Mertonian norms (Merton 1973) of communalism, universality, disinterestedness, originality and scepticism (CUDOS). It is not monolithic, has no formal hierarchy with a president or prime minister or an arch-bishop at its head. Ziman believes that many scientists have a very outdated view of this institution (which he calls the 'legend'), particularly relating to its links with the rest of society and how society views it, and he has proposed a different set of norms to characterise 'post-academic' science: proprietary, local, authoritarian, commissioned and expert (PLACE) (Ziman 2001; Ziman 2002).

Many scientists retain a belief that, ideally, science seeks objective truth through a process driven by open-mindedness, self-analysis and self-criticism, scepticism, commitment to reason, rationality and rigour, as a means of focussing our creativity and imagination in attempts to understand the natural world. However, for many physical scientists, particularly chemists (Schummer 2003), this could well be as far as their analysis takes them, especially if they rely solely on the limited (if not absent) formal content of their degree courses for a grounding on this topic.

The scientific attitude (which may not be restricted to professional scientists nor even demonstrated by all of them!) according to an early analysis (Davis 1935) shows *'a willingness to change his opinion on the basis of new evidence....search for the whole truth without prejudice....have a concept of cause and effect relationships....makes a habit of basing judgement on fact; and...have the ability to distinguish between fact and theory'*. This list of characteristics has been expanded more recently (Arons 1983).

Scientists, particularly natural scientists, believe that there is something special about a scientific way of knowing (Wolpert 1992) and, in some respects and about certain things, they may be said to 'know best'. However, they may feel the need to qualify this assertion in some way. If they do not, others will certainly do so, on their behalf!

In his essay 'What is Science?' Richard P. Feynman (Feynman 2001) exemplifies this continuing tradition but also stresses the provisional status of the theories and ideas that science develops, particularly recognising the possibility of error: *'Science is a belief in the ignorance of experts'*. Feynman also suggests something much more subtle, that *'each generation that discovers something from its experience must pass that on, but it must pass it on with a delicate balance of respect and disrespect, so that the race (....) does not inflict its errors too rigidly on its youth, but it does pass on the accumulated wisdom, plus the wisdom that it may not be wisdom'*. Alexander Pope expressed the point more poetically in his 'Essay on Man', in which he concludes that man is *'born but to die, and reasoning but to err'*. Eternal vigilance, over oneself and over others, is thus the key to avoiding the perpetuation of errors: indeed, the word 'sceptical' is derived from the Latin for 'reflective' (Shermer 2002).

Furthermore, we need to define precisely what the scientific knowledge is that we are talking about. Is this knowledge of a 'hard' science, such as physics, or 'soft', such as sociology or economics? Such a distinction might affect the precision and quantification of our knowledge or its scope of application, that is, the degree to which it can be sensibly applied to a circumstance relevant to general human experience. This may, in its turn, influence public attitudes towards science.

This belief in the special status of scientific knowledge is, however, not shared by some cultural theorists who associate the workings of science with its determination to retain a role in exercising power in society. They thus challenge the special status of science (though, somewhat paradoxically, not their own special status in mounting this challenge). While Ziman does not go along with these critics entirely, he does express a view that should be reflected upon, *viz.* *'that science is often held responsible for the bad effects of its application (and not credited with its successes)'*. He also suggests, more seriously in terms of its possible contribution to sustainability, that *'it is increasingly difficult to enlist science as a non-partisan force against obscurantism, social exploitation and folly'*.

Science may be done for many reasons: pure enquiry and curiosity; to help solve serious societal problems; to discover new technologies; to improve old ones. The term 'science' has been used by some sociologists (Irwin 1995) to encompass its technological application, whereas some humanists (Goodman, quoted in Postman 1993) make a clear distinction, suggesting that technology, with its profound social implications and choices, should be considered a branch or moral philosophy, not of science. Where scientists have close affiliations with those seeking to profit from their advances (van Kolfshooten 2002), they may feel protected from attacks by opponents who impugn their motives and question the independence of their research findings by the 'legendary' view (Hammond *et al.* 1976) that *'when scientists look for the truth and truth appears in doubt, neither scientific work nor the scientific ethic requires the investigation of the person working on the problem: instead, they require the analysis of the method by which results were produced'*.

However, the attitude of society towards science and technology has changed, with Ulrich Beck, (Beck 1992) the German sociologist, articulating a bleak and influential analysis based on the premise that individuals experience a sense of powerlessness in the face of the invisible hazards to which modern technology exposes them, the so-called 'risk society'. (Whether or not contemporary technophobia arises from a rational appreciation of the true extent of technological risk or is more governed by perceptions of risk (Starr 1969; Schulze *et al.* 1981; Hohenemser *et al.* 1983; Slovic 1987; Freudenburg 1988) is something to which I will return.) If I understand Beck's thesis correctly, the risk society, rather than basing social stratification on class (with the associated driving force *'I am hungry'*) is instead being determined by the relationship between individuals and hazardous technology (summed up by the phrase *'I am afraid'*). He suggests that the threat of catastrophic accidents associated with nuclear power, chemicals production and biotechnology are putting the advanced nations in a perpetual state of anxiety and vigilance. The treatment of these potential or perceived threats in parts of the journalistic media is seen by some as contributing to the level of concern and anxiety, sometimes on the basis of the selective or imbalanced use of scientific data. Consequently, some fundamental questions arise concerning scientific evidence and its presentation and use in affecting perceptions when we come to consider what is meant by the verb 'to know'.

However, significant to a discussion on sustainability, is to ask whether we have yet arrived at the condition summed up by the phrase *'I would rather be hungry than afraid'*; that is, a state in which the benefits of technology are knowingly rejected by the likely beneficiaries, so as to reduce the impact of humanity on the environment or to allow improved living standards for others. However, one possible outcome of such a condition could be a <u>selective</u> rejection of technology, with the consequence of denying its benefits to those in the developing world, or crudely, to a circumstance characterized by the phrase *'so <u>I</u> won't feel so afraid <u>you</u> must continue to be hungry'*. This is not an academic point, as demonstrated by the refusal of the Zambian government to feed genetically modified grain to the starving, who might, not unreasonably, assert, '<u>you</u> *may be less afraid, but* <u>I</u>*'m still hungry*'. However, while we probably accept that science and technology are part both of the problem and the solution, there are many dangers linked with the consequences of an increasingly sceptical public failing to recognise the importance of science. Here we find common ground with Beck, when he said *'scientific rationality without social rationality remains empty, but social rationality without scientific rationality remains blind'*. These are ideas, analyses and perceptions outside of science (or at least my own view of it) that inform the wider context in which issues of sustainability and the contributions of technology will be discussed. Even if we do not agree with them, we need to be aware of them.

At the heart of the matter is the distinction, in practice, to be made between truth and fiction (Hobsbawm 1993) and the issue of perception and what is believed to be true.

4   Science and the Truth

The definition of 'truth', given in my dictionary, that which is '*in accordance with fact or reality; in accordance with reason or correct principles*', begs a series of questions relating to the meaning to be attached to terms such as 'correct'.

We should also draw a distinction between 'truth' and the term 'belief' (without getting drawn into the philosophical basis for, overlap of and interaction between these two concepts). However, more recently, some sociologists of science and social and cultural theorists have put forward a 'relativist' perspective to challenge the traditional epistemic authority of science, suggesting that no one knowledge system (such as science) should have a privileged position over any other (indeed, referring, rather inelegantly, to 'knowledges' and 'understandings').

However, I agree with Steven Weinberg when he suggests (Weinberg 2001) that, because physical scientists generally do not speculate why they believe something to be true, but only why it <u>is</u> true, they have tended (with notable exceptions (Sokal *et al.* 1998; Gross *et al.* 1997; Gross *et al.* 1994)) to ignore the teachings of the cultural relativists or to dismiss them as irrelevant in the face of incontrovertible evidence of the success of science and the self-evident benefits of its application through technology. This is the territory on which the 'Science Wars' debate was fought, a topic amply analysed elsewhere. For a balanced series of essays on this subject, see Segestråle 2002 and for a readable summary of the critique of science by historians and sociologists, see Forman 1995 and Forman 1995a.

However, I believe those scientists who have ignored or simply dismissed these ideas have been wrong because, unchallenged and unqualified, they have begun to resonate in the public mind, particularly when linked with concerns over some of the disbenefits of technology and of the uncertain prospects arising from well publicised areas of scientific development, particularly in the biological sciences.

I conclude that there is a need for scientists to become more aware of the full range of claim, counter-claim and critical comment challenging science (Berry

2000; Porritt 2000; Wilson 2002) and indeed the rationality on which it and all logical discourse is based. That an article with the title 'Can Moral Norms be Rationally Justified?' (Patzig 2002) can appear in a mainstream chemistry journal suggests that this process has already begun. Scientists should also engage more fully and more energetically in continuing to justify the positive status of scientific knowledge while seeking to free it from the current taint, real or imagined, associated with its close links to industrial technology, identified by Ziman.

In summary, because of the rise of consumerism, access to information and wider concepts of accountability, the future of science will be governed by public perceptions of its discoveries and their implications, as well as of its dealings with those with power and authority in society (and how these are changing). It is in the interest of scientists to understand these trends and what lies behind them.

As Galileo experienced in a particularly extreme form, science and scientists have frequently had an uneasy relationship with authority, whether of the religious or secular type, and this is at its most ambiguous when such authority is the source of funding! The motto of the Royal Society, founded in 1660, contains an important imperative as valid as ever: *'Nullius in verba'*, 'take nothing on authority'.

Scientists should not defer to authority (including that associated with public opinion) to the extent of denying what we, as scientists, know to be true. Dryzek and Torgerson (Dryzek *et al.* 1993) have already noted that *'reason has not always been at ease with democracy'*. There exists, however, scientific knowledge that is essentially undisputed (such as the validity of the laws of thermodynamics) and, while one would hope that scientific literacy among the educated lay public would extend to an awareness of such laws and theories, one asks how many schemes for pollution prevention claiming 'zero waste' have been fully tested against the laws of thermodynamics?

We must take care to limit the scope of science to the physical world. However, while science has properly sought to avoid the territory of metaphysics, moral philosophy and what might be considered to be the realm of human relationships and human happiness, recent developments in neuroscience, the cognitive sciences (Pinker 2002) and in genetics makes this an interface whose boundaries are becoming ever more blurred. Likewise, when, as scientists, we comment on moral, ethical, cultural or political matters, it should simply be as citizens, not as representatives of science nor claiming any special privilege for our beliefs because we are scientists.

Defining the role of the scientist in the area of public discourse and political debate, then, is one that requires great care (Winterton 2003). Nevertheless, increasingly, science and other institutions are expected to be accountable to, and derive their license to operate from, the general public, as voters, consumers and as citizens – so the way the public mind is influenced is critically important. This will be discussed further in Section 6.

5 Science and Sustainability

Much has been written on the definition and description of sustainability and sustainable development. The meanings of these (and related) terms seem often to be adjusted to suit the user and the circumstance. Some academics more certain of their views on this point have been quite scathing (Stott 2002). Joseph Huber (Huber 2000) is somewhat more restrained, calling the management rules for the use of resources set down by the Brundtland Commission (WCED 1987), '*empirically empty categorical imperatives*'. I commend Roland Clift's definition of 'Sustainability' (Clift 2000): ('*a blissful state of existence, in which humanity's techno-economic skills are deployed, within the long-term ecological constraints imposed by the planet to provide resources and absorb emissions, to provide the welfare on which human society relies for an acceptable quality of life*') and its distinction from 'Sustainable Development', which '*meets the needs of the present without compromising the ability of the future to meet their own needs*' (WCED 1987; see also Merkel 1998). In addition, there are detailed descriptions set down in international agreements, the result of negotiation, debate and compromise, involving politicians, lawyers and civil servants. There have also been attempts by academics to arrive at rigorous and objective definitions (Lee *et al.* 2000; Lélé 1991).

Some see an inherent paradox (indeed, contradiction) at the heart of sustainable development and attempts to reconcile improvement in the human condition world-wide with protection of the global and local ecosystems on which life depends. However, it is possible to see sustainability as an ideal, like justice (Rawls 1972), that can be recognised in the abstract, with profound disagreement about its precise definition or the processes and priorities needed to bring it into realisation.

Nevertheless, the direct linkage between the current state of the planet with the activities of processors of materials and manufacturers of artefacts providing goods and services for a growing world population requires an acceptance of responsibility by the individual consumer of their role in the process as well as of the role to be played by science and technology in seeking to meet the challenges of sustainability. In this respect, scientists need to help to counter the argument that sustainability can be achieved without the participation of science and technology. I have dealt with these issues in more detail elsewhere (Winterton 2001; Winterton 2003) as have many others (Lesher 2002; Raven 2002; Clarke 2002; Editorial 2002; Waggoner *et al.* 2002; Wackernagel *et al.* 2002; Arrow *et al.* 1995, Waggoner 1996; Starr 1996; Mayer *et al.* 1999; Ausubel *et al.* 1999).

The question before us is 'how can this contribution of science best be made?'

First, we must recognise not simply the scientific problems (of description, observation, analysis, understanding, conclusion, validation; issues of generality, complexity, uncertainty, indeterminancy, scale, level and scope) and the social-scientific problems (differing approaches and methods of analysis of differing disciplines, their overlap and interaction; the synthesis of consensus views; the reliability and limitations of modelling, prediction, and forecasting) but also the problems associated with participation, presentation and acceptance in the wider processes of policy development and priority setting (including the selection, procurement, management and direction of research programmes).

Each one of these elements is worth a chapter in itself.

In the context of sustainability and understanding of changes in the natural world arising from man's activities a prime requirement is good monitoring data. There are obvious practical and cost difficulties associated with simultaneous, continuous, precise and accurate observation and measurement of the natural world at the level (spatial, temporal, statistical, size, complexity, phase, trophic) needed to understand its overall behaviour in the ways necessary to provide the basis for comprehensive and confident policy-making. Extrapolation, modelling and prediction necessary to define scenarios and explore potential outcomes of different interventions (or the absence of intervention), with some estimate of the likelihood of the outcome and the reliability of this estimate, bring their own challenges, dangers and uncertainties. The study of scientific uncertainty in the context of understanding and resolving environmental problems is a specialism whose learning and expertise should be extended to wider issues of sustainability (Lemons 1996; Kinzig *et al.* 2003).

Measurements will usually be incomplete and difficult to assemble retrospectively if needed to throw light on some long-term trend or provide evidence of some past change. Irwin (Irwin 1995) identifies the different types of uncertainty faced by scientists when asked to address particular controversies: *pragmatic* uncertainty (lack of facilities if asked to provide advice; short notice of request); *analytical* uncertainties (limits of detection; sampling errors (see Giles 2002, in relation to climate-change scenarios arising from atmospheric modelling)); *theoretical* uncertainties (no strong theoretical framework, especially if the topic is trans-disciplinary); *indeterminancy* (impossibility of prediction in an inherently unstable or chaotic system; see Ayres 1996, in the context of sustainability). All these are relevant to issues of sustainability. Ignorance can be reduced but indeterminancy cannot.

In the context of sustainable development, to what extent should one rely on predictions of future behaviour of such complex and inter-related physical, biological, ecological and social systems such as those that make up the earth that depend on an assumption of deterministic behaviour of the component parts? Or, may new conceptual interpretations be needed to understand the behaviour of the integrated system at the higher level, such as those that lie behind the idea of 'emergence' (Anderson 1972; Laughlin *et al.* 2000; Goldenfeld *et al.* 1999; Humphreys 1997; Vicsek 2002). Even if one does not accept this idea, there is much empirically to be learned by the study of systems at a higher level of complexity (or even 'holistically'), as the approach of geophysiology (also known as the Gaia concept (Lovelock 1995)) has demonstrated.

In addition, the problem of incommensurability will plague judgement-making and priority setting, arising from the extreme difficulty, if not impossibility, of using the same metric to compare phenomena (or possible outcomes) in different environmental domains. Attempts using money value as the metric in such comparisons are being made by environmental economists to aid processes of judgement-making on sustainability issues, particularly in such cases where comparisons are difficult (such as seeking to compare the costs and benefits of protecting biodiversity with those of organic agriculture) or next to impossible (such as the comparison of the impact of species loss with that of ozone depletion).

Experience of the Intergovernmental Panel on Climate Change is instructive and a recent paper (Siebenhüner 2002) has analysed the management of the processes and the dynamic between those involved in assessing the state of the science, considering the possible outcome of various scenarios and options for mitigation. Much of this involves a large number of different studies, using different techniques making different assumptions, modelling different scenarios leading to different predictions. The output from such work can be misleading or, worse, misused by special interests with their own agenda, unless attempts are made (Giles 2002, Allen 2003) to quantify the probabilities, to put into context outcomes with very low estimates of probability and to provide audit trails relating the evidence and its uncertainties.

Similar care will be needed to ensure robustness of the various metrics on sustainability that are being developed. For example, the Environmental Sustainability Index seeks to quantify parameters that are seen to contribute to sustainability and then to aggregate them together at the country level to provide country-to-country comparisons (World Economic Forum 2002). The issues of comparability of the values of these metrics and parameters from country to country; of clear and accessible trails leading to the raw data on which they are based, of the assessment of the sensitivity of aggregated data to input type and quality; of the time lag between data collection and aggregation and of the ability to sense changes in the metric over time will need to be addressed to generate confidence in the validity of the metrics and the uses to which they are put. However, it is much more difficult to assess precisely the relative impact of two different technological approaches to a particular problem (including, first and foremost, defining system boundaries to permit a valid comparison to be made). In the case of chemical technologies, steps forward are being made which seek to put aspects of feedstock source, reaction efficiency, utilities usage, waste treatment on a common exergy metric (see, for instance, Dewulf *et al.* 2000; Dewulf *et al.* 2002; Constable *et al.* 2002). Scientists, particularly chemists, need to be aware of these developments and to apply them in their own work.

Rayner and Malone (Malone 2001) have suggested that there are two styles of research: first: *interpretative-style* done by those seeing themselves as the centre of the environment, experiencing it from within, their involvement gaining them their knowledge; second: *descriptive-style* done by those outside of the environment, their distance allowing them to gain their knowledge. They believe that attempts to meld these two approaches are unrealistic and, instead, they should be used to complement one another to bring their knowledge to bear on global problems. They make a sound point.

Such questions of research style are relevant to the many disciplines and perspectives that are needed for an understanding of sustainability questions. Glaze, former editor of the ACS journal, *Environmental Science and Technology*, has coined a useful term, hyperdisciplinarity (Glaze 2001), to characterise the knowledge needs for sustainable development. We are all, to varying extents, prisoners of our past, as discipline boundaries have been driven by the departmental structure of academic institutions (and attempts to address multi-, trans- and interdisciplinary problems) and are not necessarily appropriate for the needs of new and complex areas of enquiry and scholarship, such as sustainability (Kates 1989; Kates 2002). My own experience of assembling, coordinating and directing multidisciplinary teams on interdisciplinary problems, both in industry and academia, suggests that the development of a shared view of the problem and how it is to be tackled is as big a challenge as the achievement of the objective of the project or collaboration itself. In the context of sustainable development, therefore, the idea of anyone 'knowing best' is problematical, though in some instances I would contend that science would know better than most, albeit imperfectly. The fact that science may not be perfect does not negate its importance. Nor does the fact that science may not be sufficient make science's contribution unnecessary (Medawar, quoted in Greenberg, 1969). In any event, most of the factual evidence on which argument and decision-making will be based will come from the physical, biological and environmental sciences. Gaining a shared detailed view on anything as complex and as value-laden as sustainability may be impossible. How then is this view to be arrived at? Will it be imposed politically or by bureaucrats? How can society express its wishes? (Vollenbroek 2002). Can (or should) scientists themselves do it? Will they be allowed to do it? Parallels may be noted between these considerations and those surrounding management of technological risk. Research on the underlying aspects of risk assessment, risk perception and risk communication and approaches to reconciling technical expertise with democratic control, all provide perspectives that may inform discussion about the role of science in sustainability.

Just as it seems sensible to bring scientists closer together, fragmentation and further specialisation into sub-disciplines drives them apart: I recently discovered the sub-discipline of psychobiogeography (Trudgill 2001). The author of this paper struggles with the dilemma of public misunderstanding of verifiable ecological facts and the 'common-sense' attitudes about the natural world held by lay people. He also suggests *'The very notion of the lay-expert divide means that people are dis-empowered by experts. Setting up this divide means that it is inevitable that the basic problem of democratising the conservation debate is then seen in terms of how to involve 'the public' who are seen as largely 'ignorant''*. We could easily replace the word 'conservation' by the term 'sustainability'. This analysis is widely held among sociologists and I will return to in a moment, as it is one that physical scientists, some of whom may choose not to include sociology among the sciences, need to confront.

For a natural scientist it is sobering enough to be confronted with the challenges from related natural science disciplines let alone from those from other disciplines that fall within the broad envelope of science (such as, economics, ecology, anthropology, sociology, geography). But the issues raised by sustainable development require us to go beyond even this and address the contribution of other areas of scholarship and intellectual study, such as the law, ethics, philosophy and politics. Again, what each of these has to say would take a working life to review, but nevertheless each provides important and valid perspectives on the challenges we face. All one can do is to reflect humbly on our profound states of individual ignorance (perhaps the proper state for any self-respecting expert!).

However, we physical scientists need to note to what extent the public agenda on matters relating to development and sustainability is being set by these other intellectual domains, particularly sociology and economics, with influential concepts such as the risk society (Beck 1992) (and the related topics of risk perception and communication), industrial ecology (Graedel *et al*, 1995; Seager *et al.* 2002; Ehrenfeld 2002) and ecological modernisation (a concept from environmental social science) (Weidner 2002; Massa *et al.* 2000; Anderson *et al.* 2000; Seipel 2000; Ashford 2002; Huber 2000; Cohen 1998; Cohen 1997; Langhelle 2000), carrying capacity (Ehrlich 1994), ecological footprints (Rees *et al.* 1994), as well as the more technologically-driven ideas such as back-casting (Weaver *et al.* 2000), ecobalances, life cycle assessment, Factor 4 and Factor 10 (Refs 4 – 7 in Robèrt *et al.* 2002), exergy (Rosen *et al.* 1997) and ecothermodynamics (Ayres 1998).

Keeping abreast of developments in any one of these topics would be difficult for the specialist, let alone a non-expert entering this territory searching for enlightenment or guidance. One is forced to conclude that the 'knowing' in relation to science and sustainability may have to take on a special collective, rather than an individual, meaning, if this is not stretching the definition of knowing too far. However, the formal institutions of science (beyond the processes of peer review and the norms concerning the proper use that scientists may make of scientific knowledge) do not exist to permit science to formulate such consensus views, though *ad hoc* groups seeking to do this are appearing (Clark 2002). We should consider the implications of the fact that other institutions in society, such as government, business and special interest groups, are doing it for us.

Science has been long been under well-argued intellectual challenge (Berry 2000), among other things, because of its so-called reductionist approach. 'Reductionism' may be seen as a means to an end, *i.e.*, is a process of seeking understanding of natural phenomena at the most fundamental level (see Chapters 2 and 10 in Weinberg 2001). In this it has been remarkably successful in addressing complex phenomena, including those underlying environmental problems. One may also see reductionism as the end in itself, that is, to seek a 'theory of everything' (Laughlin *et al.* 2000) or to bring a unity to our view of nature (*i.e.*, 'consilience' (Wilson 1998), to use E.O. Wilson's term).

For many scientists, this enhanced understanding removes neither a sense of wonder nor an aesthetic appreciation of the natural world, as some critics contend. Even less does it suggest that there are not moral, ethical and social questions to which science itself has little to say (but where scientists may have lots to say!). However, it is certainly the case not only that the technological application of much of current science has ethical questions to be resolved, but also similar questions are being raised about some aspects of scientific enquiry itself. How science and scientists meet this challenge could well govern the extent to which science can make its full contribution to the development of more sustainable technologies.

Can sustainable development's need for science be met *via* the existing disciplines or even from current knowledge or, as some have suggested, do we need a separate science, called 'sustainability science' (Kates *et al*. 2001)? My own view, pragmatic as befits a chemist (Schummer 2003), is that there will be a need for both. Just as innovative ideas for cleaner chemical processes will emerge from chemistry research whether or not it sees itself as 'green', similarly, an approach to sustainability governed solely by programmes of sustainable development (Robèrt 2000) will not, on its own, be sufficient.

On the other hand, Professor Steve Rayner, Professor of Science in Society, Saïd Business School, Oxford University and Director of the Economic and Social Research Council's science in society programme, has suggested (Rayner 2002) that we need neither 'sustainability science' nor a new style of scientific research to galvanise policy-makers into action. In his view, the major challenges to political leaders are not the paucity of scientific information but political will and ethical dilemmas. We do need the best information, but, in Rayner's view, we need sustainability politics rather than sustainability science. He has also, tongue-in-cheek, sought to call attention to his view (Macleod 2002) by calling for a science strike, to highlight his contention that further research is not a pre-requisite for sound policy action, science having already established a justification for such action. He says that the role of scientists as citizens may be more important than their role as technical experts. I have addressed this question myself, elsewhere (Winterton 2003). If we ignore Professor Rayner, what should the role of sustainability science be? How would it be different and what new purpose would it serve? Some of the core questions of sustainability science, designed to address interactions at the interfaces between nature and society and their monitoring and management, have been posed in an article in *Science* (Kates *et al*. 2001) developed further in recent articles in the *Proceedings of the National Academy of Sciences of the USA* (Clark *et al*. 2003; Kates *et al*. 2003; Parris *et al*. 2003; Turner II *et al*. 2003; Turner II *et al*. 2003a; Cash *et al*. 2003).

We are beginning to move onto some of the more difficult questions to be addressed. It is clear that this entails not just dealing with the factual evidence of changes to our surroundings, conclusions about how these come about and what should be done about them, but also with personal opinions and cultural values, the ways they may be perceived and the processes for their resolution.

6 Science and the Public

While it may be unclear who it is that knows best on matters of sustainability, ultimately, decision-making will be in the hands of the public, whether as consumers, stakeholders, voters or activists. It is also clear that society will increasingly challenge science and will expect to engage with scientists in agenda setting. Is science, and are scientists, ready?

Nevertheless, robust policy formulation should integrate input from experts, from those directly and indirectly affected and from the wider (non-affected) public

(Cvetkovich *et al.* 1992; Renn 2001). The unresolved problem is to agree on how to bring this about, the priority to be attached to the input from the different interested parties and the relative weight given to the technical input from experts and the perceptions, opinions or experience of those directly or indirectly affected. Research on risk management and associated decision-making suggests that securing a consensus is most difficult when the stakes are high, knowledge of the consequences is limited and the distribution of benefits and burdens is unequal (see, for instance, Bullard 1988; Freudenburg 2001). It is likely that many of the decisions about sustainable development fall into this category and will require '*a deeper kind of prudence*' (Freudenburg 2001) and '*a capacity to worry intelligently*' (Kates 2001).

The pertinent question for scientists is whether there is a need to distinguish between their own participation in this debate as scientists rationally seeking to assemble evidence and dispassionately draw conclusions from it or as citizens trying to win an argument using rhetoric and selective information?

The benefits and otherwise of scientific rationality have been the concern of thinkers from the renaissance, as has the role of science in the development of human culture and its contributions to wider human well-being, often taken for granted, arising from its application through technology. More generally, it is not clear how receptive members of wider society are to consider in a fully informed manner scientific controversies in general and science's role in sustainable development, in particular. How, indeed, would today's public respond to C.P. Snow's question "*Can you describe the Second Law of Thermodynamics?*", a question which Snow suggested was the equivalent in science of asking "*Have you read a work of Shakespeare's?*" (Snow 1998). To be 'cultured' might conventionally be demonstrated by a familiarity with the works of Shakespeare, whereas ignorance of something as basic, central and universal to an awareness of science as an understanding of the second law of thermodynamics seems not to provide evidence of a lack of culture. Bearing in mind the second law's dominant importance in the realisation of greater sustainability and its constraints on what is possible, Snow's test takes on a new and contemporary significance, despite the contempt his ideas elicted from some humanists.

There should thus be concern regarding the degree of understanding of the lay public of scientific issues, particularly as ignorance about science doesn't stop people from holding strong opinions on techno-scientific controversies, such as climate change, GM crops and so on, basing their opinions on information and comment from newspapers, TV and radio. While this may be self-evident, it is reinforced by the conclusions from a very recent analysis of TV, newspaper, radio and magazine content combined with an attitude survey (Radford 2002) involving 1035 people. In the case of the link between autism and the use of a triple vaccine for protection against mumps, measles and rubella (MMR), 67% of those surveyed believed that the evidence was either evenly balanced or in favour a link, whereas most published scientific evidence concludes that there is no link. This example of what has been called '*the social amplification of risk*' (Kasperson *et al.* 1988; Renn *et al.* 1992; Frewer *et al.* 2002) raises some important general questions regarding 'knowing' and the quality of that knowledge that constitutes part of the question I am seeking to address. Significantly, the attitude survey confirmed that people tend to absorb oft-repeated associations they hear in the media. This suggests that newspapers, radio and television and, increasingly, the internet, are not themselves passive transmitters of unbiased information (Davies 2001). Indeed, one is worried that there may be instances in which perceptions are manipulated in the expectation that they will become the beliefs of those exposed.

If it ever was acceptable, we certainly can no longer justify a lofty, detached or patronising attitude to public ignorance on scientific matters. There is a view (not universally held (Shamos 1995)) that public concern about technological developments (and the science that underpins them) could be allayed simply if the general public were more scientifically aware, knowledgeable or literate. In 1983, Miller (Miller 1983), formulated a stratified model of science and technology policy formulation with a small number of decision-makers and policy leaders on top, and dividing the public into *ca.* 20% of the 'attentive' and 80% of the 'non-attentive', though inattention was not necessarily because of ignorance or lack of intellectual ability. This model can be seen to be outdated in many ways, though the vetoing role of the majority, if mobilised, was certainly recognised.

This proposed 'linear' relationship between scientific knowledge and scientific attitude has been questioned recently (Cohen 1997), based on empirical analysis of public-opinion data. The relationship was, instead, found to follow more of an inverted U. Societal support for science increases with scientific knowledge only up to a certain point, after which further understanding contributes to a decline. Suggested reasons for the decline include increased educational opportunities that stimulate independent analysis leading to the propagation of alternative knowledge systems; technological scares which imply that science has compromised its reputation as a virtuous and open knowledge system.

In addition, the very idea that the deficit of understanding about science among the lay community can (or should) be addressed by relieving the non-expert of his or her ignorance implicit in the foregoing, is challenged by social scientists, who suggest that science is simply one form of knowledge among many and, while it has occupied a privileged position, it must now address challenges to its privileged status, particularly as, it is argued, the exercise of its status (particularly in controversies concerning risk, actual, potential or perceived) simply is designed to maintain the existing power structure and the position of the scientist within it.

Interestingly and perhaps significantly, a recent major study (Office of Science and Technology (OST) 2000) on public attitudes to science was aimed not at the scientific community, but at the 'science communication community' *i.e.*, a group containing social scientists as well as non-scientists. Perhaps they have given up on the average scientist as a communicator. The main conclusions of the study, however, are important to scientists in that they suggest that communication about science is usually provider-driven not consumer-driven. How much of our own science communication is put into the context of the non-specialist reader or viewer? Research Councils UK, through which much of the UK's research funding is channelled, has recently published a report 'Dialogue with the public: practical guidelines' (Research Councils UK 2002) listing several ways in which to engage in dialogue. This study also concludes that there is a need to bring public opinion more into the development of science policy. The public will indeed have views on any associated ethical and social factors and not just on the science. To express such views, it is believed, they do not need to understand much science.

Public concerns also arise out of ignorance about 'what goes on behind closed doors' (shouldn't scientists be more pro-active in opening up their laboratories to public view?) and a feeling that scientists try new things out without stopping to think about the risks.

The trustworthiness of information is an important factor and, according to the OST survey, people tend to place their trust in sources that are perceived as neutral or independent (*i.e.*, university scientists, scientists working for research charities or health campaigning groups, TV news and documentaries). Next are those seen to have a degree of vested interest, such as environmental groups, well known scientists and the popular scientific press. The least trusted are politicians and, significantly, newspapers. These issues of trust are important and I will return to them in a moment. 64% of those surveyed agreed that the media sensationalize science, with 53% agreeing that politicians are too easily swayed by the media's reaction to scientific issues. The role of the media is thus critical in this discussion (see, for example, Singer 1990; Frewer *et al.* 2002; Jauchem 1992).

An interesting example is a front page story in the UK daily newspaper, the '*The Independent*' (Morris 2001) of 25 April 2001, at the height of the foot and mouth epidemic, suggesting that some individuals showed symptoms that might be consistent with the human form of the disease. This report, written by the paper's political correspondent, appeared while tests were being done to establish whether or not the individuals had the disease or not. This may be an example of a phenomenon revealed in another survey (ESRC 2002), this time of science journalists in the UK, that suggests that they were marginalized or even ignored within their own news organisations as science-driven stories became the subjects of campaigns driven directly by senior executives, who then used general reporters, political reporters and environmental reporters as much as science reporters. '*Science is the first casualty when there is an emotive story to be had*' said one; '*scientists are useless, which is why there are armies of PR people in universities, research councils and funding agencies*' said another.

(In passing, it may be of interest to note that a further report (Science Media Centre 2002) led to the setting up of a Science Media Centre, based at the Royal Institution in London, which works '*to promote the voices, stories and views of the scientific community to the news media when science is in the headlines*' and '*to promote more balanced, accurate and rational coverage of the controversial science stories that now regularly hit the headlines*'. Very recently, the American Association for the Advancement of Science has announced that it is planning a Center for Public Engagement with Science and Technology (AAAS 2003).)

A further example arises from a press campaign undertaken by another UK daily newspaper, the '*Daily Mail*' (Marsh *et al.* 2002), promoting the suggestion that a combined inoculation for measles, mumps and rubella (MMR) is associated with increased incidence of autism. Most published scientific evidence has concluded that no such link exists (confirmed in a recent Danish study involving 0.5M children (Madsen *et al.* 2002)). However, as previously mentioned, a recent survey of public attitudes in the UK revealed that 67% of respondents believed that either the evidence was evenly balanced or supported the idea of an enhanced risk. A search of the Daily Mail website shortly after publication of the Danish study found no mention of it.

It is fair, in the face of this and other evidence (for the UK at least) (Winterton 2003), to ask whether media treatment of scientific controversy is designed to create a perception that then determines public attitudes. In other words, the media are not simply transmitters of information but seek to mould opinion (Singer 1990; Frewer *et al.* 2002). And if this is so, it is relevant to ask whether it follows in these circumstances that the public can be said to 'know best'.

The 'who knows best' dilemma (in the context of risk management) is summed up well by the following quotation (Raucher 2000): '*Whose views should determine policy- the expert's or the public's?.....if the public is really misinformed we have a dilemma as to whether we ignore the public and follow the experts or blindly follow the public even though we know they're misinformed.*'

There is perhaps a need, as one science journalist respondent in the survey suggested, to expand the research agenda to include an extended analysis of the criteria for 'expertise' and 'authority' in the work of journalists and perhaps to explore the extent to which initiatives designed to improve the legitimacy of experts – for example transparency of relationships and processes of accountability – might be extended to the media as well.

This, on a more general but still relevant level, was the overall contention of Onora O'Neill, Principal of Newnham College, Cambridge and a political philosopher, in her 2002 series of BBC Reith Lectures (O'Neill 2002), called 'A Question of Trust'. Her main thesis was that well-intentioned efforts to improve accountability of public institutions, such as local and national government, the police and legal system, health, education and social services, often had the opposite effects to those intended. However, the fifth and final lecture was addressed to the poor standards of professionalism among journalists in some sections of the UK press and the absence of the sort of rigorous scrutiny and accountability that is found in other professions. I agree with Onora O'Neill, but also wonder whether an inability to discriminate properly between fact, speculation and fiction is not simply bad journalism but also may be a reflection of the literal application of the cultural relativist viewpoint among some journalists and editors.

On a more positive note, Irwin (Irwin 1995) urges science to consider the lay public as a source of information, suggesting the relevance of what he calls 'popular epidemiology', that is, the anecdotal experience of those who may be the victim of some pollution or environmental problem. Their stories satisfy many of the criteria for 'good' journalism and can inevitably attract the interest of those with 'an agenda'. Those responsible for the problem may well use the arguments of science in their own defence. Irwin suggests that '*scientific failure at the local level will generally be diagnosed as a minor matter of inappropriate application or variance from the norm – rather than as a major area of cognitive and institutional challenge for the activities of science*'. Those who believe themselves to be damaged as a result of the application of technology are usually a statistically small proportion of those who benefit, but whose role in production or direct use of the technology may make them more vulnerable to the consequences of their exposure. Even if we are talking about 1 in a million risk, our developing and changing attitude to the individual (and the voices now raised on behalf of such individuals) provides further ethical and legal questions to be addressed (Heinzerling 2000; Ravetz 1999).

On the other hand, Irwin fails to distinguish between 'vocal' and 'passive' citizenry, nor does he address the role played by activist groups. The consequence of their involvement may be to raise the apparent importance of one perspective relating to an environmental concern against another. The 'precautionary principle' (Harremoës et al. 2002; see also Foster et al. 2000; Foster 2002) is often invoked under such circumstances particularly when scientists and others cannot claim 100% certainty in their statements or are unable to prove negative statements, such as 'this process, product or activity carries no risk'. The associated problems of risk aversion (Furedi 1997; Furedi 2001) and the concern of public institutions to limit opportunities for litigation tend to take precaution (understandably) to excessive limits. Action based on an unthinking application of the precautionary principle may itself have negative consequences (Anderson 1991). Ironically, it is those for whom 'precaution' is believed to be necessary that often are the sufferers. Elevation of what some see as a mere a rule of thumb (and, in its 'strong' form, not a very useful one) (Sustein 2003) to a rigid principle limits effective decision-making, paralyses action and misallocates scarce resources ('*the social cost of fear reduction*' (Durodié 2003)). Attempts have been made to justify the costs of false alarms (see, for instance, Cross 1996), on the grounds that the potential benefits of averting disaster are so much greater (Pacala et al. 2003). This has been disputed (Murray 2003). More constructively, there have been suggestions that the term 'principle' be replaced by 'approach' (Stirling et al. 2002) and attempts to reconcile the precautionary approach with the concepts of cost-benefit analysis (Geistfeld 2001; Commission of the European Communities 2002).

In his book, Irwin (Irwin 1995) quotes Nelkin (Nelkin 1975) who gets to the nub of the point: '*The complexity of public decisions seems to require highly specialised and esoteric knowledge, and those who control this knowledge have considerable power. Yet democratic ideology suggests that people must be able to influence policy decisions that affect their lives*'.

In a paper cited earlier (Trudgill 2001), the author concludes that '*division between expert and lay needs dissolving without dis-empowering the expert or alienating the public*'. The questions are: 'is this possible?' and 'if so, how?'.

On the other hand, we need to guard against those who agree with Mulkay (Mulkay 1991), also quoted by Irwin, who says '*I have come to see sociology's ultimate task, not as that of reporting neutrally the facts about an objective social world, but as that of engaging actively in the world in order to create the possibilities of other forms of social life*'.

Irwin's basic challenge is the one we need to address in the context of sustainable development, with its social, political and ethical dimensions: natural scientists will need to engage *directly* with citizens (and not just with each other) about highly complex and uncertain environmental and developmental issues at all levels to enable a mutually comprehensible dialogue. While this may be difficult, and will be discussed further in the next section, such a dialogue will be preferable to an exclusion of either the rationality of science or the experience of the citizen.

7 Science and Decision Making

While under certain defined circumstances science may be presumed to 'know best', it does not necessarily follow that it is science that should ultimately decide. Deciding involves determining which way society collectively wishes to go, and in this, scientists are citizens just like anyone else. Such processes will be governed by a spectrum of opinions, values, ideals and attitudes. However, with science's special knowledge comes special responsibility that requires increased public engagement.

Decision-making is a complex process of negotiation and compromise between groups and individuals, a process we call politics, that ideally seeks to reconcile conflicting interests *via* participative decision-making.

Who should be involved? Who should negotiate and compromise with whom? Who should have the final say? How should science and scientists get involved? Can we construct better processes to make such decisions more efficiently or more perfectly? Are we engaged in solving a technical problem or are we engaged in a battle between competing ideologies? Is the debate about facts or values or norms? How much central planning should be part of the process of the decision-making? Are many small decisions better than a small number of big ones? The aggregation of the purchasing decisions of individual consumers now guides the practices and policies of many big industrial and commercial enterprises. The power to be wielded under such circumstances has not been lost on activist groups and some Non-Governmental Organisations, in planning their campaigns.

Faced with these and other questions, it is sobering to learn that, until very recently, the two disciplines most intimately involved in seeking to understand the processes of public decision making about technological risk, *viz.*, that concerned with technical risk assessment and social science research on how risk is perceived, were entirely separate and did not interact. Fortunately this has changed, with issues addressed from a series of perspectives in readable papers by Renn *et al.* 1993; Renn 1998; Renn 1999; Cross 1998, Okrent *et al.* 1998; Okrent *et al.* 2001; Klinke *et al.* 2002.

Is it inevitably (as it so often appears) a contest in getting a message across between a media-savvy advocate with sincerely-held if simplistic views and a scientist inexperienced in dealing with the media giving a guarded careful presentation on a complex problem?

This is not a new problem, the contest between rhetoric and expertise as a means of determining public decision making was the subject of the Socratic dialogues (Plato [1994]) – though, importantly, in Athens there was a direct debate occurring in front of the public, without mediation.

The matrix shown in the Table 1 (based on one produced by King (King 1994)) provides four generic outcomes to the process of reconciling facts (related to knowing) with values (related to deciding), depending on whether there is agreement or disagreement.

I think that aspects of the sustainability debate may be fitted into one or other of these quadrants and they will shift from one to another over time, as more information becomes available or as values and attitudes change regarding what is acceptable and what is needed. The effort of science may need to be directed towards preventing attacks on itself in the conflict quadrant from undermining its potential for aiding progress in the other three and towards improving communication with the wider public to make the conflict quadrant smaller. As Tait (Tait 2001; Tait 2001a) has argued, negotiating resolutions in a number of areas of controversy, science and rationality may founder when differing value-systems collide.

Table 1. Facts and Values (after King 1994)

		Values (Deciding)	
		Agreement	**Disagreement**
Facts (Knowing)	**Agreement**	Consensus	Negotiation
	Disagreement	Research	Conflict

I do not presume to be able to answer these questions: all I am attempting to do is to draw attention to the fact that science's participation should not be automatically assumed *a priori* nor may it always be necessary. It will always have to be argued for. And who will articulate that argument? As science has few global formal structures this will depend ultimately on scientists.

Particularly concerning is that the current challenges to science are such that there may be a danger that society will lose the value of the expertise that some of its scientist members have, if we impugn motives or fail to see the difference between (and potential benefits arising from) 'expertise' on the one hand and 'lay' experience on the other. Do we do without the expert, or fail to value his or her contribution on some misplaced notion of elitism and the need for democratisation?

On the other hand, science must not claim too much. While science and scientists may be said to 'know' and even 'know best', their knowledge is limited to statements about the natural world. These statements may be statements of fact, generalisations, predictions and speculations, with varying types and degrees of uncertainty associated with them. Those made about sustainable development will be no different.

How will science and scientists get involved in this process of negotiation and compromise? As individual experts contributing as they have done in the past? Or through new institutions set up specially to deal with matters of sustainable development, as discussed in a recent report (Clark *et al.* 2002; Kinzig *et al.* 2003), which calls for a new contract between society and science and technology. Set against this we may cite Rayner's view that we do not need any more sustainability science, just more sustainability politics. Who will decide?

At the end of the day, science and scientists need to convince individual consumers, voters and citizens of the returns from investment in science and technology in relation to sustainable development. A variety of methods (see Chess *et al.* 1999 for a survey limited to the US and Canada) have been attempted to bring this about, though none appear to display all the necessary characteristics (Hampton 1999) for wide engagement, rational consideration of the issues and a process of producing defined outcomes in a reasonable time.

These include the following:

- Scientific advisory committees that traditionally have guided or influenced government legislation and regulation fall into one of three categories:
- *expert* ('let the facts decide') panels. such as the Intergovernmental Panel on Climate Change and others (Siebenhüner 2002; Siebenhüner 2002a)
- *democratic* ('let the people decide') structures, with a variety of forms of democratic participation: open (but how can it lead to a conclusion?); closed (but how can it be effective?)

What about theoretical potential victims? Won't this include everyone? How many potential victims should make up a veto. Should the statistical individual have legal rights? (Heinzerling 2000) How do we define the public? In the developed or in developing countries? In this context, it is pertinent to raise the issue of the use of DDT to control the malaria vector in developing countries, such as Ethiopia and the attitude of NGOs from the developed countries.

- *pragmatic* (sensible judgement) approach based on a common-sense view of what is practical, reasonable or manageable.

Other methodologies designed to aid public participation in decision making include:

- *public enquiries* (different types: preliminary; formal; special; community),
- *environmental courts* (A New Zealand approach which grew out of the public planning processes (Birdsong 2002)),
- *science courts* or *citizen juries, consensus conferences* (Purdue 1999) (A UK example of the governance of biotechnology),
- *study circles* (widely used in Sweden, though polarisation rather than consensus may be the outcome),
- *public information campaigns* (though these can be dismissed as propaganda),
- *science shops* (Irwin 1995) (science-citizen mediation, pioneered in Holland, designed to promote socially relevant R&D by providing technical assistance to members of the public who do not have the means to pay for it or to collect it themselves),
- *consultative panels, deliberative polls, focus groups, surveys.*

Renn and others have proposed methods by which experts, stakeholders and others can be engaged in an 'analytical-deliberative' procedure. Efforts to balance different interests and values in policy judgements on the commercial growing of genetically-modified (GM) crops in the UK appear to be based on such an approach. Science has contributed both a scholarly review of what has been published on the topic (GM Science Review Panel 2003) and results (published in a series of eight peer-reviewed papers in *Philosophical Transactions of the Royal Society* (Royal Society 2003)) from farm-scale trials of three different crop/pesticide combinations. Treatment of this research by advocates of the pro- and anti-GM positions appears to have borne out Tait's contention, obscuring the conclusions from the work to an extent that the President of the Royal Society was moved to complain about '*kneejerk opposition*' (May 2003). Who is to blame if science (and scientists) fails to move the debate out of the 'conflict' quadrant? We can hardly argue that the current relationship between science and citizens is satisfactory and requires no modification. We need to accept that science, to have everyday relevance, must change this relationship. In my view, failure to do the latter will inevitably lead to an even greater public onslaught on science, to the detriment of a more sustainable future.

8 Conclusions

Despite all the challenges that have been levelled at science and scientists, it is still possible to conclude that science indeed knows best about certain aspects of certain things (and certainly the ability to distinguish 'good' science from 'bad' science). However, it is from wider society that science receives its license to operate and scientists must actively argue the benefits of science to achieving sustainability.

In doing this, science needs to be careful how it defines and sees itself. It should be inclusive rather than exclusive, with membership built on attitudes of rationality, objectivity, scepticism, open-mindedness and tolerance, which are not the exclusive domain of science.

Scientists should also accept that science is a social process with a strong ethos, and seek to understand these aspects of it in collaboration with those who make it their subject of study.

While it polices infractions of its rules science must also watch the company it keeps, particularly so as not to compromise its critical role in correcting error.

We should aim for the highest standards of excellence in science that we do, whatever it is.

In recognition of the hyperdisciplinarity of the science needs of sustainable development, we should extend our awareness of other scientific disciplines and other domains of intellectual activity. We should build 'ivory bridges' (Sonnert *et al.* 2002) across the canyons that divide the disciplines and the rifts that divide science from society.

In recognition that science represents only part of the foundation for sustainable development, scientists should engage more in the communication of the nature and the role of science, generally, but also understand and reflect on the fears arising from some of the technological applications made possible by scientific advances, while challenging those that seek to misrepresent or to exploit those fears. (*e.g.*, May 2003)

We should apply sustainability precepts to our own specialities, as far as is practicable.

We should recognise where our specialities, indeed where science itself, has limited things to say (but be alert to the fallacies of some critics).

We should maintain the multiplicity and diversity of scientific approaches towards sustainability.

We should avoid cultural centralisation in our approaches to sustainable development. It will be important to cultivate the sceptics, the unorthodox and license the irritants, as there is too much we do not know to believe we have even some of the answers.

However, the absence of a formal organisation, representation or leadership to articulate the nature and methods of science results in these messages being presented incompletely, incoherently and sometimes falsely by those claiming to speak on behalf of science.

The most effective means of achieving a successful presentation of such messages is through scientists themselves, a process that can be improved only by better formal education and training in the underpinning philosophy, history and methodology of science, particularly at the degree level.

Finally, in answer to the question 'Science and sustainability: Who knows best?' I would like first to quote H.L. Mencken, as a reminder of the folly of seeking simple answers to complex problems:

'For every complex problem there is a solution that is simple, neat – and wrong'

My own answer to this question is, nevertheless, impossible to challenge: 'History and hindsight'.

References

AAAS (2003) A Monthly Newsletter for AAAS Members. American Association for the Advancement of Science, March 2003.
Allen MR (2003) Possible or probable? Nature 425:242.
Andersen MS and Massa I (2000) Ecological Modernization – Origins, Dilemmas and Future Directions. J Environ Policy Plann 2:337.
Anderson, PW (1972) More is Different: Broken Symmetry and the Nature of the Hierarchical Structure of Science. Science 177:393.
Arons AB (1983) Achieving wider scientific literacy. Daedalus 112:91.
Arrow K Bolin B Constanza R Dasgupta P Folke C Holling CS Jansson B-O Levin S Mäler K-G Perrings C and Pimentel D (1995) Economic Growth, Carrying Capacity and the Environment. Science 268:520.
Ashford NA (2002) Governmental and Environmental Innovation in Europe and North America. Am Behav Scientist 45:1417.

Ausubel JH (1999) Reasons to Worry about the Human Environment. Technol Society 21:217.
Ayres RU and Axell R (1996) Foresight as a Survival Characteristic: When (If Ever) Does the Long View Pay. Technol Forecast Soc Change 51:209.
Ayres RU (1998) Eco-thermodynamics: Economics and the Second Law. Ecol Econ 26:189.
Beck U (1992) Risk Society: Towards a New Modernity. Sage, London.
Berry W (2000) Life is a Miracle: An Essay against Modern Superstition. Counterpoint, Washington, DC.
Birdsong BC (2002) Adjudicating Sustainability: New Zealand's Environment Court. Ecol Law Quart 29:1.
Bullard CW (1988) Management and Control of Modern Technologies. Technol. Society 10:205.
Cash DW Clark WC Alcock F Dickson NM Eckley N Guston DH Jäger and Mitchell RB (2003) Knowledge systems for sustainable development. Proc. Natl. Acad. Sci. USA 100:8086.
Chess C and Purcell K (1999) Public participation and the environment: do we know what works? Environ. Sci. Technol. 33:2685.
Clark WC Buizer J Cash D Corell R Dickson N Dowdeswell E Doyle H Gallopin G Glaser G Goldfarb L Gupta AK Hall JM Hassan M Imevbore A Iwu MM Jäger J Juma C Kates R Krömker D Kurushima M Lebel L Lee YC Lucht W Mabogunje A Malpede D Matson P Moldan B Montenegro G Nakicenovic N Ooi LG O'Riordan T Pillay D Rosswall T Sarukhán J and Wakhungu J (2002) Science and Technology for Sustainable Development: Consensus Report of the Mexico City Synthesis Workshop, 20-23 May 2002, Cambridge MA: Initiative on Science and Technology for Sustainability. http://sustainabilityscience.org/ists
Clark WC and Dickson NM (2003) Sustainability science: the emerging research program. Proc. Natl. Acad. Sci. USA 100:8059.
Clarke T (2002) Wanted: Scientists for Sustainability. Nature 418:812.
Clift R (2000) Forum on Sustainability. Clean Prod Proc 2:67.
Cohen MJ (1997) Risk Society and Ecological Modernisation. Futures 29:105.
Cohen MJ (1998) Science and the environment: assessing cultural capacity for ecological modernisation. Public Understand Sci 7:149.
Commission of the European Communities (2002) Communication on the Precautionary Principle, 2 February 2000, Brussels (http://europa.eu.int/comm/off/com/health_consumer/precaution.htm)
Constable DJC Curzons AD and Cunningham VL (2002) Metrics to 'green' chemistry – which are best? Green Chem 4:521.
Cross FB (1996) Paradoxical perils of the precautionary principle. Wash. & Lee Law Rev., 53:851.
Cross FB (1998) Facts and values in risk assessment. Reliability Eng. System Safety 59:27.
Cvetkovich G and Earle TC (1992) Environmental Hazards and the Public. J. Social Issues 48:1.
Davies AR (2001) Is the Media the Message? Mass Media, Environmental Information and the Public. J Environ Policy Plann 3:319.
Davis I (1935) The measurement of scientific attitudes. Sci Educ 19:117, quoted in Miller JD (1983) Scientific literacy: a conceptual and empirical review. Daedalus 112:29.

Dewulf J Van Langenhove H Mulder J van den Berg MMD van der Kooi HJ and de Swaan Arons J (2000) Illustrations toward quantifying the sustainability of technology. Green Chem 2:108.

Dewulf J and Van Langenhove H (2002) Assessment of the Sustainability of Technology by Means of a Thermodynamically Based Life Cycle Analysis. Environ Sci Pollut Res 9:267.

Dryzek JS and Torgerson D (1993) Editorial: Democracy and the policy sciences: A progess report. Policy Research 26:127.

Durodié B (2003) The true cost of precautionary chemical regulation. Risk Anal. 23:389.

Editorial (2002) Leadership at Johannesburg. Nature 418:803.

Ehrenfeld JR (2002) Industrial Ecology: Coming of Age. Environ Sci Technol 36:281A.

Ehrlich PR (1994) Ecological Economics and the Carrying Capacity of the Earth in Jansson AM Hammer M Folke C and Constanza R (1994) eds. Investing in Natural Capital: The Ecological Economics Approach to Sustainability. Island Press, Washington, DC.

ESRC, 2002, The views of science journalists: a survey and some additional reflections, www.esrc.ac.uk/esrcccontent/PublicationsList/whom/views.html.

Feynman RP (2001) The Pleasure of Finding Things Out: The Best Short Works of Richard P. Feynman, edited by J. Robbins. Penguin Books, London, p188.

Forman P (1995) Truth and Objectivity Part 1: Irony. Science 269:565.

Forman P (1995a) Truth and Objectivity Part 1: Trust. Science 269:707.

Foster KR (2002) The Precautionary Principle – common sense or environmental extremism. IEEE Technol. Soc. Mag. Winter 2002/2003:8.

Foster KR, Vecchia P and Repacholi MH (2000) Science and the precautionary principle. Science 288:979.

Freudenburg WR (1988) Perceived risk, real risk: social science and the art of probabilistic risk assessment. Science 242:44

Freundenburg WR (2001) Risky thinking: facts, values and blind spots in societal decisions about risk. Reliability Eng. System Safety 72:125.

Frewer LJ, Miles S and Marsh R (2002) The media and genetically modified foods: evidence in support of social amplification of risk. Risk Anal. 22:701.

Furedi F (1997) Culture of Fear: Risk Taking and the Morality of Low Expectations. Cassel, London.

Furedi F (2001) The Blame Game. New Sci 171:48.

Geistfeld M (2001) Reconciling cost-benefit analysis with the principle that safety matters more than money. New York U Law Rev. 76:114.

Giles J (2002) When doubt is a sure thing. Nature 418:476.

Glaze WH (2001) Hyperdisciplinarity and Environmental Studies. Environ Sci Technol 35:471A.

GM Science Review Panel (2003) GM Science Review First Report: An open review of the science relevant to GM crops and food based on interests and concerns of the public. http://www.gmsciencedebate.org.uk.

Goldenfeld N and Kadanoff LP (1999) Simple Lessons from Complexity. Science 284:87.

Graedel TE and Allenby BR (1995) Industrial Ecology. Prentice-Hall, Englewood Cliffs, NJ.

Greenberg DS (1969) British AAS: Counterattack on Gloom about Science and Man. Science 165:1239.

Gross PR Levitt N and Lewis MW (1997) eds The Flight from Science and Reason. New York Academy of Sciences, New York.

Gross PR and Levitt N (1994) Higher Superstition: the Academic Left and its Quarrel with Science. The Johns Hopkins University Press, Baltimore, MD.

Hammond K and Adelman L (1976) Science, values and human judgement. *Science* 194:389.

Hampton G (1999) Environmental equity and public participation. Policy Sci. 32:163.

Harremoës P Gee D MacGirvan M Stirling A Keys J Wynne B and Guedes Vaz S (2002) eds. The Precautionary Principle in the 20th Century: Late Lessons from Early Warnings. Earthscan Publications Ltd, London.

Heinzerling L (2000) The Rights of Statistical People. Harvard Environ Law Rev 24:189.

Hobsbawm E (1993) The New Threat to History. New York Review of Books, 16 December 1993, p62, quoted in Sokal A and Bricmont J (1998) Intermezzo: epistemic relativism and the Philosophy of Science, in Intellectual Impostures. Profile Books, London, p95.

Hohenemser C Kates RW and Slovic P (1983) The nature of technological hazard. Science 220:378.

Huber J (2000) Towards Industrial Ecology: Sustainable Development as a Concept of Ecological Modernization. J Environ Policy Plann 2:269.

Humphreys P (1997) Emergence, not Supervenience. Philos Sci 64:S337.

Irwin A (1995) Citizen Science: A Study of People, Expertise and Sustainable Development. Routledge, London.

Jauchem JR (1992) Epidemiological studies of electric and magnetic fields and cancer: a case study of distortions by the media. J. Clin. Epidemiol. 45:1137.

Kasperson RE Renn O Slovic P Brown HS Emel J Goble R Kasperson JX and Ratick S (1988) The social amplification of risk – a conceptual framework. Risk Anal. 8:177.

Kates RW and Parris, TM (2003) Long-term trends and a sustainability transition. Proc. Natl. Acad. Sci. USA 100:8062.

Kates RW (1989) The Great Questions of Science and Society do not Fit Neatly into Single Disciplines. The Chronicle of Higher Education XXXV (36) B1, B3.

Kates RW (2002) Humboldt's Dream, Beyond Disciplines and Sustainability Science: Contested Identities in a Restructured Academy. Ann Assoc Am Geog 92:75.

Kates RW Clark WC Corell R Hall JM Jaeger CC Lowe I McCarthy JJ Schellnhuber HJ Bolin B Dickson NM Faucheux S Gallopin GG Grubler A Huntley B Jager J Jodha NS Kasperson RE Mabogunje A Matson P Mooney H Moore III B O'Riordan T and Svedin U (2001) Policy Forum: Environment and Development: Sustainability Science. Science 292:641 and subsequent correspondence: Swart R Raskin P and Robinson J (2002) Critical challenges for sustainability science. Science 297:1994; Kates R and Clark WC (2002) Response, Science 297:1994.

King DM (1994) in Jansson AM Hammer M Folke C and Constanza R (1994) eds Investing in Natural Capital: The Ecological Economics Approach to Sustainability. Island Press, Washington, DC, p325.

Kinzig AP Starrett D Arrow K Aniyar S Bolin B Dasgupta P Ehrlich P Folke C Hanemann M Heal G Hoel M Jansson AM Jansson B-O Kautsky N Levin S Lubchenko J Maler K-G Pacala SW Schneider SH Siniscalco D and Walker B (2003) Coping with Uncertainty: A Call for a New Science-Policy Forum. Ambio 32:330.

Klinke A and Renn O (2002) A new approach to risk evaluation and management: risk-based, precaution-based and discourse-based strategies. Risk Anal. 22:1071.

van Kolfschooten F (2002) Can you believe what you read? Nature 416:360.

Langhelle O (2000) Why Ecological Modernization and Sustainable Development should not be Conflated. J Environ Policy Plann 2:303.
Laughlin RB and Pines D (2000) The Theory of Everything. Proc Nat Acad Sci USA 97:28.
Laughlin RB Pines D Schmalian J Stojkovich BP and Wolynes P (2000) Proc Nat Acad Sci USA 97:32
Lee K Holland A and McNeil D (2000) eds Global Sustainable Development in the 21st Century. Edinburgh University Press, Edinburgh.
Lélé SM (1991) Sustainable Development: a Critical Review. World Develop 19:607.
Lemons J (1996) ed. Scientific Uncertainty and Environmental Problems. Blackwell Science, Oxford, UK.
Lesher A (2002) Science and Sustainability. Science 297:897.
Lovelock J (1995) Gaia: a New Look at Life on Earth. Oxford University Press, Oxford.
Macleod D (2002) Academic calls for 'strike' over environment. The Guardian, Tuesday, 27 August, 2002.
Madsen KM Hviid A Vestergaard M Schendel D Wohlfahrt J Thorsen P Olsen J and Melbye (2002) A Population-Based Study of Measles, Mumps and Rubella Vaccination and Autism. New Eng J Med 347:1477.
Malone EL and Rayner S (2001) Role of the Research Standpoint in Integrating Global-scale and Local-scale Research. Clim Res 19:173.
Marsh B and Irwin J (2002) New Alert on MMR Jab. The Daily Mail, 6 February 2002.
Massa I and Andersen MS (2000) Special issue introduction: Ecological Modernisation. J Environ Policy Plann 2:265.
May R (2003) GM warriors have killed the debate. Guardian, November 23 2003.
Mayer PS and Ausubel JH (1999) Carrying Capacity: a Model with Logistically Varying Limits. Technol Forecast Soc Change 61:209.
Merkel A (1998) The Role of Science in Sustainable Development. Science 281:336.
Merton RK (1973) The Sociology of Science. University of Chicago Press, Chicago, IL.
Miller JD (1983) Scientific Literacy: a Conceptual and Empirical Review. Daedalus 112:29.
Morris N (2001) Foot and Mouth: New Fears of Human Infection. The Independent, Wednesday, 25 April 2001.
Mulkay MJ (1995) Sociology of Science – A Sociological Pilgrimage. Open University Press, Milton Keynes, UK, p xix, quoted in Irwin A (1995) p181.
Murray I (2003) Comment: Environmental scientists must stop crying wolf. Financial Times, 17 September 2003.
Nelkin D (1975) The Political Impact of Technical Expertise. Social Studies of Science 5:37.
Office of Science and Technology (2002) Science and the Public: A Review of Science Communication and Public Attitudes to Science and Britain. A Joint Report of the Office of Science and Technology and The Wellcome Trust, October 2000 (www.wellcome.ac.uk).
Okrent D and Pidgeon N (1998) Editorial: risk perception versus risk analysis. Reliability Eng. System Safety 59:1.
Okrent D and Pidgeon N (2001) Guest editors' introduction: risk perception, policy and democracy. Reliability Eng. System Safety 72:113.
O'Neil O (2002) A Question of Trust. BBC Reith Lectures 2002, (www.bbc.co.uk/radio4/reith2002/).

Pacala SW, Bulte E, List JA and Levin SA (2003) False Alarm over Environmental False Alarms. Science 301:1187.

Parris TM and Kates RW (2003) Characterising a sustainability transition: goals, targets, trends and driving forces. Proc. Natl. Acad. Sci. USA 100:8068.

Patzig G (2002) Can Moral Norms be Rationally Justified? Angew Chem Int Edn 41:3353.

Pinker S (2002) The Blank Slate: The Modern Denial of Human Nature. Allen Lane (Penguin)/Viking, London.

Plato [1994] Gorgias: a New Translation by Robin Waterfield. Oxford World Classics, Oxford University Press, Oxford, UK.

Porritt J (2000) Playing Safe: Science and the Environment. Thames and Hudson, London.

Postman, N (1993) Technopoly: The Surrender of Culture to Technology. Vintage Books, New York.

Purdue D (1999) Experiments in Governance of Biotechnology: a Case Study of the UK National Consensus Conference. New Genetics and Society 18:79.

Radford T (2002) News Scientist. Guardian, 4 September 2002; ESRC, 2002, Media has little impact on public knowledge.

Raucher R (2000) quoted in Freeze RA (2000) The Environmental Pendulum. University of California Press, p140.

Raven PH (2002) Science, Sustainability and the Human Prospect. Science 297:954.

Ravetz RL (1999) Environmental Regulation, Cost-Benefit Analysis and the Discounting of Human Lives. Columbia Law Rev 99:941.

Rawls J (1972) A Theory of Justice. Clarendon Press, Oxford.

Rayner S (2002) We know enough. The Guardian, Monday, 2 September 2002.

Rees RE and Wackernagel M (1994) Ecological Footprints and Appropriated Carrying Capacity: Measuring the Natural Capital Requirement of the Human Economy, in Jansson AM Hammer M Folke C and Constanza R (1994) eds. Investing in Natural Capital: The Ecological Economics Approach to Sustainability. Island Press, Washington, DC.

Renn O Burns WJ Kasperson JX Kasperson RE and Slovic P (1992) The social amplification of risk: theoretical foundations and empirical applications. J. Social Issues 48:137.

Renn O Webler T Rakel H Dienel P and Johnson B (1993) Public participation in decision making: A three-step procedure. Policy Research 26:189.

Renn O (1998) The role of risk perception for risk management. Reliability Eng. System Safety 59:49.

Renn O (1999) A model for an analytic-deliberative process in risk management. Environ. Sci. Technol. 33:3049.

Renn O (2001) The need for integration: risk policies require the input from experts, stakeholders and the public at large. Reliability Eng. System Safety 72:131.

Research Councils UK (2002) Dialogue with the public: Practical guidelines. Research Councils UK August, 2002. (www.research-councils.ac.uk/guidelines/public/).

Robèrt K-H (2000) Tools and Concepts for Sustainable Development, How do they Relate to a General Framework for Sustainable Development, and to Each Other. J Cleaner Prod 8:243.

Robèrt K-H Schmidt-Bleek,B Aloisi de Larderel J Basile G Jansen JL Kuehr R Price Thomas P Suzuki M Hawken P and Wackernagel M (2002) Strategic Sustainable Development – Selection, Design and Synergies of Applied Tools. J Cleaner Prod 10:197.

Rosen MA and Dincer I (1997) On Exergy and Environmental Impact. Int JEnergy Res 21:643.

Royal Society (2003). Theme issue: The Farm Scale Evaluation of Spring-sown Genetically Modified Crops. Phil. Trans. R. Soc. Lond. B. 358:1775.
Schulze WD and Kneese AV (1981) Risk in Benefit-Cost Analysis. Risk Anal. 1:81.
Schummer J (2003) The Philosophy of Chemistry. Endeavour 27:37.
Science Media Centre (2002) Consultation Report. March 2002 (www.sciencemediacentre.org/rismc/index.jsp).
Seager TP and Theis TL (2002) A Uniform Definition and Quantitative Basis for Industrial Ecology. J Cleaner Prod 10:225.
Segerstråle U ed (2002) Beyond the Science Wars: The Missing Discourse about Science and Society. SUNY Press, Albany, NY.
Seippel Ø (2000) Ecological Modernization as a Theoretical Device: Strengths and Weaknesses. J Environ Policy Plann 2:287.
Shamos MH (1995) The Myth of Scientific Literacy. Rutgers University Press, NJ.
Shermer M. (2002) Skepticism as a Virtue. Sci Am April 2002.
Siebenhüner B (2002) How do Scientific Assessments Learn? Part 1 Conceptual Framework and Case Study of the IPCC. Environ Sci Pol 5:411.
Siebenhüner B (2002) How do Scientific Assessments Learn? Part 2 Case Study of the LRTAP Assessments and Comparative Conclusions. Environ Sci Pol 5:421.
Singer E (1990) A Question of Accuracy: How Journalists and Scientists Report Research on Hazards. J .Commun. 40:102.
Slovic P (1987) Perceptions of risk. Science 236:280.
Snow CP [1959](1998) The Two Cultures. Cambridge University Press, Cambridge.
Sokal A and Bricmont J (1998) Intellectual Impostures. Profile Books, London.
Sonnert G. with Holton G (2002) Ivory Bridges: Connecting Science and Society. MIT Press, Cambridge, Mass.
Starr C (1969) Social benefit versus technological risk. Science 165:1232.
Starr C (1996) Sustaining the Human Environment: the Next Two Hundred Years. Daedalus 125:235.
Stirling A and Gee D (2002) Science, precaution and practice. Public Health Rept. 117:521.
Stott P (2002) 'Sustainable Development' is just dangerous nonsense. The Daily Telegraph, 16 August, 2002.
Sustein CR (2003) Beyond the precautionary principle. Univ. Penn. Law Rev. 151:1003.
Tait J (2001) Faust or Frankenstein: the Role of Interests and Values in Risk Assessment, Perception and Communication. BA Festival of Science, University of Glasgow, 3-7 September 2001.
Tait J (2001) Faust or Frankenstein: the European Debate about the Precautionary Principle and Risk Regulation for Genetically Modified Crops. J Risk Res 4:175.
Thornton J (2000) Pandora's Poison: Chlorine, Health and a New Environmental Strategy . MIT Press, Cambridge, Mass.
Trudgill S (2001) Psychobiogeography: Meanings of Nature and Motivations for a Democratised Conservation Ethic. J Biogeography 28:677.
Turner II BL Kasperson RE Matson PA McCarthy JJ Corell RW Christensen L Eckley N Kasperson JX Luers A Martello ML Polsky C Pulsipher A and Schiller A (2003) A framework for vulnerability analysis in sustainability science. Proc. Natl. Acad. Sci. USA 100:8074.
Turner II BL Matson PA McCarthy JJ Corell RW Christensen L Eckley N Hovelsrud-Broda GK Kasperson JX Kasperson RE Luers A Martello ML Mathiesen S Naylor R Polsky C Pulsipher A Schiller A Selin H and Tyler N (2003) Illustrating the couples

human-environment system for vulnerability analysis: three case studies. Proc. Natl. Acad. Sci. USA 100:8080.

Vicsek T (2002) The Bigger Picture. Nature 418:131.

Vollenbroek FA (2002) Sustainable Development and the Challenge of Innovation. J Cleaner Prod 10:215.

Wackernagel M Schulz NB Deumling D Callejas Linares A Jenkins M. Kapos V Monfreda C Loh J Myers N Norgaard R and Randers J (2002) Tracking the Ecological Overshoot of the Human Economy. Proc Nat Acad Sci USA 99:9266.

Waggoner PE (1996) How Much Land can Ten Billion People Spare for Nature?. Daedalus 125:73.

Waggoner PW and Ausubel JH (2002) A framework for Sustainability Science: a Renovated IPAT Identity. Proc Nat Acad Sci USA 99:7860.

WCED (1987) World Commission on Environment and Development, Our Common Future. Oxford University Press, Oxford.

Weaver P Jansen L van Grootveld G van Spiegel E and Vergragt P (2000) Sustainable Technology Development. Greenleaf Publishing, Sheffield, UK.

Weidner H (2002) Capacity Building for Ecological Modernization: Lessons from Cross-National Research. Am Behav Scientist 45:1340.

Weinberg S (2001) Facing Up: Science and its Cultural Adversaries. Harvard University Press, Cambridge, Mass.

Wilson EO (1998) Consilience: The Unity of Knowledge. Knopf, New York.

Wilson EO (2002) The Future of Life. Little Brown, London.

Winterton N (2000). Chlorine – the only green element: towards a wider acceptance of its role in natural cycles. Green Chem 2:173.

Winterton N (2001) Science, scientists and sustainability. Clean Prod Proc 3:62.

Winterton N (2003) Sense and Sustainability: the Role of Chemistry, Green or Otherwise. Clean Technol Environ Policy 5:8.

Wolpert L (1992) The Unnatural Nature of Science. Faber and Faber, London.

World Economic Forum (2000) 2002 Environmental Sustainability Index: An Initiative of the Global Leaders of Tomorrow Environmental Task Force, World Economic Forum, Annual Meeting 2002.

Ziman J (2001) Real Science: What it is, and what it means. Cambridge University Press, Cambridge, UK.

Ziman J (2002) Postacademic Science: Constructing Knowledge with Networks and Norms in U. Segerstråle, ed., Beyond the Science Wars: The Missing Discourse about Science and Society. SUNY Press, Albany, NY, USA, Chapter 7, p135.

Sustainability: Ecological, Social, Economic, Technological, and Systems Perspectives

Heriberto Cabezas, Christopher W. Pawlowski[1], Audrey L. Mayer and N. Theresa Hoagland

U.S. Environmental Protection Agency, National Risk Management Research Laboratory, Sustainable Technology Division, Sustainable Environments Branch, 26 West Martin Luther King Drive, Cincinnati, OH 45268, USA

Abstract

Sustainability is generally associated with a definition by the World Commission on Environment and Development, 1987: "... development that meets the needs and aspirations of the present without compromising the ability to meet those of the future ..." However, there is no mathematical theory embodying these concepts, although one would be immensely valuable in humanity's efforts to manage the environment. The concept of sustainability applies to integrated systems comprising humans and the rest of nature; the structures and operation of the human component (society, economy, law, etc.) must be such that they reinforce the persistence of the structures and operation of the natural component (ecosystem trophic linkages, biodiversity, biogeochemical cycles, etc.). One of the challenges of sustainability research lies in linking measures of ecosystem functioning to the structure and operation of the associated social system. We review the nature of this complex system including its ecological, social, economic, and technological aspects, and propose an approach to assessing sustainability based on Information Theory that bridges the natural and human systems. These principles are then illustrated using a model system with an ecological food web linked to a rudimentary social system. This work is part of the efforts of a larger multidisciplinary group at the U.S. Environmental Protection Agency's National Risk Management Research Laboratory.

---

[1] Postdoctoral Research Fellow, Oak Ridge Institute for Science and Education.

1 Introduction

While it has proven difficult to develop a detailed consensus around the concept of sustainability, there is increased recognition that the current growth of human activity cannot continue without significantly overwhelming critical ecosystems. The Brundtland Commission (World Commission on Environment and Development 1987) defined sustainable development as "development that meets the needs of the present without compromising the ability of future generations to meet their own needs." This statement addresses the concern over the extent to which ecosystems can continue to provide functions and services into the future, given the activities of human societies.

The concept of sustainability applies to integrated systems comprising humans and nature. The structures and operation of the human component (in terms of society, economy, government etc.) must be such that these reinforce or promote the persistence of the structures and operation of the natural component (in terms of ecosystem trophic linkages, biodiversity, biogeochemical cycles, etc.), and vice versa. Thus, one of the challenges of sustainability research lies in linking measures of ecosystem functioning (especially those that are desired by human societies) to the structure and operation of the ecosystem itself (DeLeo and Levin 1997). Sustainability indicators for these systems are needed to improve our understanding of the nature of human demands on ecosystems, and the extent to which these can be modified. Information Theory may prove useful to the development of indicators that bridge the natural and human systems and make sense of the disparate state variables of these systems.

Changes in system regime then are important from the point of view of ecosystem sustainability, since we generally want to sustain those ecosystem regimes that provide the most useful functions and services to humans. We have developed a form of Fisher Information that characterizes order in a dynamic system's steady state. Changes in steady state variation indicate changes or disturbance in regime or changes in internal dynamics–changes in the relationship or organization of the system elements. Such information could be used with other knowledge about the system (what regime changes are possible, which if any of these are desirable or undesirable, etc.) to assess what changes are occurring and what can be done to avoid any expected negative consequences. If a change in regime is occurring due to disturbances of an anthropogenic nature, we are more likely to be able to mitigate them, precisely because they are anthropogenic.

To investigate the interaction between human activity and ecosystem components we discuss the salient aspects of sustainability and ecology, society, economy, technology, and finally systems at the simplified model level. Hence, we introduce a deterministic model of a relatively simple ecosystem with an integrated, rudimentary human culture, and we explore a series of simulated scenarios. We examine the use of Fisher Information to assess the order or organizational changes in this model system under different disturbance events. Fisher Information is used to indicate changes in the pattern of system state variables, or at least the occurrence of changes. Fisher Information collapses the many state variables into one index that measures overall system order.

2 Sustainability and ecology

Ecological sustainability inherently requires that ecosystems retain their ability to function through environmental changes and, if necessary, provide the evolutionary capacity (through genetic and species diversity) to form more appropriate ecosystem structures and functions as environmental conditions change. Ecosystems consist of species that operate in fairly characteristic trophic (feeding) levels, which engage in a wide variety of interactions, from wholly negative competition to wholly positive symbiosis. At present, higher species diversity exists in tropical areas near the Equator, particularly in about 25 areas known as "hotspots", which have high numbers of endemic species (Myers et al. 2000; Brooks et al. 2002). Little is understood about the importance of the full compliment of diversity, from one-celled bacteria to the largest mammalian predator, to the stability and productivity of these ecosystems (Tilman 1999; Bond and Chase 2002). However, due to habitat loss, over-harvesting, invasion by non-native species, and other anthropogenic pressures, many species are currently at risk of extinction (Ehrlich 1995). As species are lost from systems, the connectivity (and perhaps redundancy) of the system declines, and critical functions (such as nitrogen fixation or pollination) may no longer be provided (Kearns 1997). As more land is appropriated for human use, extinction rates have generally risen proportionately to the area of natural habitat damaged or destroyed (MacArthur and Wilson 1967; Doak and Mills 1994; Pimm and Askins 1995). As more species decline to extinction or low population levels, the evolutionary capacity of ecosystems decline, reducing their ability to adapt to changing environments.

Loss of biodiversity in ecosystems would only be sustainable if immediate restoration of these systems were possible, but this is unlikely to ever be the case. Global extinction of a species is irreversible at present, and reintroduction of species from population centers to areas of extirpation is usually unsuccessful and often very expensive (Cade 1988). Few species in the United States have been removed from the endangered species list due to recovery, while many more have gone extinct waiting for an endangered status listing (U.S. Department of Interior 1990). The restoration of ecosystems depends heavily on the order of species reintroduction to a system. Community assembly rules, resistance to invasion, and individual species ecological characteristics are all critical ecosystem components to restoration efforts, but are at best vaguely understood for a few ecosystems (Levin et al. 2001; Sakai et al. 2001; Ferenc et al. 2002).

Ecosystems are complex, dynamic systems that often display characteristic regimes of behavior dictated by their internal dynamics and the disturbances that act on them (Scheffer et al. 2001). A regime is delineated by a particular multidimensional neighborhood or set of values within which the system state, e.g., the species populations, fluctuates. Each set forms a basin of attraction about a particular steady state, meaning that, absent changes in external disturbances or random variations within the system, the system will remain within the basin and, moreover, will approach the steady state. Disturbances can range in size, intensity and frequency, and can originate from natural (e.g., lightning strike fires, floods) or anthropogenic sources (e.g., agriculture, deforestation). The size of disturbance that can be tolerated by an ecosystem before a change in regime occurs is a measure of the ecosystem's resilience to disturbances (Holling 1973, 1996; Gunderson 2000). Although ecosystems may naturally pass through many regimes, the functions and services that ecosystems can provide human societies under these can vary (Wardle et al. 2000; Portela and Rademacher 2001). Hence, in this respect, some regimes may be more desirable to humans than others.

3 Sustainability and society

Society as an element of sustainability must reflect the actual interaction between humans and the biological system. For this, a historical viewpoint is necessary. Diamond (1999) traces the development of human societies from the primitive to the modern. Essential to that development was the domestication of animals and plants, which led to an increase in human population and density. It also led to a surplus of food resulting in job specialization, centralized conflict resolution and decision-making, collection and redistribution of wealth, technology development, and amalgamation of smaller territories into larger ones. Even today, according to Diamond, of the 20,000 domesticated plant species, only 12 make up 80 percent of the world's tonnage. Six of the 12 grains are cereals and more than one half of all calories consumed are from cereals. Only five large mammals of global importance have been domesticated. In short, the human control of a few key plant and animal species in the food web has over the last 10,000 years resulted in human domination of the planet's resources and fate.

The importance of domestication in the food web cannot be underestimated. However, humans also consume non-domesticated food, such as edible wild plants, seafood and other fish and game species. There are also plant and animal species that are an important part of the food web, but humans neither domesticate nor consume them. Thus, from the human perspective, the food web can be divided into two branches–domesticated and non-domesticated. The social, legal, and economic ramifications of this division are substantial. For example, domesticated food sources today are for the most part private property, produced and traded as commodities on the open market. Plants may be private property by virtue of their fixed nature, but non-domesticated animal and fish species are either regulated or protected by the government or free for the taking (Grosse 1992). Thus, relationships within the domesticated branch of the food web are substantially affected by society's economic system. Relationships in the non-domesticated branch are more often affected by the legal and political systems or by biological interactions.

Humans are also able to appropriate resources before they become part of the food web (soil beneath asphalt is no longer available for plant growth; water consumed by industry or contaminated by pollutants is no longer available for drinking). In fact, humans *must* appropriate these resources in order to support the infrastructure needed to sustain the level of development brought about by domestication. As an example, according to Meadows et al. (1972) "to achieve a 34 percent increase in world food production from 1951–1966, agriculturists increased yearly expenditures on tractors by 63 percent, annual investment in nitrate fertilizers by 146 percent and annual use of pesticides by 300 percent. The next 34 percent increase will require even greater inputs of capital and resources." This is in addition to the massive physical infrastructure and energy needed to transport, process, and distribute the agricultural products (if we consider the food web to include consumption as well as production, we must take into account that centralized food production necessitates a distribution system in order for consumption to occur).

There is a related aspect, as well. Societies are problem-solving entities and tend to increase organizational complexity in order to solve problems. As a society evolves toward greater complexity, the support costs levied on each individual increases and the population as a whole must allocate an increasing portion of its energy and resource budget to maintaining organizational institutions. The idea of infinite substitutability–that we can always find alternatives through technology–does not apply to investment in organizational complexity because there is no substitute (Tainter 1988).

Thus, human society affects the food web and sustainability in several ways. It manipulates the food web in favor of domesticated species, potentially affecting resources available to non-domesticated species. It appropriates portions of the resource pool for physical and social infrastructure, and it raises the bar on sustainability by continually increasing its needs through population growth made possible by food surpluses brought about by manipulation of the food web.

4 Sustainability and economy

All economic activity is dependent upon natural resources provided by the environment, both as inputs and as sinks for pollution treatment. These resources can either be renewable (such as fish populations, timber stands, or the capacity of the environment to absorb pollution) or non-renewable (such as coal or petroleum) on the time scale of human existence. Renewable resources can be exhausted if harvest rates are too high (Slade 1982; Reed 1986; Pauly et al. 2002). Depletion rates of natural resource stocks are often dictated by the scarcity of the stock (which influences its price), instead of the biological limits of regeneration of the stock, which may lead to extraction and harvest rates for renewable resources that are not sustainable (Daly 1996). The identification of feedback loops between the productivity of ecosystems and economic systems has become important in many fields, spawning the discipline of ecological economics. Feedback loops between ecological and economic systems can significantly alter projected resource availability and economic productivity, and are important in determining sustainable resource use (Settle et al. 2002). Economists have also begun to internalize pollution and other negative impacts to the environment in economic analyses, which once treated these impacts as externalities (Samuelson 1954; Freeman 1984; Bird 1987).

The goods and services that ecological systems provide to human economies and societies are rarely directly valued in economic markets (Daily 1997). Several studies have attempted to make the costs of these services explicit, and estimate that the planetary value of all ecosystem services is around U.S.$16–54 trillion per year (Costanza et al. 1997). Some of these estimates are based on the cost of designing and building a technological system that could provide the same services, such as reverse osmosis or desalination technology to produce freshwater in the place of wetlands (Postel and Carpenter 1997). Other estimates are based on a variety of hedonic pricing, contingent valuation or other, more indirect methods (Farber et al. 2002). The purpose for valuing ecosystem goods and services is not necessarily to include them in regular market transactions, but rather to provide an accounting system that is recognizable to economic analysts and that can be used to monitor rapid depletion or unsustainable use of these goods and services.

In many analyses of economic sustainability, the discount rate chosen and the substitutability of environmental capital for other resources and for manmade capital can greatly influence the perceived sustainability of an activity. Higher discount rates cause firms to extract natural resources faster earlier, saving less for later, and thus leaving a reduced (or exhausted) resource base left for future generations (Costanza et al. 1997; Barbier and Markandya 1998). On the other hand, resources for which there are many substitutable sources may be more likely to be used sustainably, as economic agents can more easily shift from one resource to another when prices rise (due to increasing scarcity). However, some resources have no substitution, such as breathable air or fresh water (without these, humans and most other life forms would die). Economists are at odds regarding the degree of substitutability between natural and manmade capital. Of the economic valuation techniques described above, some assume that technology can design systems that are perfectly substitutable for natural systems in terms of the goods and services produced, although it has long been recognized that this may not be the case in many instances (Krutilla 1967).

5 Sustainability and technology

Technology has a major role to play in sustainability because it provides the means by which humans take resources from the environment and transform them to meet their needs. Further, the size of the human population practically guarantees that any technology that is widely adopted will use and transform large amounts of resources with consequent impacts on the environment. New technologies are constantly making new resources available or improving access to currently available resources. Technology, however, comes with a price: the environmental impact of the processes involved in the manufacture of the technology itself, the processes involved in accessing resources made available through the technology, or both. Moreover, technology is subject to social, political and economic forces that determine the extent of the uses to which technology is put and its penetration in the society. Any technology therefore must be judged on its environmental cost as well as its potential benefit in the social, economic and legal framework within which it resides. Depending on how it is used and marketed, technology can change the way we view nature and the degree to which we rely on its services, and thus our impacts on it.

One driver for technology development is efficiency. While efficient processes make it less expensive to manufacture products, efficiency plays a role in sustainability as well by reducing the amount of resources going into a product. With increased awareness of ecological impacts of various activities, technology also often is specifically developed to mitigate such impacts. For example, one may seek to replace older technologies with less polluting processes, or processes that use environmentally safer materials. More efficient and lower impact technologies in themselves, however, are not a guaranteed route to sustainability since overall this does not necessarily mean lower or better use of resources. If economic, social and legal systems are not carefully developed to coincide with sustainability goals, resource use and pollution could increase despite the availability of appropriate technologies

The development and adoption of technology as well as our use of resources is driven by various factors. The "best" technology is not necessarily that which is adopted–market phenomena such as lock-in, social forces and politics can impede the adoption of appropriate technologies even if these are available. Improved efficiency often means products can be offered at lower prices, increasing overall demand. Technology can also make us feel isolated or independent of the environment and thus free to do as we please, when in fact this is not the case. The impacts of technology–environmental, social, economic or otherwise–often are not fully known or not necessarily even studied. We must understand not only our technologies, but the social, political, legal and economic systems within which they operate.

6 Sustainability and systems

Perhaps the most subtle but critically important aspect of sustainability is the fact that it is not a feature of the parts of the system, but of the whole as already mentioned above. What is important for sustainability is that the operation of the elements making up the system sustain each other, in the same way that the organs that make up a body work together to sustain the person. A system cannot be sustainable with a major sub-system (economy, ecology, law, etc.) operating without regard to the rest of the system any more than a person can live with a malfunctioning major organ. We illustrate the concept with a diagram that depicts the path of a complex system in time as a trajectory in a multidimensional space (Fig. 1).

We further illustrate that within certain limits, here represented as a tunnel, the system is deemed to be sustainable, and not sustainable outside those limits. The important point is that the system can become unsustainable along any of its many dimensions. There can further be two types of deviations from sustainable conditions: a self-correcting and a non-self-correcting deviation that essentially represents a catastrophe in progress. It is obvious that the latter is far more serious than the former.

The need to understand the entire system presents a challenge for the scientific investigation of sustainability because it then becomes necessary to look for approaches that are applicable across a broad range of disciplines. This is difficult because most of the measurable variables, principles, and criteria commonly used in science are discipline specific, e.g., there is no exact economic equivalent to the second law of thermodynamics. There are some similarities between concepts among different disciplines, but the mapping across disciplines is not sufficiently accurate for it to form the basis of reliable principles. Therefore, measures and principles are needed that are applicable to the entire range of systems and sub-systems. We feel that we have found a way to address this last problem through the application of Information Theory, or more exactly Fisher Information. The reason is that it is always possible to express the content of any measurement of any variable regardless of disciplinary origin in terms of information. In this paper we present a form of Fisher Information that is a generalized measure of order in the system, and we show its application to a model system that includes rudimentary social and ecological sub-systems. We also explore several simulated scenarios of interest, and we discuss their application to sustainability.

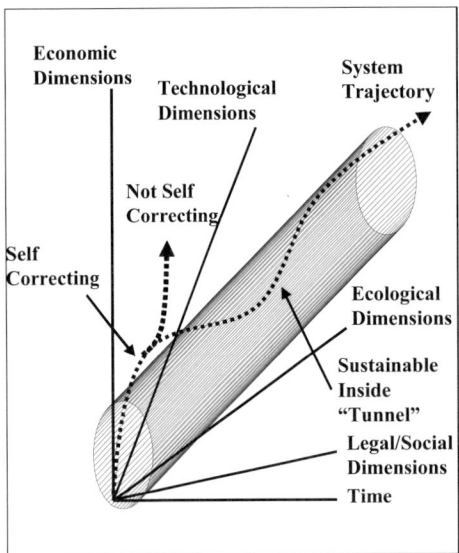

Fig. 1. Path through time for a complex eco-socio-economic system

7 Information theory

Information Theory provides a quantitative framework by which to describe processes that admit only partial knowledge. Ronald Fisher (1922) developed a statistical measure of indeterminacy now called Fisher Information. Fisher Information can be variously interpreted as a measure of the ability to estimate a parameter, as the amount of information that can be extracted from a set of measurements (the "quality" of the measurements), and, for the specific form of the Fisher Information that we use, a measure of the state of order of a system or phenomenon (Frieden 1998). This form of information is now finding its way into the scientific literature (Cabezas and Fath 2002, Fath et al. 2003). Fisher Information, I, for a single measurement of one variable is calculated as follows:

$$I = \int \frac{1}{p(\varepsilon)} \left[\frac{dp(\varepsilon)}{d\varepsilon} \right]^2 d\varepsilon \qquad (1)$$

Here $p(\varepsilon)$ is the probability density as a function of the deviation, ε, from the true value of a variable y being measured. Fisher Information takes into account the changes in the probability density shape that result from a re-ordering over the independent variable, since it involves a derivative term. This sensitivity can be useful for situations in which there is a notion of ordering. Such a situation arises in dynamic systems, for which time is a natural ordering variable.

To calculate Fisher Information, a probability density function (PDF) for the system in question must be determined. We assume that (1) the system behavior can be captured in a continuous dynamic system description; and (2) the system is in a periodic steady state. A single PDF can be identified that is based on the probability of finding the system in a given state from within a set of possible states. A state of the system is defined by a particular set of values for the state variables of the system. The longer a system is in a specific state, the more likely one is to find it in that state when sampling. The time spent in each state is calculated from the system acceleration and velocity in state space, i.e., the space defined by the state variables of the system. When normalized over the entire space of possibilities, a probability density function for the states of the system results. Fisher Information is then interpreted as a measure of the variability in the time the system spends in the various sections of its steady state trajectory (or, since time and speed are inversely related, variation in the speed of the state over its steady state trajectory). A system steady state with a uniform speed would have zero information, as one is equally likely to observe any of the states along the path. For equilibrium systems the information is infinite, since we are assured of observing the system in only one state. This relates to system order because an ordered dynamic system has biases towards specific states and a disordered system does not. A more complete derivation of the expressions given below can be found in Fath et al. (2003). The summary results below give an expression for the Fisher Information as a function of ratios of acceleration to velocity *along the path* of the system in its state space. Thus, the integral of Eq. 1 becomes:

$$I = \frac{1}{T}\int_0^T \frac{[\ddot{R}(t)]^2}{[\dot{R}(t)]^4} dt \qquad (2)$$

where $\dot{R}(t)$ and $\ddot{R}(t)$ are respectively the tangential speed and (scalar) acceleration of the system at time t, and where the integration is performed over a characteristic period of time T representing one or an integral number of cycles of the system. The Fisher Information in component vector notation is:

$$I = \frac{1}{T}\int_0^T \frac{[(\dot{y}(t))^T \ddot{y}(t)]^2}{\|\dot{y}(t)\|^6} dt \qquad (3)$$

where $\dot{y}(t)$ and $\ddot{y}(t)$ are velocity and (vector) acceleration and obey:

$$\dot{R}(t) = \|\dot{y}(t)\| \text{ and } \ddot{R}(t) = \frac{d\dot{R}(t)}{dt} = \frac{[\dot{y}(t)]^T \ddot{y}(t)}{\|\dot{y}(t)\|} \qquad (4)$$

The evaluation of Fisher Information using time series data involves estimating the state speed and acceleration along the trajectory, and performing a numerical integration over the system period. We use a central difference scheme to estimate velocity and acceleration. The acceleration must have the proper sign to account for reversals in velocity direction, i.e., when there is an instant of zero velocity but nonzero acceleration along the trajectory. Numerical estimation of velocity and acceleration is subject to noise and other data artifacts. We address these through the use of a two-point average of the calculated speed values in the numerical integration of Fisher Information. In order to preserve the original number of estimated velocity points, the very first speed data point is averaged with speed calculated as a simple difference of the first two raw data points. In addition, the integration in the expression of Fisher Information itself averages outlying values of the speed and acceleration to some extent. In addition to noise, unknown periodicity of the steady state is also a difficult issue when dealing with real-world data. One reason is that the dynamics of real systems involve multiple periodicities that are not simply related. When the integration window is not matched to the period of the system, Fisher Information will appear to fluctuate as the calculation is repeated over time even if the data are not noisy. Fisher Information also can appear "noisy" when data do not represent a periodic steady state, when the real system is non-periodic, or when the steady state changes periodicity following a disturbance.

Strictly speaking, the Fisher Information theory developed here applies to systems that have reached a periodic steady state. Thus, while the calculations described by Eqs. (1, 2, 3) can be applied to systems undergoing transient or non-periodic behavior, interpretation of the resulting values is approximate under the current development and merits further study. Nonetheless, many systems do achieve enough of a periodic regime, i.e., are close enough to being in a periodic regime, for the theory to be approximately and usefully applicable. This is an important observation since there is a very extensive literature in science and engineering on the useful application of approximate theory. For example, there are no systems truly at equilibrium anywhere in the universe. Hence, temperature, defined as that property which is equal for two bodies in thermal equilibrium, is, strictly speaking, ill defined everywhere in the universe. This does not, however, stop anyone from measuring, calculating, and using temperature for many systems, even those that are clearly not at equilibrium.

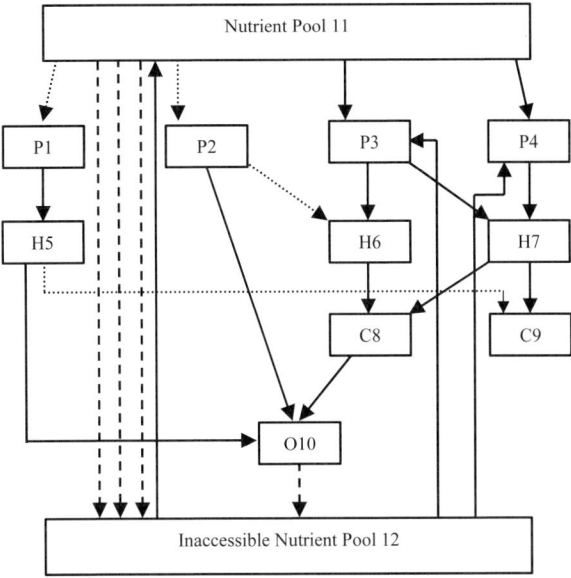

Fig. 2. Depiction of model system with four plants (P1, P2, P3, P4), three herbivores (H5, H6, H7), two carnivores (C8, C9), and a top omnivore (O10). The arrows represent mass flows. All living compartments have an implied flow back to the Nutrient Pool representing death. The plants P1 and P2, and the herbivore H5 are domesticated. The top omnivore (O10) consumes P2 and H5 as well as the non-domesticated carnivore (C8). The plants P3 and P4, the herbivores H6 and H7, and the carnivore C9 are neither domesticated nor consumed directly by top omnivore (O10). Detritivores function within the Nutrient Pool (11). The Inaccessible Nutrient Pool abstractly represents nutrients sequestered from biological accessibility by the building of infrastructure. There is a fixed, low flow from the Inaccessible Nutrient Pool to the Nutrient Pool.

Additionally, classical mechanics is known not to be exact because it ignores both quantum and relativistic effects. But it is sufficiently applicable to give useful results in many situations including the statistical mechanics of molecules where quantum effects are important (Davidson 1962). In the cases of temperature and classical mechanics there is ample evidence and experience to indicate when and where the theory is or is not applicable. For a new theory such as our work with Fisher Information, such a body of evidence and experience does not exists, and it is, therefore, necessary to proceed cautiously. We are in the process of exploring and establishing the limitations of the theory, but we suspect that the theory will likely give useful results in situations outside the strict periodic state boundaries outlined above. In what follows, we calculate Fisher Information continuously as the system shifts from one periodic attractor to another. We are most interested in detecting the transition and establishing some qualitative assessment of the system behavior during the transition. The Fisher Information that we have developed appears to be useful in this capacity.

8 Model system

We have developed a 12-compartment model system to mimic a general ecosystem with a very rudimentary social system that regulates the flows of mass according to its own criteria. The model system, which is depicted in Fig. 2, represents the flow of "mass" (which is meant to loosely characterize biomass, nutrients, water and other resources) within a closed system (i.e., the cumulative sum of masses of all the system compartments is constant). We do not claim that the model is an accurate representation of any real ecosystem or society, but treat it as a caricature. We use the model here only to study the salient features of our approach to building a theory of sustainable systems. However, this closed system may also display some of the types of economic and ecological interactions that are characteristic of a "steady state economy," envisioned by Daly (1996).

In Fig. 2, the solid arrows represent transfers of mass due solely to biological or geological drivers, with no interference from the top omnivore. The dashed arrows represent transfers of mass from the nutrient pool to the inaccessible nutrient pool that occur as a by-product of top omnivore activities in the domesticated branch, e.g., land filling, paving, etc. necessary to support domesticated agriculture and product distribution. The dotted arrows represent mass transfers that the top omnivore population can increase or decrease according its own criteria, e.g., increasing the flow of nutrients to plants (P1 and P2) to feed a growing top omnivore population, and decreasing the flow of mass from a herbivore (H5) to a carnivore (C9) thus reserving the herbivore to feed the top omnivore population. The model is, therefore, meant to mimic some of the rudimentary interactions that exist in a food web where one species represents humans with a level of management control. Table 1 gives the principal role of each compartment in the model ecosystem.

The ecosystem model has horizontal and vertical organization. The model is divided horizontally into two characteristic branches: domesticated (representing agricultural and livestock activities) and non-domesticated (representing species hunted, gathered, and species not consumed by humans). Vertically, the model has four trophic levels (plants, herbivores, carnivores, and a top omnivore) and two resource pools, one of which (an "inaccessible" nutrient pool) has a much slower rate of mass moving out of it than the other (Table 1). Although the top omnivore gets no mass directly from non-domesticated species that it does not consume, these supply ecosystem functions that are critical to the survival of the system as a whole. These non-domesticated, non-consumed species "recycle" mass from the inaccessible nutrient pool back into the rest of the system. If the mass in these non-domesticated, non-consumed compartments falls to zero, the only way for mass to get from the inaccessible to the accessible nutrient pool is through the slow transfer of mass from the inaccessible to accessible nutrient pool ascribed to "geologic processes." If this rate is too slow, most of the compartments in the domesticated and non-domesticated branches will collapse to zero or near zero populations.

Table 1. Description of compartment roles by trophic level

Trophic level	Compartment	Role
Plants	P1	Domesticated plant eaten by H5
	P2	Domesticated plant eaten by O10 and H6
	P3	Non-domesticated plant eaten by H6 and H7
	P4	Non-domesticated plant eaten by H7
Herbivores	H5	Domesticated herbivore eaten by O10 and C9
	H6	Non-domesticated herbivore species, eaten by C8
	H7	Non-domesticated herbivore eaten by C9 and C8
Carnivores	C8	Non-domesticated carnivore, eaten by O10
	C9	Non-domesticated carnivore species
Top omnivore	O10	Species at the top of the food chain having some behaviors typical of a human population
Nutrient pools	NP11	"Accessible" nutrient pool, from which all of the plants must get mass, and to which all living compartments contribute mass through death (this compartment includes detritivores).
	INP12	"Inaccessible" nutrient pool, representing the "mass" locked up (e.g., landfills, paving, etc.) due to top omnivore activity. As domestication activity increases, contributions of mass to this compartment increase.

9 Mathematical Model

A mathematical model representing the flow of mass between the 12 compartments in Fig. 2 was constructed as a series of differential Lotka–Volterra type equations. Although we recognize that Lotka–Volterra models do not necessarily provide the most accurate representation of ecological behavior, this class of models is adequate for our purposes. The arrows in Fig. 2 indicate flow of mass from (origination) one compartment to (termination) another compartment which are dictated by the following relations:

$$\dot{y}_1 = y_1(G_1(y_{10})y_{11} - g_5 y_5 - m_1) \tag{5a}$$

$$\dot{y}_2 = y_2(G_2(y_{10})y_{11} - G_{26}(y_{10})y_6 - g_{102} y_{10} - m_2) \tag{5b}$$

$$\dot{y}_3 = y_3(g_3 y_{11} + r_3 y_{12} - g_{37} y_7 - g_6 y_6 - m_3) \tag{5c}$$

$$\dot{y}_4 = y_4(g_4 y_{11} + r_4 y_{12} - g_7 y_7 - m_4) \tag{5d}$$

$$\dot{y}_5 = y_5(g_5 y_1 - G_{59}(y_{10}) y_9 - g_{105} y_{10} - m_5) \tag{5e}$$

$$\dot{y}_6 = y_6(g_6 y_3 + G_{26}(y_{10}) y_2 - g_8 y_8 - m_6) \tag{5f}$$

$$\dot{y}_7 = y_7(g_7 y_4 + g_{37} y_3 - g_{78} y_8 - g_9 y_9 - m_7) \tag{5g}$$

$$\dot{y}_8 = y_8(g_8 y_6 + g_{78} y_7 - g_{108} y_{10} - m_8) \tag{5h}$$

$$\dot{y}_9 = y_9(g_9 y_7 + G_{59}(y_{10}) y_5 - m_9) \tag{5i}$$

$$\dot{y}_{10} = y_{10}(g_{102} y_2 + g_{105} y_5 + g_{108} y_8 - g_{12} - m_{10}) \tag{5j}$$

$$\dot{y}_{11} = m_1 y_1 + m_2 y_2 + m_3 y_3 + m_4 y_4 + m_5 y_5 + \\ m_6 y_6 + m_7 y_7 + m_8 y_8 + m_9 y_9 + m_{10} y_{10} + m_{12} y_{12} - \\ y_{11}(G_1(y_{10}) y_1 + G_2(y_{10}) y_2 + g_3 y_3 + g_4 y_4 + \\ W_1(y_{10}) + W_2(y_{10}) + W_5(y_{10})) \tag{5k}$$

$$\dot{y}_{12} = g_{12} y_{10} + y_{11}(W_1(y_{10}) + W_2(y_{10}) + W_5(y_{10})) - \\ y_{12}(r_3 y_3 + r_4 y_4 + m_{12}) \tag{5l}$$

where y_i represents the mass in compartment i, g_i is a parameter representing the growth rate for compartment i, m_i is a parameter representing the mortality rate of compartment i, g_{ij} is a parameter representing the transfer of mass from compartment j to compartment i, g_{12} represents the proportion of mass transferred from the human compartment to the inaccessible nutrient pool. Hence, g_{102}, g_{105} and g_{108} represent transfers from compartments 2, 5 and 8 to the top omnivore, g_{37} is the proportion of mass transfer from compartment 3 to 7 (a "crosslink"), and g_{78} the proportion of mass transfer between compartments 7 and 8. The term m_{12} is the proportion of mass transferred from the inaccessible nutrient pool to the "accessible" nutrient pool by natural non-biological processes, e.g., erosion. Finally, the terms r_3 and r_4 represent the proportion of mass in the inaccessible nutrient pool that is recycled by P3 and P4, respectively. This allows the non-domesticated compartments to recycle nutrients out of the inaccessible nutrient pool and back into the rest of the system. Note that the domesticated branch cannot recover inaccessible mass. Also note that most transfers of mass are also proportional to mass in the compartments involved.

The symbols G_i and W_i represent variable growth and waste generation functions further discussed here. If the inter-compartment mass transfer (functions G_1, G_2, G_{26}, G_{59}, W_1, W_2, W_5 in Eq. 4) are constant, then this compartment functions essentially as a top omnivore animal, which simply hunts and gathers already existing food but does not manipulate the system. When these functions are conditional on the mass of top omnivore (compartment O10), this compartment then functions as a rudimentary human society that encourages growth of some plants and herbivores, builds "barriers" between some compartments (H5 to C9 and P2 to H6) to regulate mass flow between one branch and another, and changes the efficiency of production methods and consumption. These relationships are indicated by dotted arrows that represent the mass transfers that are functions of the mass in the top omnivore population according to:

$$G_1(y_{10}) = g_1 y_{10} \tag{6a}$$

$$G_2(y_{10}) = g_2 y_{10} \tag{6b}$$

where g_i is a positive constant. Hence, as the mass in the top omnivore increases, the growth rates of these plants increase. This represents an agricultural activity such that, as the human population grows, more labor is available for farming. "Crosslinks" between the branches are also indicated by dotted arrows (as these transfers are also a function of the mass in the top omnivore compartment):

$$G_{26}(y_{10}) = \frac{g_{26}}{1 + y_{10}} \tag{6c}$$

where G_{26} represents the rate at which H6 consumes P2 under cultivation and simply constant (g_{26}) without cultivation. Similarly, G_{59} represents the rate at which C9 consumes H5, a domesticated species, according to:

$$G_{59}(y_{10}) = \frac{g_{59}}{1 + y_{10}} \tag{6d}$$

which is a function of y_{10} under "protective domestication" and simply constant (g_{59}) without. The function in Eqs. 6c and 6d includes the fact that with a larger top omnivore population, there is more opportunity to invest in barriers, to hunt, or otherwise reduce the impact of herbivore 6 and carnivore 9 on the species P2 and H5 that are directly consumed by the top omnivore.

The dashed arrows in Fig. 2 represent direct transfers from the accessible nutrient pool to the inaccessible nutrient pool. These represent the degree of "wastefulness" or "overhead" necessary in sustaining the human population under current. These are functions of the mass in the top omnivore compartment:

$$W_1(y_{10}) = w_1 y_{10} \tag{6e}$$

$$W_2(y_{10}) = w_2 y_{10} \tag{6f}$$

$$W_5(y_{10}) = w_5 y_{10} \tag{6g}$$

where w_i is a positive proportionality constant representing the rate of transfer of mass from the accessible to the inaccessible nutrient pool in proportion to the mass in compartment i and the mass in the top omnivore compartment y_{10}. The larger the population of top omnivores, O10, the larger the necessary infrastructure and consequent waste produced.

In addition to equations 5 through 6, there is an implicit and non-limiting flow of energy through the system. Hence, the system is open to energy but closed to matter. The formal mass closure expression is:

$$\sum_{i=1}^{12} \dot{y}_i = 0 \tag{7}$$

All four plants (y_1, y_2, y_3, and y_4) are subject to a cyclic forcing function to represent seasonal variation in their intrinsic growth rate. This forcing is the source of the limit cycle behavior for the model. The expression for the forcing function is:

$$g_i = f_i(t)\left[1 + \frac{1}{3}\sin\left(\frac{2\pi t}{500} - \frac{\pi}{2}\right)\right]^2 \quad i = 1,2,3,4 \tag{8}$$

where t is time, and the function $f_i(t)$ is used to simulate the effect of a gradual change in prevailing conditions such as an environmental climatic change that affects the growth and death of plants. This same function, however, is also used to simulate: (1) changes in the mortality rate of the top omnivore such as those due to medical advances or plagues, (2) a change in the rate of waste production such as those from improving the efficiency of agriculture or poor agricultural practices, and (3) a change in the ability to keep non-domesticated species from consuming domesticated species such as would occur from building up or decaying infrastructure. The function is given by:

$$f_i(t) = f_i^l \quad 0 \le t \le t^l \tag{9a}$$

$$f_i(t) = f_i^l + \frac{f_i^h - f_i^l}{t^h - t^l}(t - t^l) \quad t^l \le t \le t^h \tag{9b}$$

$$f_i(t) = f_i^h \quad t^l \le t \le \infty \tag{9c}$$

where f_i^l is the constant value of the function before the perturbation starts, f_i^h is the constant value of the function after the perturbation ends, t^l is the point in time when the linear perturbation starts, and t^h is the point in time when the perturbation ends. For the studies below, we used $t^l = 2000$ and $t^h = 2500$. Equations (5a), (5b), (5c), (5d), (5e), (5f), (5g), (5h), (5i), (5j), (5k), (5l), (6a), (6b), (6c), (6d), (6e), (6f), (6g), (7), (8), (9a), (9b) and (9c) form the complete mathematical model representing the system depicted in Fig. 2. Uncalibrated values were assigned to the model parameters, and a computer simulation using the MATLAB[2] software was developed to explore several hypothetical cases. The parameter values used are given in the Appendix.

Table 2. Scenarios simulating desirable, undesirable, and degree of wastefulness events

Category	Scenario	Parameters changed
Desirable	1	Increase 2X growth rate of P1 and P2 (fertilization/agriculture)
		Decrease 10X mortality of O10 (enhanced medicine)
		Decrease 2X rate of mass transfer to INP (12) (increased production efficiency)
		Decrease 10X rate of consumption wastefulness (product reuse)
		Decrease 2X rate of mass transfer from P2 to H6 and H5 to C9 (barriers)
Undesirable	2	Decrease 10/7 X growth rate of P1 and P2 (bad growing conditions)
		Increase 120X mortality of O10 (epidemic)
		Increase 1.1X rate of mass transfer to INP(12) (loss of production efficiency)
		Increase 1.5X rate of consumption wastefulness (no product reuse)
		Increase 1.5X rate of consumption wastefulness
Wastefulness	3	All parameters at baseline, production wastefulness increased 1.2X, consumption wastefulness decreased 4X. (inefficient production but efficient consumption)
Wastefulness	4	All parameters at baseline, production wastefulness decreased 2X, consumption wastefulness increased 1.5X. (efficient production but inefficient consumption)

10 Simulation results

Using the aforementioned mathematical model and computer simulations, a baseline case with all the compartments stable at non-zero masses, and four perturbed

---

[2] Citation is for clarity only and does not imply approval or endorsement by the U.S. Environmental Protection Agency.

scenarios listed in Table 2 were explored. The four perturbed scenarios are: (1) a perturbed system involving five desirable (from the point of view of the top omnivore/humans) changes to the system dynamics, (2) a perturbed system with five undesirable changes to the system dynamics, (3) a perturbed system with wasteful agricultural production but efficient consumption, and (4) a perturbed system with efficient agricultural production but wasteful consumption. The associated system population dynamics and the Fisher Information calculations for these scenarios are shown and discussed below.

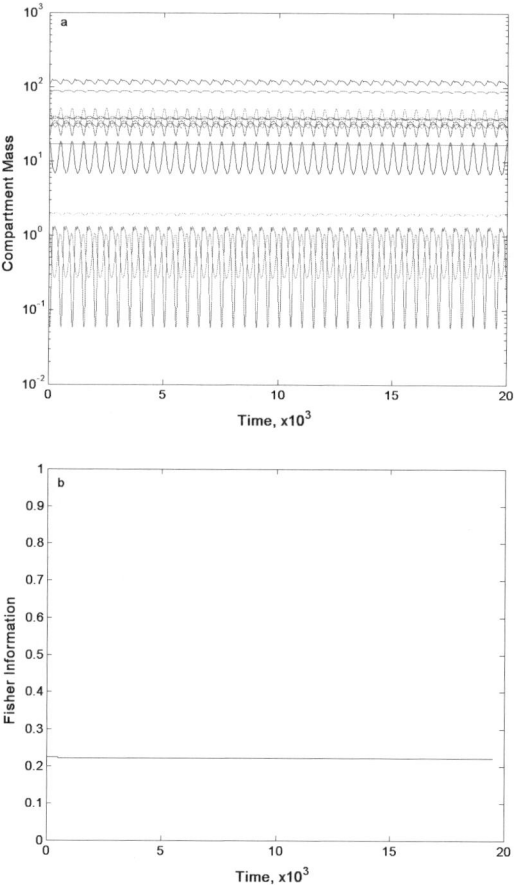

Fig. 3. System response (a) and Fisher Information (b) for the system in a stable, periodic steady state (baseline scenario)

The Fisher Information calculations are based on the ten species compartments (P1, P2, P3, P4, H5, H6, H7, C8, C9, and O10) but not the nutrient pools. The reason is that the masses of the nutrient pool and the inaccessible nutrient pool are not generally measurable quantities. The baseline and the aforementioned scenarios were simulated to test the effect of: (1) changing the mass transfer from the nutrient pool (11) to the domesticated plants (P1 and P2), (2) advances or losses in the efficiency of agricultural technology; (3) changes in the death rate of the top omnivore brought about by advances in medicine or by epidemics, (4) changes in mass transfer rates to the inaccessible nutrient pool (12) to simulate different degrees of wastefulness in top omnivore activities; and (5) changes in the transfer rates from P2 to H6 and H5 to C9 due to the effectiveness of "barriers" put between these compartments.

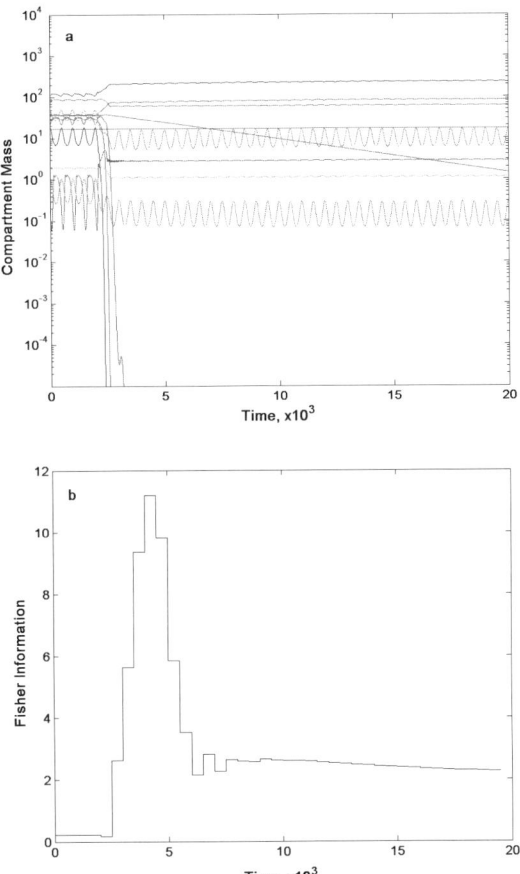

Fig. 4. System response (a) and Fisher Information (b) for the system undergoing desirable changes. Note that the mass in compartments P2, H7, C8, and C9 drops to zero

The baseline scenario with the system in a stable periodic steady state is shown in Fig. 3. Fig. 3a shows the mass in all 12 compartments cycling about non-zero values as a function of time. Fig. 3b shows the Fisher Information calculated repeatedly in blocks of 500 time steps (one period) for the ten species compartments

(P1, P2, P3, P4, H5, H6, H7, C8, C9, and O10). Note that the Fisher Information of a system in a periodic steady state is constant with time.

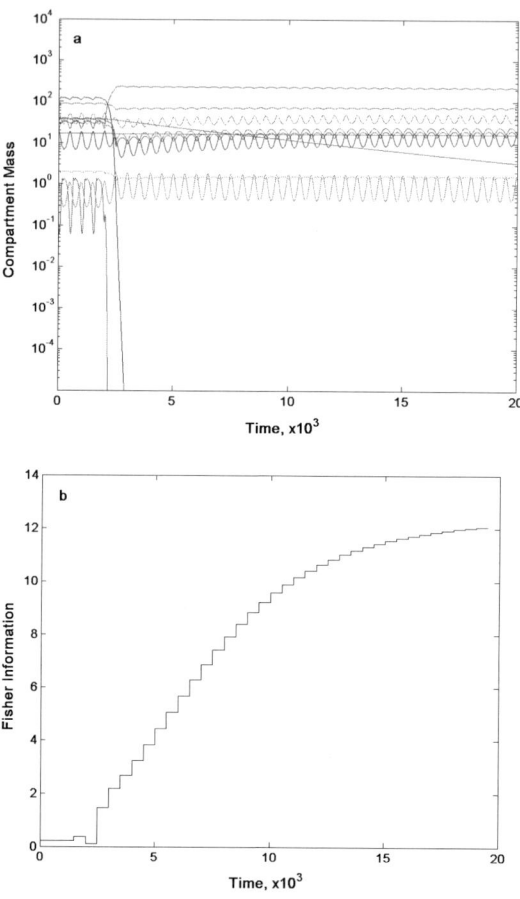

Fig. 5. System response (a) and Fisher Information (b) for the system undergoing undesirable changes. Note that the mass in compartments P1, H5, and C9 drops to zero

The case where all of the desirable changes are implemented is shown in Fig. 4. Note that the system starts in the same steady state as the baseline case of Fig. 3. Parameter changes are then gradually made using Eqs. 9a, 9b and 9c starting at time step 2,000 and ending at time step 2,500. The changes lead to the eventual extinction (compartment mass of zero) for carnivore C9 (Fig. 4a), plant P2, herbivore H7 and carnivore C8. Fisher Information for the system rises, indicating an increased variability in the time the system state spends in certain portions of its trajectory relative to others, i.e., an increase in system preference of certain states (Fig. 4b). Note that the system has changed its organization by eliminating several species. While the loss of species is generally not desirable, if such a change results in a system that is more robust to disturbances, is easier or less expensive to manage, etc., this could be beneficial to the top omnivore.

Note that variability as measured by Fisher Information does not refer to the range over which the state variables vary, but to the degree of non-uniformity in the time spent over of the steady state trajectory in a period (i.e., variation in speed over the steady state trajectory). As seen in the figure, the system had not completely arrived at a new steady state when the simulation ended. While our theory for computing the Fisher Information does not support rigorous interpretation in the transition between steady states, it is clear that a major change in the system variability has occurred in the transition.

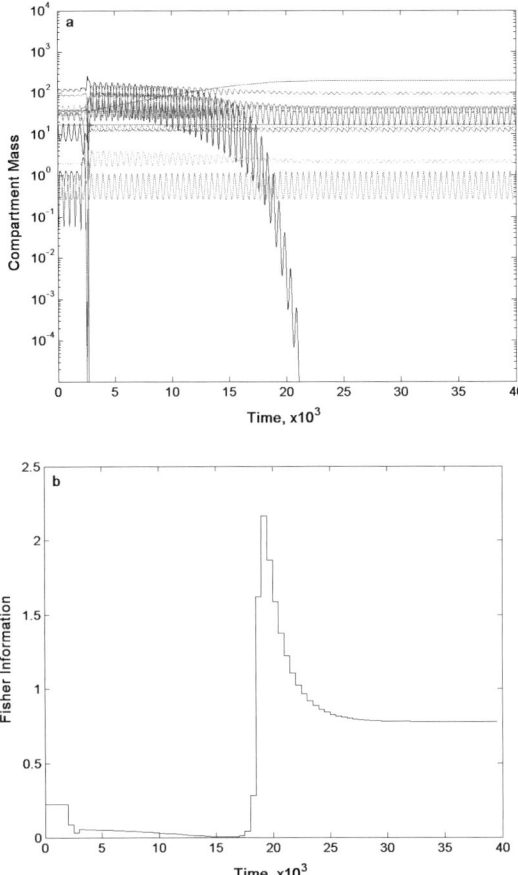

Fig. 6. System response (a) and Fisher Information (b) for the system at baseline with production wastefulness increased and consumption wastefulness decreased. Note that the mass in compartments P1, H5, and C8 drops to zero

The case where all of the undesirable changes are implemented is shown in Fig. 5. The system starts in the baseline steady state of Fig. 3, and then the parameter changes are implemented using Eqs. 9a, 9b and 9c starting at time step 2,000 and ending at time step 2,500. The changes lead to extinction (compartment mass of zero) for domesticated plant P1 and domesticated herbivore H5 (Fig. 5a), as well as the eventual extinction of carnivore C9. This leaves domesticated plant P2 and non-domesticated carnivore C8 as the only sources of food for the top omnivore population. The Fisher Information for the system again settles to a higher value (Fig. 5b). While the level of Fisher Information is higher than the baseline, the fact that there are fewer food species available for humans makes this steady state generally less desirable.

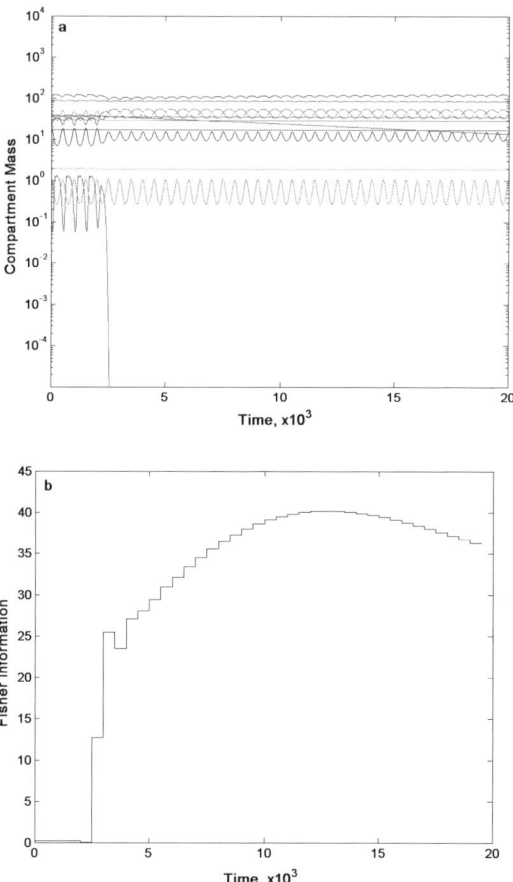

Fig. 7. System response (a) and Fisher Information (b) for the system at baseline with production wastefulness decreased and consumption wastefulness increased. Note that the mass in compartments H5 and C9 drops to zero

The third and fourth cases studied involve changes to the wastefulness parameters. With the other parameters at baseline throughout the simulation, we first increase production wastefulness and decrease consumption wastefulness between time steps 2,000 and 2,500. The result is shown in Fig. 6. Plant P1, herbivore H5 and carnivore C8 go extinct, with P1 taking roughly 20,000 simulated years to go extinct.

The response of Fisher Information (Fig. 6b) in this case is especially interesting, showing details of the transient behavior that are not evident from the time series. Two significant changes are evident, one following the loss of C8 and H5, and the other occurring as P1 goes extinct, with a relatively steady level of Fisher Information in between. Fig. 7 shows the system response and Fisher Information for the case in which production wastefulness is decreased relative to the baseline and consumption wastefulness is increased. Herbivore H5 and carnivore C9 eventually go extinct. As in previous cases, Fisher Information has gone up relative to the baseline.

11 Discussion

The Fisher Information is a measure of the level of variability in the system regime or order for the system. While the specific form that we have developed strictly applies only to periodic steady state regimes, the previous results demonstrate the sensitivity of the calculation to changes in steady state behavior when it is applied repeatedly to time series data, and, as has been previously discussed, the theory is likely to give useful insights outside its strict range of applicability in the manner of various other theories in science. The results that we obtain from the series of simulations where desirable changes are implemented (Fig. 4) indicate that the changes lead to a functioning system with fewer species. The Fisher Information calculation provides a clear signal that the steady state regime is changing, and changing rapidly (Fig. 4b). In the case of real systems, this may not be clear from field data, especially when large numbers of variables are involved. The usefulness of the Fisher Information calculation in identifying transients in system behavior is especially evident in the results of Fig. 6, where a complex transient occurs. Fisher Information shows a slow and steady drop over the 20,000 simulated years it takes to see the extinction of plant P1, and then a relatively sudden increase as the system settles once P1 is eliminated.

While an increase in Fisher Information may indicate a system that has entered a regime characterized by a greater amount of time spent in certain states along its trajectory, i.e., greater order, there is no set pattern between species loss and the level of information at steady state. The scenarios examined here exhibited higher Fisher Information with a loss of species (Figs. 4, 5, 6 and 7). However, this is not the case in all changes, and we have seen many other scenarios where Fisher Information drops with species loss. It is possible that, had we examined more scenarios, we would have found more cases in which Fisher Information did not change or declined concurrently with the loss of species from the system. While Fisher Information appears to be a good indicator of the onset of transient behavior, additional knowledge concerning the type of disturbance being encountered and the possible system steady states is necessary in order to determine the ultimate fate of the system and repercussions on human sustainability.

The relatively simple model depicted in Fig. 2 is a caricature of an actual ecological–social system, and it is not an entirely accurate representation of the complexity of such a system. It is, however, possibly the simplest system that can still exhibit the most elementary behavior of a real ecological–social system. The model simply represents a combination of an ecological food web with a rudimentary social system that is able to manipulate the food web according to its own rules. Thus, the model allows us to explore the most elementary behavior of the real system at a reasonable level of complexity. There is ample precedence in science for using simplified model systems. For example, the basic laws of mechanics are first studied using a pendulum rather than a complex machine; however, the laws of physics extracted from studying pendulums are applicable to even the most complex of machines. In biology, simpler model organisms, e.g., rats, insects, and worms are frequently used in experimental studies of toxic responses, nerve function, and reproduction, and the results routinely extrapolated to other organisms such as human beings. It is, therefore, appropriate to use simplified models that are still complex enough for purposes of this study where we attempt to understand some of the underlying principles of sustainable systems.

If we assume that highly ordered and well-functioning complex systems have characteristic biases towards specific states within the trajectory of the system, the measurable variables of such systems also will have characteristic values with relatively narrow distributions that indicate that the system is biased towards certain states. The Fisher Information computed from the theory developed here is a very sensitive composite measure of the bias of a system towards a particular set of states. The more ordered a system, the higher the bias to specific states and the higher the Fisher Information. The less ordered a system is, the less the bias to specific states and the lower the Fisher Information. Biological systems, in fact all complex systems, must maintain their internal order in order to function. The Fisher Information computed here, because it is sensitive to change in order, can perhaps be used to assess the functionality of complex dynamic systems.

Our theory is not based on any specific view of the trends or directionality underlying the development of systems such as given by ascendancy (Ulanowicz 1997) or orientors (Müller and Leupelt 1998). Rather, we have concerned ourselves with the detection of change that is outside of the steady state pattern (in time) in the state variables of an established complex system. This may involve a change in the system's structure or function consistent with such theories, or simply be a transient response to disturbances. If resilience is defined by the size of perturbation that a system can withstand before fundamentally changing in function and structure (Gunderson 2000), then we would expect some system measures to remain unchanged for resilient systems under perturbation. A Fisher Information calculation over these measures then would also remain near the same value for resilient systems. Resilience, or other similar characteristics of systems (e.g., resistance, stability, etc.), may not always exhibit a positive relationship with diversity (Naeem 2002) and neither does Fisher Information. The resilience of a system may not change (or may even increase) with the loss of one or several species, due to the functional redundancy of species (different species that perform the same function, such as nitrogen fixation among legumes), or due to changed relationships between individual species and species composition of the system as a whole (Johnson 2000; Flower 2001). In fact, some disturbances may increase the resilience of the system by eliminating species that are sensitive to disturbance, leaving behind a community of more resilient species (Balmford 1996). However, the systems consisting of species resilient to human disturbances may not be desirable from a human perspective, and we therefore would conclude that these systems are not sustainable. Fisher Information must therefore be used with a combination of other indices to measure the sustainability of a system.

12 Conclusions

We have presented a model system that is minimally complex but exhibits some of the characteristic behavior of a real ecological–social system. Results show that Fisher Information is a very sensitive measure of the variability in the steady state regime of a dynamic system. Such changes to the dynamic regime, whether through internal reorganization or external perturbation, may not be apparent from time series plots of the system state variables. For the case of very complex systems having hundreds of state variables such as those necessary for sustainability studies, a theoretically sound index such as Fisher Information may well be indispensable. Although we recognize that the subject of sustainability is complex, our work leads us to speculate that a necessary, but certainly not sufficient, condition for sustainability is that systems in a sustainable dynamic regime must maintain constant Fisher Information with time. If the system migrates to another dynamic regime with higher Fisher Information, we would say that the system has increased the time it spends in certain states, i.e., its degree of order, and that it is functional, although this new regime may or may not be desirable. If the system migrates to a dynamic regime with lower Fisher Information, we would say that it is less likely to be found in any particular state, i.e., it has lost order, and, possibly, might be less functional from a human perspective. One limitation of the theory in its present state is that it is strictly speaking applicable only to dynamic systems at cyclic steady state, i.e., the universe of states is limited to those on the system's recurrent trajectory. This limitation, however, may not preclude the useful application of the theory to some situations outside these strict boundaries as already discussed. Work is currently underway to explore these limitations and to extend the theory to non-cyclic, non-steady state regimes. However, by repeatedly calculating the Fisher Information, some indication of changes to steady state can be identified, although they cannot be quantified. The interpretation of the Fisher Information in terms of real ecosystems needs further work. Regardless of these caveats and limitations, however, the present results are encouraging.

Acknowledgments

CWP is a Postdoctoral Research Fellow with the Oak Ridge Institute for Science and Education in residence at the National Risk Management Research Laboratory. The authors acknowledge helpful discussions with Dr. B. Roy Frieden with the University of Arizona.

References

Barbier EB (1998) The conditions for achieving environmentally sustainable development. In: The economics of environment and development. Edward Elgar, Cheltenham, pp 43–53
Bird PJWN (1987) The transferability and depletability of externalities. J Environ Econ Manage 14:54–57
Bond EM, Chase JM (2002) Biodiversity and ecosystem functioning at local and regional spatial scales. Ecol Lett 5:467–470
Brooks TM, Mittermeier RA, Mittermeier CG, de Fonseca GAB, Rylands AB, Konstant WR, Flick P, Pilgrim J, Oldfield S, Magin G, Hilton-Taylor C (2002) Habitat loss and extinction in the hotspots of biodiversity. Conserv Biol 16:909–923
Colubi A (1996) Comparative studies of two diversity indices based on entropy measures: Gini-Simpson's vs Shannon's. In: de Andalicia J (ed) Information processing and management of uncertainty in knowledge-based systems, IPMU '96, vol 2, pp 675–80
Costanza R, Cumberland J, Daly H, Goodland R, Norgaard R (1997) An introduction to ecological economics. St. Lucie Press, Boca Raton, Fla., 275 pp
Costanza R, D'Arge R, de Groot R, Farber S, Grasso M, Hannon B, Limburg K, Naeem S, O'Neill RV, Paruelo J, Raskin RG, Sutton P, van den Belt M (1997) The value of the world's ecosystem services and natural capital. Nature 387:253–260
Daily GC (1997) Nature's services: societal dependence on natural ecosystems. Island Press, Washington D.C. 392 pp
Daly HE (1996) Beyond growth: the economics of sustainable development. Beacon Press, Boston, Mass., 253 pp
Davidson N (1962) Statistical mechanics. McGraw-Hill, New York
De Leo GA, Levin S (1997) The multifaceted aspects of ecosystem integrity. Conserv Ecol [online] 1(1): 3. URL: http://www.consecol.org/vol1/iss1/art3
Diamond J (1999) Guns, germs, and steel: the fates of human societies. WW Norton, New York, 480 pp
Doak DF, Mills LS (1994) A useful role for theory in conservation. Ecology 75:615–626
Ehrlich PR (1995) The scale of the human enterprise and biodiversity loss. In: Lawton JH, May RM (eds) Extinction rates. Oxford University Press, Oxford, pp 215–226
Farber SC, Costanza R, Wilson MA (2002) Economic and ecological concepts for valuing ecosystem services. Ecol Econ 41:375–392
Fath DB, Cabezas H, Pawlowski CW (2003) Regime changes in ecological systems: an information theory approach. J Theor Biol, in press
Ferenc J, Istvan S, Gabor V (2002) Species positions and extinction dynamics in simple food webs. J Theor Biol 215:441–448

Fisher RA (1922) Philos Trans R Soc London 222:309
Flower RJ (2001) Change, stress, sustainability, and aquatic ecosystem resilience in North African wetland lakes during the 20th century: an introduction to integrated biodiversity studies within the CASSARINA project. Aq Ecol 35:261–280
Freeman AM (1984) Depletable externalities and Pigouvian taxation. J Environ Econ Manage 11:173–179
Frieden BR (1998) Physics from Fisher Information: a unification. Cambridge University Press, Cambridge, 319 pp
Frieden BR, Plastino A, Soffer BH (2001) Population genetics from an information perspective. J Theor Biol 208:49–64
Grosse WJ (1992) The protection and management of our natural resources, wildlife, and habitat. Oceana Publications, Dobbs Ferry, NY, 353 pp
Gunderson LH (2000) Ecological resilience–in theory and application. Annu Rev Ecol Systematics 31:425–439
Holling CS (1973) Resilience and stability of ecological systems. Annu Rev Ecol Systematics 4:1–23
Holling CS (1996) Engineering resilience versus ecological resilience. In: Shultze P (ed) Engineering within ecological constraints National Academy Press, Washington, D.C., pp 31–44
Johnson KH (2000) Trophic-dynamic considerations in relating species diversity to ecosystem resilience. Biol Rev (Cambridge) 75:347–376
Kearns CA (1997) Pollinators, flowering plants, and conservation biology. BioScience 47:297–307
Krutilla J (1967) Conservation reconsidered. Am Econ Rev 57:777–786
Levin SA, Dushoff J, Keymer JE (2001) Community assembly and the emergence of ecosystem pattern. Sci Mar 65[Suppl 2]:171–179
MacArthur RH, Wilson EO (1967) The theory of island biogeography. Monographs in population biology. Princeton University Press, Princeton, N.J., 203 pp
McCann KS (2000) The diversity–stability debate. Nature 405:228–233
Meadows DH, Meadows DL, Randers J, Behrens WW (1972) The limits to growth. Universe Books, New York, p 53
Müller F, Leupelt M (eds) (1998) Eco targets, goal functions, and orientors. Springer, Berlin Heidelberg New York
Myers N, Mittermeier RA, Mittermeier CG, Fonseca GAB, Kent J (2000) Biodiversity hotspots for conservation priorities. Nature 403:853–858
Naeem S (2002) Biodiversity equals instability? Nature 416:23–24
Pauly D, Christensen V, Guénette S, Pitcher TJ, Sumaila UR, Walters CJ, Watson R, Zeller D (2002) Towards sustainability in world fisheries. Nature 418:689–695
Pimm SL, Askins RA (1995) Forest losses predict bird extinctions in eastern North America. Proc Natl Acad Sci USA 92:9343–9347
Portela R, Rademacher I (2001) A dynamic model of patterns of deforestation and their effect on the ability of the Brazilian Amazonia to provide ecosystem services. Ecol Model 143:115–146
Postel S, Carpenter S (1997) Freshwater ecosystem services. In: Nature's services: societal dependence on natural ecosystems. Island Press, Washington, D.C., pp 195–214
Reed WJ (1986) Optimal harvesting models in forest management–a survey. Nat Resour Model 1:55–80

Sakai AK, Allendorf FW, Holt JS (2001) The population biology of invasive species. Annu Rev Ecol Systematics 32:305–332
Samuelson P (1954) The pure theory of public expenditure. Rev Econ Stat 36:387–389
Scheffer M, Carpenter S, Foley JA, Folke C, Walker B (2001) Catastrophic shifts in ecosystems. Nature 413:591–596
Settle C, Crocker TD, Shogren JF (2002) On the joint determination of biological and ecological systems. Ecol Econ 42:301–311
Shannon C, Weaver W (1949) The mathematical theory of communication. University of Illinois Press, Champaign, Ill., 125 pp
Tainter JA (1988) The collapse of complex societies. Cambridge University Press, Cambridge, 250 pp
Tilman D (1999) The ecological consequences of changes in biodiversity: A search for general principles. Ecology 80:1455–1474
Ulanowicz RE (1997) Ecology, the ascendent perspective. Columbia University Press, New York
US Department of the Interior (1990) The endangered species program: Audit Report. Report No 90-98. US Department of Interior, Office of Inspector General, Washington, D.C., 27 pp
Wardle DA, Bonner KI, Barker GM (2000) Stability of ecosystem properties in response to above-ground functional group richness and composition. Oikos 89:11–23
World Commission on Environment and Development (1987) Our common future. Oxford University Press, Oxford

Appendix

Table 3 shows all of the model parameter values used in simulating the scenarios previously shown. It should be noted that these parameter values have no particular physical or biological significance other than the fact that they generate the dynamics of a model system which is: (1) in a stable cyclic steady state and (2) relative masses in the different trophic levels that are not grossly inappropriate.

Table 3. Model parameter values

Variable	Value	Variable	Value	Variable	Value
g1	0.0030742	g59	2.5056e-003	m6	0.86282
g2	0.014371	g37	2.9554e-007	m7	0.067135
g3	0.0057021	g102	0.0062406	m8	6.8072e-005
g4	0.046056	g105	0.010037	m9	2.299524e-4
g5	0.0040908	g108	0.0080011	m10	0.0089928
g6	0.050993	g78	0.0098707	m12	1.1351e-007
g7	0.0047921	m1	0.053185	r3	0.0015156
g8	1.517e-007	m2	0.02416	r4	0.0049866
g9	5.9494e-006	m3	0.037033	w1	0.15846
g12	0.27566	m4	0.31467	w2	0.10646
g26	0.00025391	m5	.1	w5	0.10733

US EPA/Academia Collaboration for a Green Engineering Textbook For Chemical Engineering

David R. Shonnard[1], David T. Allen[2], Sharon Austin[3], Nhan Nguyen[4]

[1] Department of Chemical Engineering, Michigan Technological University, 1400 Townsend Drive, Houghton, MI 49931, Tel: 906-487-3468, Fax: 906-487-3213, e-mail: drshonna@mtu.edu

[2] Department of Chemical Engineering, Campus Mail Code: C0400, University of Texas, Austin, TX 78712, Tel: 512-471-0049, Fax: 512-475-7842, e-mail: allen@che.utexas.edu

[3] Sharon Austin, U.S. Environmental Protection Agency, 1200 Pennsylvania Ave., Mail Stop 7406M, Washington, DC 20460, Tel: 202-564-8523, Fax: 202-564-8528, e-mail: austin.sharon@epamail.epa.gov

[4] Nhan Nguyen, U.S. Environmental Protection Agency, 1200 Pennsylvania Ave., Mail Stop 7406M, Washington, DC 20460, Tel: 202-564-8526, Fax: 202-560-6953, e-mail: nguyen.nhan@epamail.epa.gov

Abstract

This paper describes the collaborative efforts between the Chemical Engineering Branch of the US EPA's Office of Pollution Prevention and Toxics (OPPT) and academia to develop a textbook titled "Green Engineering: Environmentally Conscious Design of Chemical Processes". Development of the GE textbook was initiated in the Fall of 1998 as part of the Green Engineering Program. An important goal of the Green Engineering Program is to incorporate "green" or environmentally conscious thinking and approaches in the academic and industrial communities regarding the design, commercialisation, and use of processes and products. Initially, the Green Engineering Program focused on the academic community, with the aim to produce the next generation of engineers with knowledge to design environmentally beneficial processes. The flagship of the program is the Green Engineering textbook, published in the fall of 2001, that applies classroom initiatives to real world engineering problems. To date, it is estimated that up to 50 engineering departments, both domestically and abroad, are using the textbook and/or incorporating the materials into their curricula. Schools using Green Engineering are achieving Accreditation Board for Engineering and Technology (ABET) criteria 2000, which includes having a working knowledge of the environment.

1 Motivation for a Green Engineering (GE) Textbook

Environmental issues are gaining prominence within engineering education due to a number of recent developments. As industrialised economies continue to grow, the increased output of goods and services creates mounting pressure to efficiently use resources and reduce environmental impacts of products and industrial processes. The traditional approaches of pollution control at the end-of-pipe are being seen as less desirable in the face of more stringent environmental regulations and the escalating costs of waste management. One answer to the dual needs of sustained economic growth <u>and</u> a healthy environment is pollution prevention at the source of waste generation, that is, within the manufacturing processes themselves. But, designing manufacturing processes and products to have lower environmental impacts requires a different approach than traditional design. Green Engineering involves integration of pollution prevention with environmental and health considerations into the engineering design of processes. Taking these considerations into account in design encourages the adoption of P2 activities that eliminate or significantly reduce potential environmental and health impacts.

The changing nature of engineering education provides additional motivation for including environmentally-conscious design and production in engineering curricula. The Accreditation Board for Engineering and Technology, Inc. (ABET, 2002) in the United States has stipulated in Criteria 3 (Program Outcomes and Assessment) that students must 1) demonstrate an understanding of the impacts of engineering solutions in a global and societal context, 2) have a knowledge of contemporary issues, and 3) understand professional and ethical responsibility. Furthermore, several professional engineering associations have made specific calls to include environmental aspects of engineering activities. For example, the American Institute of Chemical Engineers (AIChE), in their Program Criteria for chemical engineering education, states that graduates must demonstrate a "working knowledge, including safety and environmental aspects, of ..." chemical engineering practice. The key question therefore is *what is the nature of environmental effects that engineering students need to learn in order to perform their professional duties with a minimum of environmental impacts*? Risk assessment and Green Engineering concepts could be important answers to this question because these approaches require the engineering student to incorporate environmental issues into all aspects of design and operation.

Even more motivation is provided by the availability of environmentally-conscious methodologies and computer-aided tools, which allow for a systematic design synthesis and assessment, with the goal of decreasing environmental impacts. Many of these new methods and tools have been or are being developed by the U.S. Environmental Protection Agency (EPA) (US EPA, 2003a,b). For example, EPA's New Chemicals Program, located in the Office of Pollution Prevention and Toxics (OPPT), was established to help manage the risks from chemicals introduced into the marketplace. A series of systematic assessments are performed on a manufacturer's or an importer's Premanufacturing Notification (PMN) using these methodologies and software tools. As shown in Figure 1, several assessment steps are required and a number of software tools are used in the evaluation of each PMN application. Using only the molecular structure of the chemical and by applying structure activity relationships (SAR), the EPI Suite software estimates environmental physical-chemical and fate properties. Similarly, the ECOSAR tool predicts the toxicity of the chemical to aquatic organisms, while OncoLogic estimates potential carcinogenic effects. Next, releases of the chemical from specific process units are estimated using emission factors from the Air CHIEF software and using correlations in the TANKS program. In addition to air release estimation, the ChemSTEER software also predicts occupational exposure using workplace mass balance models. The assessments for the PMN process shown in Figure 1 are performed systematically and sequentially, starting with chemical structure and molecular-level properties, then process-level exposure assessment, and finally environmental and general population exposure. The models and concepts embedded within the software tools mentioned above could be very effective in teaching Green Engineering.

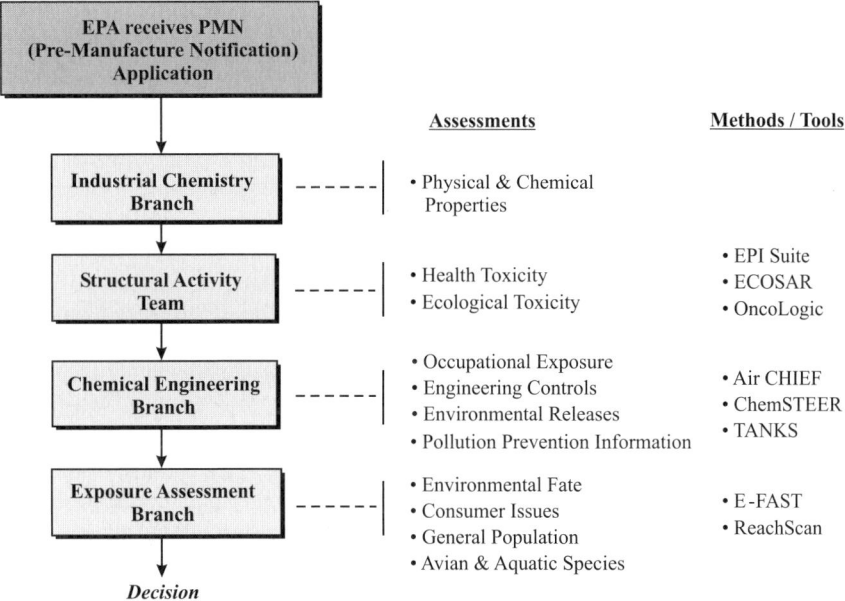

Fig. 1. Assessment activities for the PMN process at the U.S. EPA OPPT

The purpose of this article is to summarise the evolution of a collaborative effort between members of academia and the EPA to define the scope and to write a Green Engineering textbook for Chemical Engineering (Allen and Shonnard, 2002). An outline of the textbook chapters will be presented, which includes a systematic and hierarchical approach for integrating environmental impacts and risk into chemical process design. Finally, a description is presented of engineering education outreach activities sponsored by the EPA and implemented through the American Association for Engineering Education (ASEE) and the AIChE.

2 Overview of the Green Engineering Textbook

The intent of this textbook is to describe environmentally preferable or "green" approaches to the design and development of chemical processes and products. The idea of writing this textbook was conceived in 1997 by the staff of the Chemical Engineering Branch (CEB), Economics, Exposure and Technology Division (EETD), Office of Pollution Prevention and Toxics (OPPT) of the EPA. In 1997, OPPT staff found that, although there was a growing technical literature describing "green" approaches to chemical product and process design (Allen & Rosselot, 1997; Billatos and Basaly, 1997; Bishop, 2000; Dorf, 2001; Mulholland & Dyer, 1999; Rubin & Davidson, 2001; Theodore & McGuinn, 1992), and a growing number of University courses on the subject, there was no standard textbook on the subject area of Green Engineering. In addition, the university courses being taught at that time tended to focus on pollution control or prevention, mostly without emphasis on risk concepts or systematic design approaches. Therefore, in early 1998, OPPT initiated the Green Engineering Project with the initial goal of producing a text describing "green" design methods suitable for inclusion in the chemical engineering curriculum.

Several years of work, involving interaction between chemical engineering educators and EPA staff, have resulted in this text (Allen and Shonnard, 2002), which is intended for senior level or graduate level chemical engineering students. As shown in Table 1, the text begins (Chapters 1-4) with an introduction to environmental issues, risk assessment, and environmental regulations. With a focus on the chemical industry, this background material identifies the types of wastes and their impacts, introduces risk concepts, covers the fundamentals of environmental regulations and the regulatory process, and discusses the roles and responsibilities of chemical engineers for occupational safety and the environment protection.

Once the environmental issues have been defined, the design of processes, with the intent to reduce these environmental impacts, begins. Chapters 5-12 describe tools for systematically assessing and improving the environmental performance of chemical processes. The structure of these chapters follows a hierarchy of design activities that is commonly used in textbooks to create a chemical process designs (Douglas, 1988, 1992; Smith and Petela, 1991; Seider et al. 1999; Turton et al. 1998). To this hierarchy is added environmental assessment activities, beginning with screening methods and tools for evaluating environmental hazards of chemicals, continuing through unit operation and flow-sheet environmental impact analysis, and concluding with environmental costs of the design. A hierarchy of process environmental assessment activities that is employed in the textbook is shown in Table 2. Each design stage has a corresponding set of environmental assessment methodologies and specific design methods to prevent pollution and improve the environmental performance. The final section of the text (Chapters 13 and 14) describes tools for improving product stewardship through Life-Cycle Assessment and increasing the level of integration between different chemical manufacturing facilities and, potentially, between other industry sectors (Industrial Ecology). The textbook also contains an appendix with a number of sections including; a summary of prominent federal environmental regulations, data for estimating emissions for process units, tables of chemical-specific environmental physical and toxicological property data, and references to downloadable software to be used in the analysis and solution of end-of-chapter problems.

Table 1. Chapter Outline for "Green Engineering: Environmentally Conscious Design of Chemical Processes"

Green Engineering - the environmentally conscious design and commercialisation of processes and products
Textbook Outline
Part I: A Chemical Engineer's Guide to Environmental Issues and Regulations:
This section provides an overview of major environmental issues, and an introduction to environmental legislation, risk management and risk assessment.
1. An Introduction to Environmental Issues
- The Role of Chemical Process and Chemical Products
- An Overview of Major Environmental Issues
- Water Quality Issues
- Ecology

2. Risk Concepts
- Description of Risk
- Value of Risk Assessment in the Engineering Profession
- Risk-Based Environmental Law
- Risk Assessment Concepts

3. Environmental Law and Regulations: from End-of-Pipe to Pollution Prevention
- Nine Prominent Federal Environmental Statutes
- Evolution of Regulatory and Voluntary Prog-rams from End-of-pipe to Pollution Prevention
- Pollution Prevention Concepts and Terminology

4. The Roles and Responsibilities of Chemical Engineers
- Responsibilities for Chemical Process Safety
- Responsibilities for Environmental Protection
- Further Reading in Green Engineering Ethics

Part II: Environmental Risk Reduction for Chemical Processes: This section will describe a variety of analysis tools for assessing and improving the environmental performance of chemical processes. The group of chapters will begin at the molecular level, then proceed to an analysis of process flowsheets.

5. Evaluating Environmental Fate: Approaches based on Chemical Structure
- Chemical & Physical Property Estimation
- Estimating Environmental Persistent
- Estimating Ecosystem Risks
- Using Property Estimates to Estimate Environmental Fate and Exposure

6. Evaluating Exposures
- Occupational Exposures: Recognition, Evaluation, and Control
- Exposure Assessment for Chemicals in the Ambient Environment
- Designing Safer Chemicals

7. Green Chemistry
- Green Chemistry Methodologies
- Quantitative/Optimization Based Frameworks for the Design of Green Chemical Syntheses Pathways
- Case Studies

8. Evaluating Environmental Performance During Process Synthesis
- Tier I Environmental Performance Tools
- Tier II Environmental Performance Tools

9. Unit Operations and Pollution Prevention
- Pollution Prevention in Material Selection for Unit Operations
- Pollution Prevention for Chemical Reactors
- Pollution Prevention for Separation Devices
- Pollution Prevention Applications Separative Reactors
- Pollution Prevention in Storage Tanks and Fugitive Sources
- Pollution Prevention Assessment Integrated with Haz-Op Analysis
- Integrating Risk Assessment with Process Design

10. Flowsheet Analysis for Pollution Prevention
- Process Energy Integration
- Process Mass Integration
- Case Study of a Process Flowsheet
- Summary

11. Evaluating the Environmental Performance of a Flowsheet
- Estimation of Environmental Fates of Emissions and Wastes
- Tier III Metrics for Environmental Risk Evaluation of Process Designs

12. Environmental Costs Accounting
- Definitions
- Magnitudes of Environmental Costs
- A Framework for Evaluating Environmental Costs
- Hidden Environmental Costs
- Liability Costs
- Internal Intangible Costs
- External Intangible Costs

PART III: Moving Beyond the Plant Boundary: This section will describe tools for improving product stewardship and improving the level of integration between chemical processes and other material processing operations.

13. Life Cycle Concepts, Product Stewardship and Green Engineering
- Introduction to Product Life Cycle Concepts
- Life Cycle Assessments
- Streamlined Life Cycle Assessments
- Uses of Life Cycle Studies
- Summary

14. Industrial Ecology
- Material Flows in Chemical Manufacturing
- Eco-Industrial Parks
- Assessing Opportunities for Waste Exchanges and Byproduct Synergies
- Conclusion

Series of Appendices:
Addendum Overview / Software Addendum
Case Studies /Publication / Reference Addendum
Glossary / Index

Table 2. Hierarchical Approach for Environmental Assessment and Improvement of Chemical Process Designs

Design Stage/Issues	Environmental Impact Assessment	Tools For Pollution Prevention	Textbook Chapters
1. Conceptual Design • reaction pathways • materials selection • solvent selection	Tier 1 • envir. properties • chemical toxicity • stoichiometry	Green Chemistry • alternative routes • atom efficiency • mass efficiency	5, 7 and 8
2. Flowsheet Synthesis • define process units • separating agents • units configuration	Tier 2 • release estimation • productivity index • targeted releases	Green Engineering Choices • process technologies • modes of operation • process configurations	8 and 9
3. Detailed Design • process integration • process parameter evaluation	Tier 3 • multimedia model of environmental fate and transport • impact assessment	Process Integration Analysis • heat / mass integration	10 and 11

3 University/EPA Collaboration in the GE Textbook Project

This textbook project has been a unique experience in many respects. As mentioned in the previous section, the textbook was a result of several years of interaction between the textbook authors and the EPA staff. The initial drive to create this textbook was with the Chemical Engineering Branch of the OPPT of U.S. EPA. The primary authors, Allen and Shonnard, were chosen based on prior experience in textbook writing and because of their strong interests in environmental-related education and research in chemical engineering. The outline for the textbook was developed after several meetings between the EPA staff and the textbook authors. The outline is an attempt to merge the strengths of the well-known hierarchical process design approach with environmental exposure and assessment expertise of the EPA staff.

Because of the broad range of technical expertise needed to merge these two disparate fields into a single text, several contributing authors were recruited from the EPA staff and from industry for certain chapters. Risk Assessment is a very specialized field which combines elements of toxicology and environmental / occupational exposure assessment. Three experts in the field of risk assessment from the EPA, Gail Froiman, Fred Arnold, and John Blouin, wrote the introductory chapter on risk assessment, Chapter 2. Other U.S. EPA scientists and engineers from the OPPT contributed expertise in areas of pollutant release estimation from process units, occupational exposure modelling and environmental dispersion. Dr. Fred Arnold has developed several occupational and environmental exposure models for the PMN process (Figure 1) and Chapter 6 contains descriptions of modelling approaches for estimating exposure to chemicals released in the workplace and emitted to the environment. Mr. Scott Prothero has developed software to estimate air emissions from process units in support of the PMN process. He contributed significantly to Chapter 8. Dr. Paul Anastas from the White House Office of Science and Technology Policy, a recognized leader and pioneer in Green Chemistry worldwide, co-authored Chapter 7 on Green Chemistry. Ms. Kirsten Sinclair Rosselot, owner of Process Profiles, a consultant firm specializing in environmental planning and management, co-authored Chapters 10, 12, and 13.

Prior to writing the textbook, the primary author, Dr. David Allen, visited EPA and met with a number of EPA staff with expertise in various risk assessment topics and tools. EPA staff provided Dr. Allen with documentation on the various methods and tools used by EPA. Some of the risk-based tools and approaches shown in Figure 1 are incorporated into the Green Engineering textbook (EPI Suite, Air CHIEF, TANKS). The contributing authors reviewed the relevant literature and provided the necessary details and example problems in each of the chapters. Sharon Austin was the work assignment manager for the project. She provided overall project management, keeping authors on schedule to meet deadlines and organized the educator workshop activities described next. In addition, she contributed greatly to defining the format of individual chapters during the writing phase. Nhan Nguyen, Chief of the Chemical Engineering Branch, initiated the Green Engineering Program and originally conceived the GE textbook project, helped define content in several chapters, and provided leadership and valuable guidance throughout the project.

4 Outreach Activities of the Green Engineering Program at US EPA

The goals of the Green Engineering Program at the EPA are 1) to incorporate "green" or environmentally-conscious thinking and approaches in the academic and industrial communities regarding the design, commercialization and use of processes and products and 2) to promote and foster development and commercialization of green approaches and technologies. Thus, the two most important targets for GE outreach activities are academia and industry.

Over the past several years, the GE Program has worked with universities and the ASEE's Chemical Engineering Division to develop GE education materials, to provide GE training for professors, and to incorporate GE into chemical engineering curricula. For example, a number of "Green Engineering Educator" workshops were conducted between 1999 and 2002 in association with ASEE. These workshops, which were conducted as the textbook was being created and also after its completion, provided participants with lecture modules, problem sets and solutions, and software tools on CD to aid in teaching environmentally-conscious process design and course materials. Figure 2 shows the locations of these workshops and also the locations of some of the 90 or so institutions of the approximately 200 professors who have attended. Draft textbook chapters were disseminated at these workshops to allow early feedback on the textbook.

Fig. 2. "Green Engineering Educator" workshop locations in the United States. ● = workshop locations; ✱ = attendee institution locations. Not shown are the approximately 10 foreign institutions.

In addition to these workshops, other outreach programs have been initiated. At the AIChE Annual Meeting, a Green Engineering poster contest for undergraduate and graduate students has been established to encourage and reward chemical engineering projects that demonstrate innovative applications of Green Engineering concepts or novel uses of Green Engineering education. In addition, at these AIChE meetings awards are given to students for the best papers describing original research into Green Engineering topics, such as process optimization using environmental objectives, process integration to achieve waste minimization, environmental reaction engineering, green chemistry, and many other topics. The GE program, over the next few years, will work with AIChE and others professional engineering societies to convert the GE Textbook and lecture materials into a format for training practicing engineers.

The GE program completed and published, via Prentice Hall in September 2001, a textbook titled "Green Engineering: Environmentally Conscious Design of Chemical Processes". This textbook is being used and/or incorporated into a number of chemical engineering departments domestically and abroad. It is estimated that up to 50 engineering departments in the US and abroad have used textbook in their course offerings ((e.g.U. of Iowa, U. of MO, W. VA University, U. of Notre Dame, U. of Nevada, U. of OH, and Rowan University). The GE program is also working with other groups (including industry) to incorporate GE into their activities. For example, the AIChE's Chemical Center for Process Safety (CCPS) has agreed to incorporate GE into process safety training. Further information on GE Program and outreach activities is found on the GE website at www.epa.gov/oppt/greenengineering

5 Concluding remarks

This textbook project has been a unique experience in many respects. It provided an opportunity for EPA to transfer knowledge and methods in the area of risk assessment and to contribute to undergraduate engineering education, by emphasizing the environmental impacts of chemical process designs. A textbook has been developed that stresses a systematic and hierarchical approach for evaluating and improving the environmental performance of chemical process designs. A number of student-oriented activities have been established to encourage participation by undergraduate and graduate students. Outreach programs to academia and industry is and will continue to enhance the acceptance of Green Engineering concepts and approaches. In addition, course specific modules are being developed to incorporate GE into core textbooks and courses of chemical engineering. EPA is committed to the continued development of green engineering curriculum for engineering.

References

ABET - Accreditation Board for Engineering and Technology, Inc., (2002) Baltimore, MD, http://www.abet.org/

Allen D T, Rosselot K S (1997) Pollution Prevention for Chemical Processes. John Wiley and Sons, New York

Allen D T, Shonnard D R (2002) Green Engineering: Environmentally-Conscious Design of Chemical Processes. Prentice Hall, Upper Saddle River, NJ

Billatos S B, Basaly N A (1997) Green Technology and Design for the Environment. Taylor & Francis, Washington, DC

Bishop P L (2000) Pollution Prevention: Fundamentals and Practice. McGraw-Hill, Boston MA

Dorf R C (2001) Technology, Humans, and Society: Toward a Sustainable World. Academic Press, San Diego, CA

Douglas J M (1988) Conceptual Design of Chemical Processes. McGraw-Hill, New York

Douglas J M (1992) Process synthesis for waste minimization. Industrial and Engineering Chemistry Research, 31, pp. 238-243

Mulholland K L, Dyer J A (1999) Pollution Prevention: Methodologies, Technologies, and Practice. American Institute of Chemical Engineers, New York, NY

Rubin E S, Davidson C I (2001) Introduction to Engineering and the Environment. McGraw-Hill, Boston, MA

Seider W D, Seader J D, Lewin D R (1999) Process Design Principles: Synthesis, Analysis, and Evaluation. John Wiley & Sons, New York

Smith R, Petela E (1991) Waste minimization in the process industries. Part 1: The problem. Chemical Engineering, 31, pp. 24-25

Theodore L, McGuinn Y C (1992) Pollution Prevention. Van Nostrand Reinhold, New York, NY

Turton R, Bailie R C, Whiting W B, Shaeiwitz J A (1998) Analysis, Synthesis, and Design of Chemical Processes. Prentice Hall, Upper Saddle River, NJ, p. 814

US EPA (2003a) U.S. Environmental Protection Agency, Office of Pollution Prevention and Toxics, Green Engineering Program website, http:www.epa.gov/oppt/greenengineering.

US EPA (2003b) U.S. Environmental Protection Agency, Office of Pollution Prevention and Toxics, Exposure Assessment Tools and Models website, http:www.epa.gov/oppt/exposure.

Innovative Industrial Ecology Education Can Guide Us to Sustainable Paths

Kristan Cockerill

Secretary, International Society for Industrial Ecology, P.O. Box 7731, Albuquerque, NM 87194, E-mail address: kristan5@unm.edu

Abstract

Many activities labeled "industrial ecology" are as ancient as human society (e.g. reusing materials, using waste from one process to fuel another). The idea, however, that industrial ecology is the "science of sustainability" has gained prominence only in recent decades. Within this landscape, industrial ecology is becoming more formalized – there is a journal, an international society and increasing numbers of educational efforts dedicated to the topic. While industrial ecology has become a fairly common reference in various types of literature, its inclusive nature makes it difficult to define – much like the concept it strives to support – sustainability. Hence, while there is general agreement among practitioners that education dedicated to industrial ecology is important, there is not agreement on the specific direction this should take. The metaphor —applying ecological principles to industrial systems — is *de facto* interdisciplinary. This creates philosophical and administrative conflict when designing courses and programs. There are several approaches evolving that employ the industrial ecology concept, but each has quite distinct foci. A simplified delineation of these approaches might include: 1) Focus on developing innovative technology/models; 2) Focus on quantifying processes and identifying "best" technologies and/or best uses for technology/models; 3) Focus on societal factors (economic, behavioral, paradigmatic) to find alternative ways to do things using existing technologies. The ideal approach is likely some combination of these three. However, in designing a formal curriculum, it is not feasible (nor necessarily desirable) to cover all three in depth. For industry leaders (and policy-makers) understanding the values and limitations in each is important. If industrial ecology is to promote sustainability, then decisions about how it is taught will greatly influence efforts to define and reach sustainability. Understanding the tradeoffs and opportunities inherent in the diverse directions that industrial ecology education is moving is important if we wish to continue to identify and clarify pathways to sustainability.

1 Introduction

Contemporary literature, both academic and popular, provides a plethora of articles about sustainability and its importance to the future of life on planet Earth. There are numerous organizations, public and private, dedicated to promoting sustainability and/or the idea of sustainable development. There are also numerous publications criticizing the sustainability concept as ill defined and not situated appropriately to affect real change. As Tarlock (2001) has noted, sustainable development and the more specific Environmentally Sustainable Development (ESD), are appealing ideas (who would oppose a clean environment?) without the institutional support to realize their potential. Contention surrounds the very language used because many believe that "sustainable development" is oxymoronic. While debate continues about what qualifies as sustainable and what a sustainable approach to development might be, there is little doubt that the terminology surrounding sustainability has entered the mainstream and environmentally sustainable development has become a catch phrase for attempts to prevent further travel down what are perceived to be unsustainable paths.

Within the ongoing discussions about sustainability, is another ESD – Education for Sustainable Development which has contributed research efforts and program ideas for teaching students at all levels about sustainability. Of course, this ESD is also fraught with debate as to what to teach, to whom, and to what purpose. Yet, there is a strong, perhaps even intuitive, sense that education is the single most promising vehicle for eventually reaching the basic sustainability goal set out by the Brundtland Commission of meeting our needs without jeopardizing future generations' ability to meet theirs. This perhaps reflects a commonality between sustainability and education. Both are advanced through trail and error and trial and success approaches. There is no obvious or precise mechanism to guarantee success in either.[1] Into this flow of ideas we also have ecological principles merging with industrial practices. Industrial ecology (IE) is becoming a noticeable presence in education in the United States and throughout the world, largely because it is perceived to be an applied effort that can provide direction for environmentally sustainable development.

The connections between IE and sustainable development are obvious. The philosophy driving IE applications, such as reusing materials, designing resource efficient products and processes, using waste from one process to fuel another, is not new. In fact, many waste eliminating ideas have direct counterparts in pre-industrial practices. In post-industrial society, however, we are facing the need to consider the ramifications from decades of development that disregarded ecosystem impacts and natural resource limitations. Rejuvenating old ideas and developing new methods are inherent in IE, explicitly and implicitly. The tools and practices within any industrial ecology rubric are geared to help societies develop without increasing environmental damage. The contemporary concept of IE in the United States traces its beginnings to an article in a 1989 special issue of <u>Scientific</u>

---

[1] Thank you to a reviewer for providing this insight.

American bearing the cover title, "Managing Planet Earth." This publication was one of many in the late 1980s and early 1990s focused on planetary environmental and social concerns, reflecting reaction to multiple events around the world. Within a relatively short time the world witnessed the ozone hole, worldwide contamination emanating from the hot and cold wars, the Exxon Valdez oil spill, numerous negative reports concerning biodiversity, global warming, and poverty. In the US, Time Magazine did not select a "Man of the Year" in 1988, but instead highlighted Earth as "Planet of the Year." This emphasis on Earth and its inhabitants followed the trail that the Brundtland Commission blazed in 1987 with its call to appreciate Our Common Future. From within this fomentation sustainability arose as a potential product, process, and paradigm to address the rather negative news about planet Earth's condition. There also came calls for industry to change common practices. Hence, in 1989 we find two researchers from industry proposing that we might use ecosystems as a metaphor for designing industrial operations to enable us continue traditional development, but with fewer negative impacts (Frosch and Gallopoulos 1989).

In the ensuing decade IE began maturing and is currently being professionalized. There now exist the Journal of Industrial Ecology and the International Society for Industrial Ecology (ISIE). There have also been a solid number of books published about IE. As IE has grown, it has engendered debate about what it does or should encompass and whether it represents the beginnings of a paradigm shift or is simply refurbishing the status quo. There is evidence that like the term "sustainability," the phrase "industrial ecology" is becoming a buzzword within higher education with diverse definitions and applications. Internet searches reveal myriad references to industrial ecology as a general concept, as a course title, as a research focus, as an assigned reading topic and as a conference session topic. A cynical person might view this as "green wash" – attempts to garner students, funding, and recognition by invoking popular ideas but without actually generating any substantive change in the educational content or process. A more optimistic interpretation is that the prevalence of the phrase "industrial ecology" within higher education implies a growing recognition of the interconnections between technology and the environment, between industry-driven lifestyles and ecological principles, and between education and environmentally sustainable development. It also may reflect a concerted effort to seek answers to the negative ramifications of many of these relationships. Industrial ecology has been called the science of sustainability[2] and this paper focuses on IE and its evolving role in creating a roadmap to guide individuals and organizations toward a more sustainable existence. I argue that the lack of consensus about what IE is or should be grants it power to become a force for change and that its growing prevalence in higher edu-

---

[2] Industrial ecology is not alone in making this claim. The Columbia University Earth Institute states that the programs at the Biosphere 2 Center teach the "science of sustainability." The Bija Vidyapeeth Education for Earth Citizenship program says that its students have the opportunity to practice the "art and science of sustainability." Neither of these programs focuses on IE. Like the general idea of sustainability, the "science of sustainability" remains a fluid concept.

cation has the potential to either promote this force or stifle it. The research reported here focused on identifying education programs that invoke the IE moniker and analyzing the potential impact these programs may have on actually moving society toward sustainability.

2 Method

To identify IE programs in higher education, I requested information from faculty and researchers at universities in the US and throughout the world. These individuals are ISIE members or attended the inaugural ISIE meeting in 2001 in the Netherlands. Additionally, in 2002 I conducted an Internet search of the schools listed on the University of Texas, Austin's Web Central, which provides links to regionally-accredited US universities. For every university web page with an internal search engine, I entered the phrase "industrial ecology." For schools without an internal search engine, I looked at academic programs and searched for the words "industrial ecology" within course listings and program descriptions in business, engineering, environmental studies/science and other programs. Many of these searches uncovered syllabi that focused entirely on IE, or featured IE as part of a course. While these individual representations are relevant and reflect IE's growing popularity, this project focused on more expansive efforts and only includes programs where students can earn a degree with an IE-relevant focus or can conduct IE-related research as part of an institute or research center. A spreadsheet summarizing information about the various programs as is available through the ISIE web site: http://www.yale.edu/is4ie

This is by no means an exhaustive report on all programs that might have IE-relevant degrees and/or research. There are likely many universities offering programs that fit within the IE paradigm that were not uncovered in this search because they do not use the specific phrase "industrial ecology." For example, there are numerous "green chemistry" programs available, but this search did not reveal them if they do not self-describe as being "industrial ecology." Additionally, programs are continuously being created and modified to reflect changing knowledge bases and societal desires. While not exhaustive, this project reviewed more than 1000 universities and therefore does provide solid data about where and how IE is evolving within higher education.

3 Results

Table 1 provides a consolidated view of where programs that use the phrase "industrial ecology" in describing themselves appeared in higher education as of late 2002. This project identified 69 different universities with 87 different programs and centers. The disciplinary focus for degree-granting programs was determined based on both the name of the department and the degree(s) awarded. For research centers, the types of disciplines featured and the kinds of research being conducted determined which disciplinary focus was appropriate. Therefore, the Engineering/Technical focus includes all types of engineering, as well as architecture and construction programs. The Environment group includes environmental science/studies and natural science departments. Business/Economics and Policy/Planning are self-explanatory. The Multi- /Interdisciplinary groups featured some combination of the previous disciplines and the Other category represents individual programs in public health, human ecology, environmental history, and one degree program called industrial ecology. While this program is technically based, it is the only one completely self-referenced as IE and hence it seemed more appropriate to put it with Other than with the Engineering/Technical group. Many universities offer IE-related research opportunities to students through institutes or research centers. These are distinguished from degree-granting programs in Table 1.

Table 1. Educational Programs Featuring Industrial Ecology

Disciplinary Focus	Total	Degree Programs	Centers
Engineering/Technical	26	21	5
Environmental	12	12	0
Business/Economics	13	13	0
Policy/Planning	5	4	1
Multi-/interdisciplinary	24	14	10
Other	7	7	0
TOTAL	87	71	16

In addition to the disciplinary distinctions, the education efforts that include industrial ecology as a focal point can be grouped into three broad emphases: 1) designing innovative technologies and processes; 2) quantifying processes and identifying "best" technologies and/or best uses for technology and/or models; 3) assessing societal factors (economic, behavioral, paradigmatic) and the relationships between human aspects and technological applications.

The educational programs in the first emphasis area include basic pollution prevention concepts often found in engineering and chemistry departments as well as some design for environment programs and other more advanced IE applications featured in engineering, architecture and other disciplines. Because so many of the programs highlighting IE are technically based, it is not surprising that many of them emphasize creating new technology that is more energy efficient and less polluting. As the contemporary IE concept emanated from within industry, it is also fits that IE is comfortable in the technical realm and comfortable with a focus on new technology.

The second area with significant IE-relevant research and education is in employing models and activities to identify "best" practices or best technologies for a particular issue. Researchers in this area try to quantify current processes, such as resource use, which can then highlight ways to reduce resource use. Both technically orientated and social science departments feature these types of activities.

Rapidly expanding efforts in life-cycle assessment and material flow analysis fit within this grouping, as do many eco-park and industrial symbiosis projects. Several multidisciplinary programs fit in this category as well.

The third type of education related to industrial ecology and sustainability encompasses social factors. Educators and researchers in some social sciences and a few in the humanities have also recognized IE as a powerful tool for moving toward sustainability. Economics are obviously a key human factor and economic ideas have been a consistent presence as IE has evolved. In recent years, some business schools have added courses and concentrations to provide students with information about the connections between industry and environmental and societal impacts. Additionally, a few policy and planning programs have included IE principles in their curricula to connect decisions about development and other human activity with environmental impacts. The search for IE education programs found efforts in diverse departments as well as in multidisciplinary programs intended to combine technical knowledge with social science and sometimes humanities disciplines. This includes several science and technology studies programs, which explicitly emphasize the connections among science and technology and society.

Of course, the three categories I have created to describe the types of efforts currently available in higher education are artificial constructs. The lines between the categories are fuzzy. Developing innovative technologies flows into quantifying processes, which merges with social factors. Design for environment courses and programs, for example, can legitimately highlight one, two or all three emphasis areas. Employing the three categories defined here is instructive, however, because they highlight that programmatic emphases can play a significant role in how IE is perceived both within and outside academia and influence where IE is likely to appear on campus. My research suggests that many programs offering degrees related to IE are being molded to fit within an existing departmental paradigm. Therefore, depending on the university and the department where a design for environment specialty arises (to continue that example) the courses and research opportunities may have very different foci. Like sustainability, the phrase "industrial ecology" is being applied to diverse activities and knowledge. There are researchers and practitioners from within each of the three categories who claim that their emphasis area is equivalent to industrial ecology. At one university, the civil engineering department may include industrial ecology as a key tenet of their program and emphasize creating energy efficient technology while another university features IE in the business department and emphasizes "green accounting" techniques. (This could also happen within a single university). The courses taught and research conducted are quite different for these two departments, yet both claim to be providing students with experience related to industrial ecology. In general, because IE is being overlain onto existing, discipline-based education, there is the potential for individuals within any given program to not recognize that IE currently operates within multiple paradigms and is not limited to the scope within their discipline. This potentially has far-reaching implications for the route that IE may take in the quest for sustainability. If, for example, people with decision-making authority (likely social science or humanities types) equate IE as simply a twist on "clean technology" it may not receive the attention it deserves in determining funding and/or actual implementation. This potential situation is (or should be) of interest to IE practitioners, as this paper discusses.

The three approaches identified here, if combined, should encompass the mix of ideas that may allow IE to help lead human society toward a more sustainable existence. When examined more closely, however, gaps appear and questions arise as to whether IE is a force for change or is contributing to maintaining status quo. For the social scientist or humanities person examining the existing educational offerings, one gap becomes immediately apparent. While there is tremendous diversity in the disciplines represented in IE education, the focus continues to grant primacy to science and engineering approaches. There is still a strong emphasis on developing new technology and refining existing technology by quantifying various parameters. There is significantly less attention being paid to the human aspects of an industrial society and the role that human-based disciplines might play in contributing to the broad concept of industrial ecology and developing its role in sustainability. Further, human factors that are emphasized, such as traditional economic theory, have been implicated as deeply as technology in contributing to our current unsustainability (Ruckelshaus 1989; Foster 2001).

4 Discussion

For decades researchers and pontificators have suggested that our increasing technological capability will not provide what is required to reach a sustainability goal. At the outset of the modern environmental movement, White's (1967) seminal article on The Historical Roots of Our Ecological Crisis called not for new technology, but for new philosophy, for new attitudes about the relationship between man and nature. Tibbs (1999) thoroughly discusses the conundrum that technology poses by noting that it provides "greater ability to solve existing environmental problems, but also the potential to make them much worse if future technology is used without social and ecological discipline" (71). Tibbs notes that changes in beliefs, values and behavior are necessary to fully harness the power of technology that is necessary for a sustainable future. Turning concepts (IE or sustainability) into actions does not hinge on technology or on a more environmentally aware business community, but on rethinking entire systems that perpetuate the status quo. More than a decade ago, William Ruckelshaus (1989) in the same publication that evinced the contemporary ideas about "industrial ecosystems" noted that moving toward sustainability would require societal-level modifications at scales equivalent to the agricultural and industrial revolutions. He called for government policy (informed by science and with access to innovative technology) to lead the charge toward this new revolution. He wrote, "in creating the consciousness of advanced sustainability, we shall have to redefine our concepts of political and economic feasibility. These concepts are, after all, simply human constructs; they were different in the past, and they will surely change in the future" (174).

Although they are important, developing new technologies or promoting less damaging corporate practices are not revolutionary forces. Being sustainable will require most human societies to revise their ideas about and feelings toward the planet and natural resources. This is an all-encompassing endeavor that includes redefining politics and economics, but reaches even further. And this is perhaps the most powerful reason that education is the most promising vehicle to carry us to a sustainability revolution. Good education provides the opportunity to challenge existing paradigms, to encourage people to see issues from a novel perspective. Just as IE actually embraces pre-industrial ideas, so too considering pre-industrial education may be appropriate for IE and for sustainability. As Foster (2001) has noted, history, philosophy and literature, the tenets of liberal arts education, may actually offer better fuel for our attempts to educate as a way to reach sustainability. Including these disciplines when we talk about industrial ecology may enable IE to become a truly revolutionary tool. As an example, philosophy is embedded in IE as researchers debate the appropriate metaphors and analogies to use in framing IE (Isenmann 2002; Ehrenfeld 2003). Because IE is not clearly defined, it offers an incredible opportunity for educators and practitioners to develop unique and innovative approaches to thinking about industrial society and incorporating new ideas into disciplines with long histories and entrenched paradigms. The power that IE may bring into education and into broader social-cultural venues is to enable all disciplines, from within their unique perspectives, to think about "industry" as a symbol for contemporary global society and its relation with ecology. IE can encourage all disciplines to question: how did industry come to be in its present state? How did my discipline contribute to this process/development? How can my discipline contribute energy toward effective change? Conversations with my engineering friends reveal that these questions are typically not welcome within the technical academic realm. These, however, are precisely the types of broad, far-reaching questions that humanities academics are encouraged to explore.

Much research on how to incorporate environmental sustainability into education concludes that the principles within sustainability must somehow be integrated into all subject areas at all grade levels (Haury 1998). Similarly, an ideal approach to IE education would encompass all three of the categories that I have identified. This, however, implies a Renaissance education with individuals becoming masters of multiple disciplines. Because our knowledge base is significantly broader and deeper than during the Renaissance and because of existing formal education structures, this is not currently realistic. There are ways, however, to get the revolutionary vehicle in gear using existing academic systems. We need to continue to develop disciplinary strength while finding a way for students to recognize and appreciate the information and ideas being generated in other disciplines.

The concept of holistic management provides one intriguing model for beginning to apply information from the diverse disciplines that currently have or are developing IE-focused programs (Savory 1999). Figures 1-3 reflect evolving approaches for addressing societal concerns and issues. We have long recognized that the model shown in Figure 1 is not appropriate. No single discipline can possibly address the "whole" of any issue. The shift to multi- and/or interdisciplinary education and applications as shown in Figure 2, reflected an attempt to address the ineffectiveness of the single discipline approach. But, because they rarely recognize or accept the "whole" as the driving force, multi- and/or interdisciplinary teams have also not been able to provide sustainable solutions to many persistent problems. What has the potential to be more successful is to ensure that students in all disciplines develop skills to allow them to see what a particular issue demands from a variety of perspectives and to recognize what their discipline and other disciplines can provide to effectively address the issue. As Figure 3 shows, allowing the "whole" to tap the most relevant information is likely to be a more effective (and perhaps efficient) way to reach some desired endpoint. Of course, because all disciplines have their own paradigms which guide pedagogy and practices, there will be incredible dissension as to what constitutes the "whole," what the desired endpoint is, as well as what the appropriate role is for any specific discipline.

To help address this, I propose that all students pursuing education related to IE and/or sustainability should receive formal training in communication and teamwork. This builds on recent movements toward more integrative approaches to policy research and utilizing collaborative processes in making policy decisions (Susskind et al. 2001; Claussen 2001). With improved skill in these areas, research and decision-making teams will more readily recognize the need to identify the "whole" of an issue and to then apply their individual specialties in a more productive fashion and to recognize when the traditional approach from any discipline may not be effective or appropriate. This model allows students and society to continue to benefit from specialized education, which has provided increased knowledge in all disciplines. Students should be encouraged to "go with their strengths" and to follow their passions. Within formal education, they need to be able to focus the majority of their time on the content within their discipline so that they become fluent in civil engineering, art, microbiology, literature, geology, business administration, psychology or whatever they choose to do. All disciplines, however, must also realize that a negative consequence of extremely specialized education is that it is very easy to lose perspective on just how narrow our educational experience becomes. The classic joke about engineers believing that a multidisciplinary team includes an electrical engineer, a civil engineer, a mechanical engineer, and a chemical engineer exemplifies this.

All specialists, including those with multi- or interdisciplinary "specialties," must recognize that their discipline alone does not have the potential to drive industrial ecology (or any other approach) all the way to sustainability. Additionally, all specialists must be open to the idea that the prevailing paradigms of their discipline may contribute to perpetuating non-sustainable activities and attitudes. Therefore, an educational model that provides students with opportunities to exchange ideas and to identify ways to apply very diverse, but very specific disciplinary training to sustainability concerns may be effective. Examining an issue from diverse perspectives and using tools from diverse disciplines to identify information and to analyze information creates a synergistic effect, which may provide the fuel necessary to propel a sustainability revolution. I am not suggesting that improved communication and teamwork will magically provide the answers for achieving sustainability. Rather, it is a first and crucial step toward finding ways to shift our philosophy that is so closely linked to our ideas about technology and development and social order. These are key to generating the revolution that Ruckelshaus (1989) says is required.

Many environmental studies and environmental science programs attempt to provide multidisciplinary opportunities for students and the research presented here show that multi-disciplinary programs are the second most common place to find IE in higher education. However, there is ongoing debate about whether existing multidisciplinary curricula are effective (Luke 1996; Soule and Press 1998). Many environmental education programs use a simplistic and ineffective model suggesting that once the science is understood, students can simply apply social science knowledge and find a solution to a problem (McKeown-Ice and Dendinger 2000). There are severe institutional barriers to establishing truly cross-disciplinary efforts in most universities. Tenure requirements, accreditation concerns, grant giving, and power struggles within and among departments make it difficult to fully collaborate across campus. Even when there are opportunities, they are typically not truly integrative. Simply taking courses from various departments or taking courses taught by professors from different disciplines does not promote the kind of interaction that I suggest is necessary. Additionally, these programs are not necessarily appropriate for students who are truly passionate about being a chemist or an historian but who want to apply their skills to IE and hence to sustainability issues. There are programs available that understand this and have been organized to address it. The University of Michigan's Certificate Program in Industrial Ecology is one example. Students overlay the IE-relevant coursework as a complement to their traditional degrees in business, engineering, natural resources, environmental health sciences or public policy. The Norwegian University of Science and Technology applies a similar approach in their multidisciplinary program. Students from various departments (natural science, engineering, social science, humanities) specialize in industrial ecology from within their home department. These types of programs reduce the potential for students to see IE as somehow disciplinary specific as they have colleagues from various departments who are also learning about IE.

Additionally, there is evidence that some research efforts are taking a much more holistic approach. Emerging IE research is addressing the connections between resource use, technology and the human factors that will combine to determine what technology is acceptable and how resources are actually used. One effort that will contribute to better understanding the role that behavior plays in determining "best" practices is a project to identify practices that seem logical, but may not be advantageous in all circumstances. For example, it seems to be common sense that developing secondary markets for goods will lessen resource use and hence reduce environmental damage and promote sustainability. This is part of the mantra: reduce, reuse, recycle. Yet, recent research is revealing that for some products, having a strong secondary market actually encourages resource use as it increases demand for "new" products (Thomas 2002). The "common sense" idea is not necessarily a uniform truth. There are also research projects emerging that couple models from social science concerning human motivation and behavior with IE-relevant tools such as life-cycle assessment and material flow analysis (Hofstetter 1998; Binder 2002). These prototype research efforts reflect a generous move toward helping IE become a more effective sustainability vehicle by better understanding people and their relationships with materials, resources, and technology. The researchers are moving well outside their disciplinary boundaries and are working with colleagues from diverse disciplines as an attempt to merge their expertise with other expertise. Encouraging students to engage in similar activities and thought processes as they pursue IE-relevant education seems prudent.

In searching for IE programs I found that many efforts are evolving within research centers or institutes. This is likely related to funding strategies as well as a result of the difficulties in implementing change in traditional departments, especially to introduce a subject that is not rigidly defined or part of any accreditation criteria. While they often suffer from lack of institutional support, centers and institutes can be an effective model for promoting IE and its role in sustainability within existing educational structures. Additionally, if a research center is successful, it can serve as a catalyst for developing new departments and degree programs. Allowing students and faculty from various disciplines (ideally from diverse schools throughout the campus) to conduct research together through an institute provides an excellent opportunity for the individuals to see an issue from diverse perspectives and to make connections among various disciplines. The approach proposed here to highlight communication and teamwork skills could be addressed within an institute as well as within a classroom setting. Such an education will prepare students for the challenges of conducting research and working with industry and/or the public once they graduate.

In fact, this approach can help meet employer expectations and redress deficiencies in current educational efforts that produce environmental managers. Thomas and Nicita (2003) found in their surveys of Australian employers that "the ability to work in a team" was the single most important attribute that they expected from environmental program graduates. Communication skills (written and oral) were also extremely important and ranked higher than research skills. Similarly, Benton and Cottle (2000) surveyed corporate and government organizations about their experiences in hiring students to work in environmental affairs and they found that among non-computer skills, 48% of respondents said that the students lacked "integrative skills." This was the most common response. Presentation/communication skills were noted by 43.4% of respondents and writing skills by 42.2%, the number two and three responses, respectively. Clearly, the programs in existence purporting to train people for careers in environmental fields are not adequately addressing these key themes.

This is not news in some technical fields. For example, professionals with the National Academy of Engineering note that a strictly technical education is no longer sufficient to prepare engineers for what they will face when they enter the workplace. Wulf and Fisher (2002) write that, "As the world becomes more complex, engineers must appreciate more than ever the human dimensions of technology, have a grasp of the panoply of global issues, be sensitive to cultural diversity, and know how to communicate effectively" (36). McLellan (2000) agrees, "To be effective in the policy world, young scientists need to learn analytical and communications skills that are relevant to that world" (40). This is especially salient to the IE community. Attendees at the 2nd International Society for Industrial Ecology conference in June 2003 repeatedly emphasized that for IE to make a difference in promoting sustainability, it is time for its ideas to move from research labs to decision-makers' desks.

The model depicted in Figure 3 reflects one way to ensure that future employer surveys no longer identify teamwork, integration, and communication skills as deficient and helps to ensure that freshly minted IE specialists know how to integrate diverse information and to better communicate across disciplines. The expanding IE presence in curricula and in research institutes provides engineering and other technical programs an opening to address growing concerns that technical education needs to be revised to adapt to changing social and cultural dynamics. Additionally, it fits with broader societal demands for improved communication between technical experts and public that includes both technical information and the values and emotions related to various issues (Waddell 1995). Just as no single discipline can address an issue, it is no longer sufficient or appropriate to employ a one-way communication process whereby technical experts simply "inform" the public and/or decision-makers. To embark on a more stable ride toward sustainability, embracing a more complex approach to communication will be paramount – especially as cultural differences are increasingly important in designing sustainable programs and practices in non-industrialized nations. Therefore technically trained individuals need exposure to non-technical perspectives, including humanities-based perspectives, and need to be comfortable communicating with other disciplinary experts about all aspects of an issue. Of course, this works the other way as well. Students in humanities and social sciences do need a better understanding of technical issues if they are to contribute to a sustainable society. However, as the data here reveal, the majority of efforts related to IE are not in the humanities or social sciences and there is a strong possibility for this to create a roadblock to our efforts to reach sustainability.

Improving communication and expanding IE education to include a broader set of social science and humanities perspectives will likely increase the nebulous nature of IE. For many, the lack of a consistent and uniform definition for sustainability and for industrial ecology is troubling. I contend that the flux in these concepts is allowing diversity to flourish and encouraging researchers and practitioners to creatively apply their own perspectives and ideas. Because sustainability will require not doing things the same way we have always done them, this creativity can be a powerful force. IE cannot be a driving force for sustainability if it limits itself to one, or a few disciplinary tracks. In contemporary educational settings, creating a static definition may lead to confining IE to a particular place on campus. The evidence presented here suggests that this is already happening by default. As technical based programs expand, the perception that IE is strictly a technical program may preclude experts in other areas, especially the humanities, from exploring the potential for all disciplines that resides in the philosophy driving IE. Being the "science of sustainability" is a nice motto, but emphasizing science and technology based programs at the expense of other perspectives will not provide the quickest route to sustainability. In fact, attempts to insulate science and technology from other, less quantifiable, human factors will likely impede efforts toward sustainability. Perhaps the revolutionary power in IE and sustainability is in NOT unifying our ideas and approaches but continuing to remain open to new ways of thinking and being. The rapidly growing interest in IE is encouraging, but there is still much work to do to ensure that the educational processes that teach tomorrow's researchers and decision-makers how to "do" industrial ecology, are not promoting a very narrow and limited view of what it will take for IE to chauffeur us along the path of sustainability.

References

Benton R Jr., Cottle SE (2000) How Well Do Universities Prepare New Environmental Managers? Environmental Practice 2 (3): 247-258

Claussen E (2001) Making Collaboration a Matter of Course: A New Approach to Environmental Policy Making. Environmental Practice 3 (4): 202-205

Ehrenfeld J (2003) Putting a Spotlight on Metaphors and Analogies in Industrial Ecology. Journal of Industrial Ecology 7(1): 1-4

Foster J (2001) *Education* as *Sustainability*. Environmental Education Research 7(2): 153-165

Frosch RA, Gallopoulos NE (1989) Strategies for Manufacturing. Scientific American 261 (3): 144-152

Haury DL (1998) Education for Environmental Sustainability. ERIC Clearinghouse for Science Mathematics and Environmental Education, ERIC Digest ED433194

Hofstetter P (1998) Perspectives in Life Cycle Impact Assessment: A Structured Approach to Combine Models of the Technosphere, Ecosphere and Valuesphere. Kluwer Academic Publishers

Isenmann R (2002) Further Efforts to Clarify Industrial Ecology's Hidden Philosophy of Nature. Journal of Industrial Ecology 6(3-4): 27-48

Luke TW (1996) Generating Green Governmentality: A Cultural Critique of Environmental Studies as a Power/Knowledge Formation. http://www.cddc.vt.edu/tim/papers.html

McKeown-Ice R, Dendinger R (2000) Socio-Political-Cultural Foundations of Environmental Education. The Journal of Environmental Education 31 (4): 37-45

McLellan E (2000) What Johnny Really Needs to Know: A View From the Hill. Geotimes, October

Ruckelshaus WD (1989) Toward a Sustainable World. Scientific American 261 (3): 166+

Savory A, Butterworth J (1999) Holistic Management: A New Framework for Decision Making. Island Press

Soulé M, Press D (1998) What is environmental studies? BioScience 48: 397–405

Susskind LE, Jain RK, Martyniuk AO (2001) Better Environmental Policy Studies: How to Design and Conduct More Effective Analyses. Washington: Island Press

Tarlock AD (2001) Ideas Without Institutions: The Paradox of Sustainable Development. Indiana Journal of Global Legal Studies 9:35-49

Thomas I, Nicita J (2003) Employers' Expectations of Graduates of Environmental Programs: An Australian Experience. Applied Environmental Education and Communication 2 (1): 49-59

Thomas V (2002) Demand Impacts of Second-Hand Markets. Unpublished draft

Tibbs H (1999) Sustainability. Deeper News 10(1), Publication of the Global Business Network

Waddell C (1995) Defining Sustainable Development: A Case Study in Environmental Communication. Technical Communication Quarterly 4(2): 201-216

White L Jr. (1967) The Historical Roots of Our Ecological Crisis. Science 155(3767): 1203-1207

Wulf WA, Fisher GMC (2002) A Makeover for Engineering Education. Issues in Science and Technology 18(3): 35-39

Binder C (2003) Personal conversation.

The sustainable industrial development: reality and vision

Jurgis Staniskis, Valdas Arbaciauskas

Institute of Environmental Engineering, Kaunas University of Technology, K. Donelaicio str. 20, LT-3000, Kaunas, Lithuania, Tel: +370 7 300764, Fax: +370 7 209 372, E-mail addresses: Jurgis.Staniskis@ktu.lt (J. Staniskis), Valdas.Arbaciauskas@ktu.lt (V. Arbaciauskas)

Abstract

The Institute of Environmental Engineering (APINI), Kaunas University of Technology developed a National Programme on Sustainable Industrial Development in Lithuania. This work was supported by the Ministry of Economy of Lithuania. The main objective of the programme is to help industry in coping with challenges and in utilizing opportunities of sustainable industrial development, i.e.: (i) to increase competitiveness of Lithuanian industry; (ii) to reduce negative process and product impact to the environment, (iii) to use energy and natural resources more rationally; and (iv) to improve work conditions and promote establishment of new workplaces. This paper presents key elements of the programme and an overview of progress in implementing sustainable industrial development measures in Lithuania over the past few years.

1 Sustainable Industrial Development and Business Competition

Most of the environmental problems that are linked to economic development and business conditions are becoming more and more complicated due to increasing environmental concerns. Sustainable development opens new business opportunities through the introduction of more environmentally friendly products and processes. Additionally, application of preventive measures, such as cleaner production, helps to improve environmental and economic performance, which, in turn, provides a competitive advantage.

The practical application of sustainable development principles in activities of industrial enterprises is becoming an important aspect of business competition.

Consequently the role of different stakeholders is becoming increasingly important (Staniskis et al. 2002):
- Customers require products and services provided by environmentally and socially responsible enterprises. Demand for products having minimal impact on the environment is increasing.
- Investors and banks evaluate activities on enterprises and increasingly consider environmental risks.
- Suppliers and clients, having environmental and quality systems, require their partners to implement such systems while attaining a particular level of environmental performance.
- Employees of enterprises are increasingly concerned about their work conditions.
- Awareness of society in the area of sustainable industrial development is increasing. In the future, society will not tolerate enterprises that do not implement measures aimed at reducing environmental impact.
- Governmental institutions constantly develop new and stricter regulations related to industrial activities. Enterprises implementing proactive measures of sustainable industrial development are prepared for such changes and, in a long run, are more competitive.

Industry's role of in the process of sustainable industrial development relates to changes in the production processes and products aimed at the reduction of impact to the environment with respect to the entire life cycle while improving the environmental and social performance of enterprises. To ensure sustainable industrial development, systematic application of the following measures is needed (Ministry of Economy 2003):
- *Cleaner production*. This strategy is based on rational use of energy and natural resources and minimization of pollution/ waste at the source where it is generated. Cleaner production helps to reduce enterprises' impact on the environment, increase economic performance and in some cases – improve work conditions. It is particularly important for enterprises to assess full costs associated with environmental protection, by using environmental accounting, as it motivates enterprises to implement innovations and new management methods.
- *Industrial Ecology*. This concept requires that an industrial system be viewed not in isolation from its surrounding system, but in concert with them. It is a systems view in which one seeks to optimise the total material cycle, particularly on the regional level. Factors to be optimised include resources, energy, and capital.
- *Integrated management systems*. These systems develop a cycle for continuous improvement with involvement of all employees. To ensure continuous improvement of environmental, economic and social performance, enterprises have to implement integrated management systems that cover environmental, quality and health and safety aspects.

- *Product related measures of sustainable industrial development (e.g. eco-design, life cycle assessment, extended producer responsibility, environmental labeling).* These measures help to reduce the negative impact on the environment during the entire life cycle of products.
- *Sustainability reporting.* This activity contributes to rising of public awareness in sustainable development. Participation in international initiatives such as the Global Reporting Initiative is particularly important. Development of sustainability reports motivates enterprises to analyse their activities and helps to identify opportunities for improvement of environmental, economic and social performance.

2 Overview of Progress in Implementing Sustainable Industrial Development Measures in Lithuanian Industry

The number of industrial enterprises in Lithuania implementing sustainable industrial development measures is increasing (fig. 1). More than one hundred out of the most advanced Lithuanian enterprises participated in various programmes and projects relating to sustainable industrial development. However, most of the implemented measures were in the areas of cleaner production and environmental/ quality management systems. Other areas such as product related measures of sustainable industrial development and sustainability reporting, generally, have not been used by Lithuanian enterprises.

Cleaner production

The Institute of Environmental Engineering, Kaunas University of Technology initiated the implementation of cleaner production in Lithuania and is the main institution in the country working in this area. In 1992-2002, with support provided by different donors and in co-operation with different foreign and local partners, APINI supported more than 100 Lithuanian companies to implement cleaner production projects. Most active in this regards were textile (21 companies) and food industry (19 companies). Cleaner production measures mainly have been used to reduce energy consumption (93 projects) and water consumption (70 projects). The number of projects in the areas of water pollution reduction, waste minimization and air emission reduction was almost the same – 34, 36 and 37 projects (Arbaciauskas and Staniskis 2002). In terms of innovation type, most investments have been used in process optimization and technology change (Fig. 2).

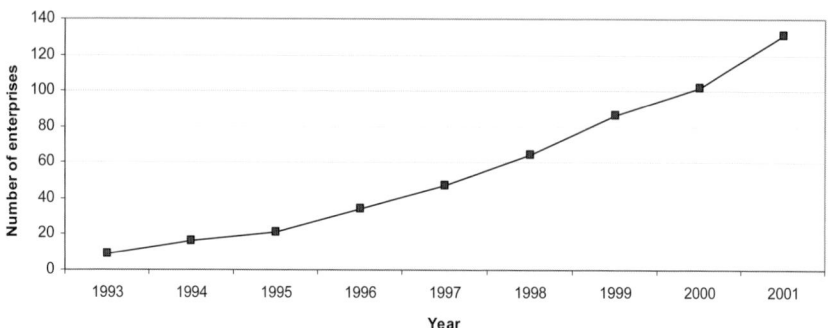

Fig. 1. Number of enterprises in Lithuania implementing sustainable industrial development measures in 1993 – 2001 (Ministry of Economy 2002)

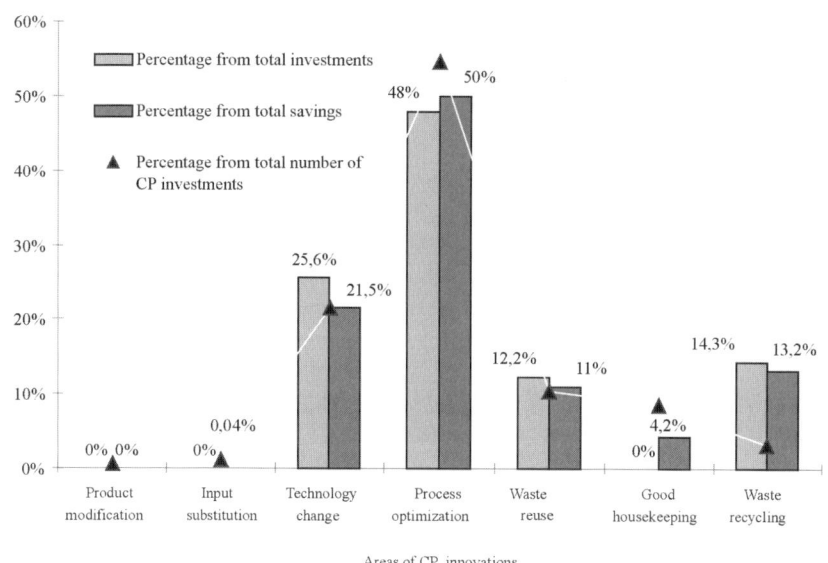

Fig. 2. Distribution of investments and savings in implemented CP innovations in 1993-2002 (Staniskis et al. 2002)

Implementation of cleaner production projects in many cases does not require large investments. Pay-back period of such projects is usually short and investments are possible from enterprises' own sources. However, implementation of larger projects with higher investments in Lithuania is often problematic. Problems exist on both demand and supply sides.

To solve such problems in Lithuania, Nordic Environmental Finance Corporation (NEFCO) established a special Revolving Facility to finance priority investments projects in the area of cleaner production (NEFCO 1997). Institute of Environmental Engineering plays a very important role in the process of project identification, feasibility analysis, development and implementation, as well as, project monitoring. Up through 2002, financing was approved for 35 investments projects.

Environmental and quality management systems

To build capacity in the area of environmental management systems, the Institute of Environmental Engineering, in co-operation with Norwegian company Det Norske Veritas and Ministry of Environment of Lithuania implemented a project, which was financed by the Norwegian government. During this project, 19 environmental auditors have been trained.

By December 2001, there were 18 companies in Lithuania that implemented certified environmental management systems in accordance to international standard ISO 14001.

Implementation of quality management systems in Lithuania started earlier and by December 2001, there were 176 companies that implemented certified quality management systems in accordance to ISO 9000 series.

To date, only foreign organizations provided certification services to Lithuanian enterprises. For some companies, particularly small and medium sized companies, such services are often too expensive. Therefore, financial support provided by Lithuanian government for companies to cover part of certification costs is very important in promoting implementation of environmental and quality management systems.

3   National Programme on Sustainable Industrial Development in Lithuania

The Institute of Environmental Engineering, Kaunas University of Technology developed a National Programme on Sustainable Industrial Development in Lithuania. This work was supported by the Ministry of Economy of Lithuania. The main objective of the programme is to help industry in coping with challenges and in utilizing opportunities of sustainable industrial development, i.e.: (i) to increase competitiveness of Lithuanian industry; (ii) to reduce negative process and product impact on the environment, (iii) to use energy and natural resources more rationally; and (iv) to improve work conditions and promote establishment of new workplaces (Ministry of Economy 2003).

Key elements of the programme are:
- establishment of framework conditions promoting the implementation of sustainable industrial development measures in industrial enterprises;

- capacity building and support to enterprises in implementing sustainable industrial development measures;
- promoting and supporting research and development;
- development and implementation of a monitoring system to evaluate implementation of the programme and monitor progress in the process of sustainable industrial development.

3.1 Framework Conditions

Regulatory and economic instruments

The role of governmental institutions is to periodically analyze regulatory and economic instruments related to sustainable industrial development and to make necessary changes. Regulations and economic instruments should promote implementation of sustainable industrial development measures in enterprises and increase their environmental performance. (OECD 1996) Currently in Lithuania, emphasis is on wastewater treatment and waste management. More attention should be given for implementation of preventive measures.

Informational measures and strengthening co-operation among different stakeholders

There is a need to develop and disseminate informational materials about sustainable industrial development measures and their practical implementation and to organize informational seminars on this topic. Informational measures should be also used to disseminate information about international initiatives related to sustainable industrial development. This would promote participation of Lithuanian organizations in these initiatives.

Information about enterprises' environmental performance should be accessible to the public. Activities on sustainability reporting should be initiated in Lithuania. There is a need to build necessary capacity in the country.

Using the experience from different award systems used in Lithuania for companies with high performance in particular areas (quality prize, prize for the best product of the year, etc.), it is proposed to establish an award system for companies that achieved tangible results in the area of sustainable industrial development.

To ensure open dialogue and co-operation among different stakeholders, an establishment of a public forum is needed. The forum would be used to exchange information and experience among industry, governmental institutions, academia and the public.

3.2 Capacity Building and Support to Enterprises

Training programmes and development of training materials

There is a need to increase capacity that would enable broader implementation of sustainable industrial development measures in enterprises, industrial associations, governmental institutions, universities and consulting companies. Two types of training programmes could be used: educational programmes in universities and special training programmes for specialists. (UNEP and APINI 2001)

There is a need to continue cleaner production training programmes using methodology developed in Lithuanian – Norwegian Cleaner Production Programme. Experience in other countries also demonstrated that long-term training programmes including both theoretical and practical training are the most effective way for capacity building (Dobes 1997). To reduce costs, regional programmes could be organized. In addition to cleaner production, the contents of such programmes should cover other aspects of sustainable industrial development. Training materials in the areas of eco-design, life cycle assessment and sustainability reporting must be developed in the Lithuanian language. Implementation of demonstration projects should also continue with more emphasis on their follow-up (Rodhe and Lindhqvist 1996).

For top managers from enterprises, representatives from governmental institutions and other stakeholders, 1-2 day training sessions could be organized with the aim to familiarize participants with sustainable industrial development measures and their practical implementation (OECD 1999).

There is a need to introduce new education modules in the university curricula. Establishment of the M.Sc. programme in cleaner production and environmental management at Kaunas University of Technology is particularly important in this regard.

Financing of investments in cleaner technologies

In cases when financing of investments is not possible from the enterprises' own sources, access to other financing sources should be ensured. Possibilities for cleaner production investment financing from NEFCO Revolving Facility should be better utilized. For this purpose, project development should be supported, particularly in small and medium sized enterprises. It should be ensured that operation of Lithuanian Environmental Investment Fund be more oriented to finance cleaner technologies.

Technical assistance to enterprises

Enterprises, particularly small and medium ones, very often require technical assistance in implementation of sustainable industrial development measures, including assistance in development of investment project documentation (OECD 1998). A special technical assistance system should be developed in the country.

Operations of such system could be co-ordinated by National Cleaner Production Centre.

3.3 Research and Development

The total volume of research activities does not have a direct link to the performance level of enterprises. Effective co-operation between enterprises and research organizations is more important in this regard. There is a need to establish conditions supporting such co-operation and promoting contacts between different partners in search for mutual benefits. Governmental support for applied research activities is an economically viable long-term policy, which ensures economically, environmentally and socially sound business development. Therefore, government should facilitate investments in research and development activities made by industry. Partial financial support could be provided by the government for projects implemented jointly by industry and research organizations.

3.4 Monitoring System

To evaluate the progress in the process of sustainable industrial development and to ensure the effective implementation of the Programme on Sustainable Industrial Development, there is a need to develop a monitoring system.

In terms of the Programme's objectives and goals, a set of quantitative indicators has to be developed, e.g. a fraction of enterprises that implemented certified management systems, a fraction of environmentally labelled products, a fraction of enterprises publishing sustainability reporting. To evaluate the progress of industry in the process of sustainable development, indicators such as use of energy and natural resources for a product unit and a number of work hours lost due to accidents could be used.

The main objective of the monitoring system should be to support the decision-making process. Therefore, one of the key challenges is to ensure quality of the monitoring system. Clear procedures for collecting and processing information must be established.

4 Conclusions

A number of programmes and projects in the area of sustainable industrial development have been implemented in Lithuania during the last decade. These projects resulted in a number of cleaner production measures implemented in industrial enterprises. A number of companies implemented effective environmental management systems based on cleaner production approach. Lithuanian experience in financing cleaner production investments is particularly valuable and could be used in other countries.

To strengthen ongoing activities and to ensure systematic application of sustainable industrial measures in industrial enterprises, the National Programme on Sustainable Industrial Development was developed in Lithuania. The Programme will help to create the framework conditions that promote sustainable industrial development and will create capacity to enable effective implementation of sustainable industrial development measures, particularly in the areas where not many activities have been undertaken in the country so far, e.g. in product related and sustainability reporting areas.

References

Arbaciauskas V, Staniskis J (2002) Sustainable Industrial Development in Lithuania. In "The Survey of Lithuanian Economy", No. 1. Ministry of Economy, Department of Statistics.
Dobes V (1997) Five Key Factors for Ensuring Sustainability of Cleaner Production Programmes within a Country. In Proceedings of 4th European Roundtable on Cleaner Production, Oslo.
Ministry of Economy Lithuania (2003) Programme on Sustainable Industrial Development. Vilnius.
Ministry of Environment (2002) National report on Sustainable Development, Vilnius.
NEFCO (1997) "NEFCO Revolving Facility for Cleaner Production Investments". In Proceedings of the 4th European Roundtable on Cleaner Production, Oslo.
OECD (1996) Which Policies, Which Tools? Washington Waste Minimisation Workshop. Volume II.
OECD, EAP Task Force (1998) Policy Statement on Environmental Management in Enterprises in CEEC/NIS. Paris.
OECD, EAP Task Force (1999) Cleaner Production Centres in Central and Eastern Europe and the New Independent Sates. Paris.
Rodhe H Lindhqvist T (1996) Danish Cleaner Production Projects in Central and Eastern Europe 1991-1995. Danish Environmental Protection Agency, Copenhagen.
Staniskis JK, Stasiskiene Z, Kliopova I (2002) Cleaner Production: Systems Approac (In Lithuanian language). Kaunas, Technologija.
UNEP, APINI (2001) Introduction to Cleaner Production Concepts and Practice. Kaunas, Technologija.

Sustainable Pathways

Clean technologies for wastewater management in seafood canning industries

Francisco Omil, Elena García-Sandá, Ramón Méndez and Juan M. Lema

Department of Chemical Engineering.
School of Engineering. University of Santiago de Compostela.
Campus Sur s/n. 15782 Santiago de Compostela, Galicia, Spain.
Tel: (34) 981.563.100 ext. 16021 Fax: (34) 981.547.168
e-mail address: eqomil@usc.es

Abstract

Fish canning industries generate a large number of different wastewaters that usually are treated together in a huge and complex plant, which implies high installation and maintenance costs (personnel, electricity, supplies, etc.). This paper shows that a better wastewater treatment management can be achieved if the different streams are treated as a function of their main characteristics (high organic content, low organic content, high solids content). Two scenarios are compared: i) one single plant to treat the overall effluent and, ii) a plant comprising different lines to treat the three streams resulting from the collection of similar wastewaters. In terms of investment and operational costs and, especially, removal efficiency, this latter strategy is clearly more favourable.

An additional advantage for the second scenario is that some of the main polluted effluents generated by this industrial sector can be further processed in order to obtain valuable subproducts, with the additional consequence of a substantial reduction of the organic load and volume to be treated in the wastewaters treatment plant. The production of fish-meal has been traditionally the only measure followed in this sense, in order to recover part of the rests of fish rejected during manufacturing processes. This paper shows three examples of this strategy: i) the potential reuse of the water discharged during the sterilisation step; ii) the development of a process to obtain valuable oil from the floating matter sludges separated by flotation devices; iii) the studies focussed on the processing of tuna cooking effluents in order to obtain a protein concentrate as a complement for animal foodstuff.

1 Introduction

Waste management is becoming one of the key problems of the modern world, an international issue that is intensified by the volume and complexity of waste discarded by society's domestic and industrial waste. Better practice and safer solutions than those followed in the past are required. Not only is there a need for more research on current treatment and disposal methods such as biological and physic-chemical effluent treatment, incineration, landfills, etc., but also on recycling, waste minimization, clean technologies, waste monitoring, public and corporate awareness and general education.

In spite of all the improvements achieved in industry regarding several major-polluting substances during the last few decades, industrial production processes are still the main source of pollution in Europe. Following the IPPC regulations, the permits must be based on the concept of Best Available Techniques (BAT), which can be described as those technologies which allow the highest level of preventive measures against pollution to be taken, with regard to what is technically feasible.

One of the main points of the current legislation of the European Union is the need for zero-emission production processes, which may be developed by minimising the discharge of pollutants from industrial activities. In order to achieve this objective, the first priority would be the reduction of the discharge of pollutants from each processing unit in a factory, and the second priority would focus on the recycling and treatment of wastewaters and other waste materials.

Old manufacturing processes (Fig. 1A) were based on a considerable consumption of clean water, resulting in a large quantity of effluents to be treated in a complex wastewater treatment plant (WWTP). In order to reduce water requirements, new processes are based on specific on site treatment plants for the wastewaters generated in each process. This concept implies small and specific treatment units, which can provide reused water for each process. Only the waste effluent generated from these units would be treated in a general WWTP (Fig. 1B).

The following priority would be the optimal selection and operation of the wastewater treatment process in order to combine both the achievement of a high quality effluent and the minimisation of energy consumption. The reduction of wastewaters in a manufacturing process results very frequently in the generation of a smaller, but very concentrated flow to be discharged, which may be more difficult to treat. The final layout could be the result of the appropriate combination of various wastewater treatment methods, such as biological (anaerobic, aerobic, anoxic, etc.) or physic-chemical methods (coagulation-flocculation, flotation, etc.). Therefore, the development of advanced technologies for the treatment of wastewaters and wastes that could accomplish the objectives previously mentioned (high quality, less energy, recycling) is highly necessary.

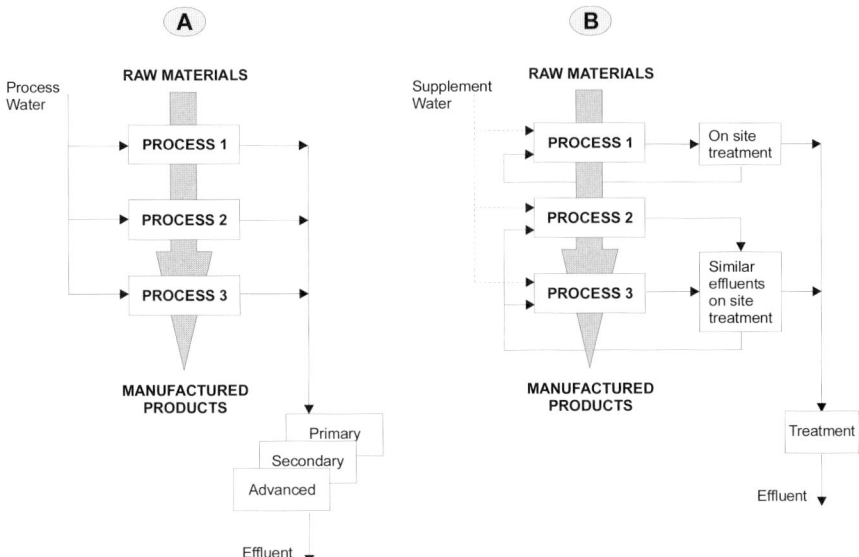

Fig. 1. Water treatment for old and new manufacturing processes (adapted from Hu et al. 1999)

1.1 Clean wastewater technologies

The combination of tighter restrictions on sludge disposal site location, air pollution, hazardous waste disposal, odour control, in addition to other factors, has had a substantial impact on the applicability of conventional treatment plants based on the activated sludges or coagulation-flocculation processes for the treatment of industrial wastewaters. In order to manage a successful scheme for the treatment of wastewaters, the development of processes combining both a high efficiency and low construction and maintenance costs has become a major priority. In this context, anaerobic wastewater treatment (AWT) is becoming increasingly popular worldwide (Lema and Omil 2001).

A first and very important issue of anaerobic technology is the significantly reduced production of excess sludges (5-20%), when compared with aerobic-based processes or coagulation-flocculation units. In the present situation in Europe, with landfills for organic wastes near the point of closure, with more limitations for agricultural applications, etc., technologies producing smaller amounts of waste sludges will be in a better situation. Sludges are often considered as toxic or dangerous wastes, especially those obtained from physic-chemical processes being therefore necessary to manage them in specialised plants. Anaerobic systems are also characterised by the possibility of applying higher loading rates, commonly varying from 5-20 kg COD/m^3·d, whereas the usual loads to aerobic systems are around 0.5-3 kg/m^3·d. This implies a substantial reduction of the reactor volume and the available space required and, therefore, lower installation costs. On the other hand, nowadays modern industries are achieving a high degree of re-use and recycling of process water, which results in reduced flows of the final liquid effluent to be discharged, but with higher pollutant concentrations (Fernández et al. 1995). Therefore, processes based on anaerobic technologies are especially indicated for the treatment of smaller flows of wastewaters highly polluted with organic matter. Besides, anaerobic treatment produces energy in the form of biogas ($12.7 \cdot 10^6$ J/kg COD converted) while aerobic treatment usually requires between 1.8 and $7.2 \cdot 10^6$ J/kg COD, depending on the technology applied (Speece 1996). This characteristic makes energy conservation and the consequent ecological and economic benefits possible. Part of the energy may be used to heat the digester, while the excess of energy can be converted into electricity ($10.5 \cdot 10^6$ J required per $3.6 \cdot 10^6$ J of electricity generated).

When physic-chemical processes are used, such as coagulation-flocculation the management of the resulting waste sludges becomes crucial. Usually these wastes have to be sent to factories specialised in the treatment of hazardous waste at a high cost, even in the case that only biodegradable matter is removed because of the nature of the additives used. In these cases, a cleaner procedure could be developed from the study of the feasibility of the use of other coagulants and flocculants with similar characteristics of biodegradation (Guerrero et al. 1998) or at least, compatible with the common practices used for organic sludges (anaerobic digestion, land application, landfills, etc.).

2 Sea food canning industries

In the European context, the fish and shellfish canning industry is mostly concentrated in Galicia (NW of Spain), with around 65% of the total Spanish production, 45% of the factories and 67% of the jobs. The industrial structure of this sector in Galicia can be divided into three groups: five large factories with around 58.6% of the total turnover; medium-sized factories, with 32% of the total turnover; and the rest is divided into a large number of small factories (more than 55% in number).

Galician cannery industries manufacture a high variety of raw materials. Tuna account for more than half of the total production. Mussel processing is also very high, and at lower extent, sardine. Other species used in smaller quantities are octopus, squids, mackerel, etc. The case of mussel is illustrative of the natural richness of the estuarine systems known as the Rías Baixas, characteristics of this region and where approximately 250 000 tons of mussel are produced, 60% of which are processed in coastal plants. This productivity is among the highest in the world (Fraga 1976). On the other hand, most of these factories are located along the two broadest estuaries where environmental problems have been detected because of the large volume of discharged wastewaters (Omil et al. 1996).

The main processes present in sea food canning plants are defrosting, cooking, canning and cleaning. Table 1 summarizes the main steps present in a representative factory with different manufacturing lines. The process for tuna comprises the following stages: defrosting, peeling, cooking, canning and the finishing operations, which are similar to all lines (sauce filling, washing of cans, sterilization and packing). In the case of mussel, the main steps are washing, trimming, cooking, size classifying, dehydrating and finishing. Sardine processing includes scraping, cutting, conveying, canning and finishing. The operation followed with other species (octopus, squids, mackerel, etc.) is similar. A more detailed description of all these manufacturing lines has been previously published (Veiga et al. 1994).

Key resources used by the sea food processing industry include raw materials, water and energy. Raw materials used have been previously quoted. Some species, like mussels, are cultured near the sea shore, whereas tuna is fished all over the oceans. Anyway, most of these factories are located near the sea shore, usually close to port facilities. Traditionally, this industrial sector has been a large water user. Water is used as an ingredient, an initial and intermediate cleaning source, an efficient transportation conveyor of raw materials, and the principal agent used in sanitizing plant machinery and areas. Although water use will always be a part of the whole process, it has become the principal target for pollution prevention, source reduction practices. A special particularity of this sector is that both fresh and sea water are used in large amounts. In the case of energy use, compared to other industries, for example, metal fabrication or pulp and paper, this industry is not considered energy-intensive. Facilities usually require electrical power, which is supplied by local utilities, to run processing machinery. Steam production has been usually done by using fossil fuel, although modern cogeneration plants using natural gas are nowadays being adopted.

2.1 Environmental problems

The main environmental problems of seafood cannery industries are related to the emission of large volume wastewaters. Emissions to the atmosphere, solid wastes classified as urban, as well as inert and hazardous wastes, although significant, not such important (IHOBE 1999).

Atmospheric emissions are mainly related to the use of fuel-oil to generate steam, although in recent years many factories have switched to natural gas in new cogeneration plants, a much cleaner technology. Emissions of odours, especially due to the H_2S generated by uncontrolled fermentation of organic wastes, is also common.

The solids wastes generated in fish canning industries are important. In fact, during the processing of species, the amount of raw products converted into waste can reach up to 50% by weight (Aguilar and Sant'Anna 1988), with a high organic content. The production of fish-meal has been traditionally the only measure followed to recover part of the rests of fishes rejected. Other solid wastes with similar nature of urban wastes are cardboard, wood and plastics. Besides, deteriorated cans comprise a type of residue completely inert. On the other hand, the amount of hazardous wastes produced by these industries is small. These wastes are fundamentally marking inks and lubricating oils, which can be easily managed.

Table 1. Main lines of production and unit operations that generate liquid effluents in sea food canning factories

Production lines	Effluents generated	
	Code	Description
Tuna	T1	Defrosting
	T2	Peeling
	T3	Washing (peeling)
	T4	Fish washing
	T5.1	Cooking (first discharge)
	T5.2	Cooking (second discharge)
	T6	Grills washing
Mussel	M1	Trimming
	M2	Washing
	M3	Cooking
	M4	Dehydrating
	M5	Conveying
Sardine	S1	Scraping
	S2	Conveying
	S3	Cutting
	S4	Grills washing
Other waters	OW	Sterilisation, general washing, etc.

As previously mentioned, there is no doubt that the main environmental impact caused by these factories is occasioned by the discharge of wastewaters. Wastewaters generated in seafood processing factories have quite a high organic load (mainly from protein, carbohydrates and fat) and also high levels of salinity. The impact caused on the environment is especially relevant when they are discharged into sea waters with low renewal capacity, as occurs in the estuaries of Galicia, where the largest number and highest density of these factories in Europe are located.

3 Characterisation of sea food canning wastewaters

Sea food-processing wastewater can be characterized as non toxic, with a high and variable content of organic matter both as suspended or dissolved solids. Fish and shellfish canning industries generate different liquid effluents that present a wide range of characteristics according to the raw material processed (tuna, mussel, sardine, etc). Cooking effluents contain the highest organic matter load. The high salinity (Na^+, Cl^-, SO_4^{2-}) is caused both by the raw material (fish and shellfish) and the seawater used in the whole factory. Table 2 summarizes the main characteristics of the streams present in one of the biggest representative factories in Galicia. A high variation in flows and composition can be observed, which makes very difficult to design a single treatment plant for all of them.

Table 2. Characteristics of the different wastewaters generated during tuna, mussel and sardine manufacturing processes and a representative overall effluent (flow in m^3/d, temperature in °C, concentrations in g/l) (Soto et al. 1990)

	Flow	pH	TSS	VSS	COD_T	Cl^-	T
T1, defrosting	28	6.48	0.53	0.39	2.55	19.72	4
T2, peeling	21	10.39	5.04	3.76	8.37	18.34	15
T3, washing (peeling)	44	8.86	0.33	0.20	0.75	19.12	15
T4, fish washing	48	7.99	0.52	0.33	1.70	19.34	15
T5.1, cooking (first discharge)	7	6.18	5.85	5.22	46.50	11.39	90
T5.2, cooking (second discharge)	44	6.18	1.31	1.19	6.93	17.65	35
T6, grills washing	3	12.40	3.38	2.09	16.78	0.30	15
T, total tuna	195	7.82	1.38	1.09	5.35	18.28	21
M1, trimming	1600	7.97	1.59	0.69	1.20	19.21	15
M2, washing	1600	8.06	0.34	0.15	0.23	19.32	15
M3, cooking	96	6.95	1.30	1.06	16.90	13.66	100
M4, dehydrating	12	6.47	1.32	1.02	27.13	17.79	100
M5, conveying	320	8.11	0.21	0.06	0.15	19.21	15
M, total mussel	3628	7.99	0.91	0.41	1.18	19.11	18
S1, scraping	4	6.70	1.03	0.65	3.12	18.23	15
S2, conveying	300	7.89	0.18	0.08	0.20	19.94	15
S3, cutting	64	6.75	1.61	1.47	4.10	19.57	15
S4, grills washing	70	9.15	0.45	0.22	0.81	18.88	15
S, total sardine	438	7.91	0.44	0.31	0.89	19.70	15
OW, Other waters	800	7.91	0.90	0.60	1.57	16.71	15
OVERALL EFFLUENT	5222	7.96	0.88	0.46	1.38	18.75	17

Another peculiar characteristic of these factories is the frequent change of raw materials to be treated (Soto et al. 1990), which implies changes in wastewater characteristics, and also the existence of stand-by periods, which imply periodical restart-up phases of the wastewater treatment plant (Balslev-Olesen et al. 1990). In addition, the volume and characteristics of the final effluent change significantly through the day, depending on the streams that are being discharged: cooking wastewaters are usually discharged in the morning; hosing down effluents, which contain large amount of suspended solids, are generated at the end of the working day; the effluent generated during the sterilization of cans, which is carried out in the afternoon, contains hardly any suspended solids or organic matter.

The adoption of measures to achieve a reduction of wastewaters results very frequently in the generation of a smaller, but very concentrated flow to be discharged. This trend was also observed in the factory studied in this work since during the 90s all the manufacturing processes were subjected to a deep modernisation, which allowed to the company to increment drastically their production and a better water management. It can be observed that the concentration of the main polluted effluents (cooking steps) was highly increased (Table 3). In the case of mussel manufacturing, the consequence of the expansion of production was not a higher effluent flow after cooking but a dramatic increase in concentration of organic matter, from 16.90 g/l up to 107.1 g/l. In the case of tuna, flows were significantly higher due to the enlargement of production. The increase in COD concentration was observed for the second discharge. On the other hand, the effluent resulting from the grills washing was less concentrated because of the better management of water and raw materials in the whole manufacturing process.

Table 3. Comparison of the main characteristics of some of the most polluted wastewaters generated in a representative factory after its modernisation and increase of production from year 1990 to 1999 (flow in m^3/d, concentrations in g/l)

	1990			1999		
	Flow	COD$_T$	TSS	Flow	COD$_T$	TSS
M3, cooking	96	16.90	1.30	60	107.1	1.68
M4, dehydrating	12	27.13	1.32	24	68.06	1.53
T5.1, cooking (first discharge)	7	46.50	5.85	54	43.90	1.84
T5.2, cooking (second discharge)	44	6.93	1.31	192	34.20	1.66
T6, grills washing	3	16.78	3.38	75	7.85	1.38

Thus, the current trend of concentration observed in this industrial sector towards the construction of fewer but larger plants, commonly with a high level of automation and more efficient equipment leads to the generation of large quantities of extremely polluted effluents, with a potential high environmental impact.

4 Overall wastewaters management: a case study

Two general alternative treatment strategies can be considered: 1) the global treatment of the effluent volume, using one or several methods in series; 2) the application of different methods to each effluent or group of effluents in parallel, depending on their characteristics. In general, the viability or convenience of applying a method of treatment depends directly on the characteristics of the effluents and on their volume and strength.

Strategy (1) is suitable for global effluents composed of individual effluents with similar characteristics, and also when the global volume is small, because a duplication of the installations, in this case, would be unfavourable. Strategy (2) should be considered whenever there is a considerable difference between the contribution of each individual effluent discharged to the final effluent, respective to their flow or their organic load. This strategy is also recommended in the case of having different factories located close to each other and that are interested in treating their effluents jointly at a local level (Veiga et al. 1994).

In the case of the factory considered in this work, its magnitude made recommendable to follow the latter strategy. Starting from the complete wastewaters characterisation reported in 1990 for this factory (Table 2), in the following paragraphs different treatment strategies will be discussed.

4.1 Treatment of the whole effluent

The overall effluent amounts to 5222 m^3/d, which represent a huge volume to treat in a single plant. However, if this solution would be considered a large and very expensive installation should be needed. According with its composition both a physic-chemical unit or a biological treatment plant based on activated sludges could be considered, separately or in combination.

Although the installation costs of the physic-chemical unit would not be extremely expensive, the cost of additives (coagulants, flocculants) for such a large effluent would be prohibitive, as well as the management costs of the sludges generated.

On the other hand, taking into account its COD content, a common process based on activated sludges could be directly applied. Of course, the magnitude of the different units will be very big and also the amount of excess of sludges. Besides, taking into account the high salinity of this effluent, the COD removal efficiency will be significantly lower than usual.

Table 4 shows the range of the main parameters to be considered for such an activated sludge plant. Due to the high salinity of the effluent a relatively low COD removal efficiency has been considered, 70% (Omil 1993), but even if this percentage could be achieved the air needed (considering a common transfer efficiency of 8%) would be more than 200 000 m^3/d. Besides, the volume of excess of sludges would be also very high, 40 m^3/d as minimum. The plant needed would be also really large, and the sum of the dimensions of the main units accounts for more than 1200 m² constructed, with a volume higher than 4000 m³. Based on this design an overall TSS and COD removal efficiencies of 75 and 78%, respectively, could be achieved, which implies a final TSS and COD concentration around 200 and 300 mg/l, respectively.

Table 4. Approximate characteristics (minimum design values) of an activated sludge based plant to treat the overall effluent (volume in m³, surface in m², flow in m³/d)

	Dimensions		Reductions	
	Volume	Surface	TSS	COD_T
Primary sedimentation	650	150	50%	-
Biological reactor	2600	900	-	70%
Secondary sedimentation	870	150	50%	-
	Flow			
Excess of sludges	40			
Design air needed	200 000			

From these values it appears clearly that the application of this treatment strategy, although beneficial for the environment because of the reduction of pollution of wastewaters, implies a high cost of investment and maintenance, a high consumption of energy and also high amounts of wastes that need a further treatment.

4.2 Segregation of effluents according with their characteristics

This option was in fact the one followed by the factory considered in this work, not only because of the high costs that would imply the adoption of a common treatment for the overall effluent, but because it has no sense to mix effluents with so different characteristics, especially M1 and M2 with the rest.

The separation criterion was made taking into account the percentages of organic load and flow that each stream contributes to the global effluent (Veiga et al. 1994). This criterion gives information about the relative importance of each stream and permits to minimize the separation of effluents. In this way, the different effluents were collected into four main streams: High Organic Concentration (HOC), Low Organic Concentration (LOC), High Solids Concentration (HSC) and Clean Waters (CW).

Table 5 shows the composition of each of these resulting streams after applying this criterion to the effluents shown in Table 2. It can be observed that the HOC effluent represents more than 38% of total organic load with only 3% of total volume. On the other hand, more than 31% of the overall flow corresponds to almost clean waters. Thus, a relatively small plant designed especially for the treatment of highly polluted wastewaters could be an alternative to treat the effluent HOC, whereas simple separation techniques should be used to convert the HSC effluent in waters with similar characteristics to CW.

Table 5. Characteristics of the four streams obtained after mixing the different wastewaters (Table 2) according with their relative contribution of organic load and flow (flow in m^3/d, concentrations in g/l, CO in kg COD/d)

	Flow	pH	TSS	VSS	COD$_T$	Cl$^-$	CO
HOC (T5.1, T5.2, M3, M4, T6)	162	6.77	1.54	1.29	16.23	14.70	2629
LOC	1699	7.98	0.67	0.44	1.21	18.16	2057
HSC (M1)	1600	7.97	1.59	0.69	1.20	19.21	1920
CW (M2)	1600	8.06	0.34	0.15	0.23	19.32	368
Overall effluent	5222	7.96	0.88	0.46	1.38	18.75	6974

4.2.1 Treatment of HOC wastewaters

The application of biological treatment methods based on the anaerobic digestion process has been widely considered during the last decades for wastewaters highly polluted in organic matter (Macarie 2000; Omil et al. 2003). For the treatment of high organic strength waste waters they present the advantages of a very low sludge generation and production of energy that can be used to cover part of the energetic needs of the factory. Taking into account the characteristics of the fish processing waste water (possible presence of high concentrations of salts and ammonium, and large volumes of effluents) high rate bioreactors should be used, that allow accumulation of high biomass concentrations and the use of relatively low hydraulic retention times (Hickey et al. 1991). This method is suitable and recommendable for wastewaters with a COD concentration higher than 5 g/l.

As a rough guide, the volume and dimensions of the anaerobic reactor needed can be made based on the characteristics shown in Table 5 and in previous works reported (Omil et al. 1996; Puñal and Lema 1999). Since the presence of organic solids and fats, oils and grease (FOGs) is important, the influent to this reactor should be pretreated in a small flotation unit. Table 6 shows the results achieved. It can be seen that a high reduction in organic matter and suspended solids can be achieved, and the dimension of the required units is small (380 m^3 and less than 75 m^2). Furthermore, there is a production of biogas that could be used to produce energy.

4.2.2 Case study

In fact, this approach has been followed in the factory considered in this work for the treatment of the HOC effluent. The only difference is that no flotation unit was considered, and a rotatory sieve device was installed prior to the entrance of the anaerobic reactor. The flow diagram is shown in Fig. 2. The main elements of this plant are: tank to collect the different wastewaters (T-03), tank of homogenisation

(T-04), tank of preconditioning and preacidification (T-05), anaerobic reactor (R-01), flexible collection tank for biogas storage (G-01), torch for biogas combustion (FL-01). The plant also comprises systems for temperature and pH control.

Table 6. Approximation of the characteristics of the HOC streams after an specific treatment comprising a flotation unit and an anaerobic reactor

Composition	Flow	pH	TSS	VSS	COD_T	Cl^-
	m^3/d		g/l	g/l	g/l	g/l
HOC influent	162	6.77	1.54	1.29	16.23	14.70
Flotation unit effluent	162	6.77	0.77	0.65	15.26	14.70
Anaerobic reactor effluent	162	7.50	0.77	0.65	1.53	14.70

Dimensions and efficiencies	Volume	Surface	HRT	Red (%)		
	m^3	m^2	d	TSS	VSS	COD_T
Flotation unit	50	16	2	40	40	60
Anaerobic reactor	320	55	2	-	-	90

Other streams	Biogas	FOGs
	m^3/d	m^3/d
Flotation unit	-	0.5
Anaerobic reactor	1300	-

From the receiving tank (T-03), the wastewaters are pumped to T-04 after passing through a rotating sieve (RS-01). This latter tank has a volume of 250 m³ which means a hydraulic residence time around 1 day, time considered enough for homogenisation of the different streams and not very long to promote an excessive acidification. Air-liquid diffusers are used to maintain mixed conditions. Tank T-05 is used for pH and temperature control, being its content in continuous recirculation with the anaerobic reactor. This latter unit is an UASB reactor of 380 m³ and is currently fed into the bottom through pipes connected from the upper part, which makes their substitution simpler in case of obstruction. The biogas produced is stored in a flexible tank where a fixed pressure is maintained before sending it to a torch.

Due to the high content of salinity due to the presence of sea water in these effluents, the biogas produced has an important content of hydrogen sulphide (up to 2-3%). In order to proceed to the further use of this gas, a scrubber unit was designed to reduce this content below 0.1%. Once this step will be implemented, the biogas will be used to produce both heat and electricity in a small cogeneration unit.

The sludges produced by this unit were not significant compared with the ones collected from the other lines. In fact, after 5 years of operation purges were not necessary and COD removal efficiencies of 70-90% were maintained at Organic Loading Rates (OLR) of 6-8 kg COD/m^3·d (Puñal and Lema 1999).

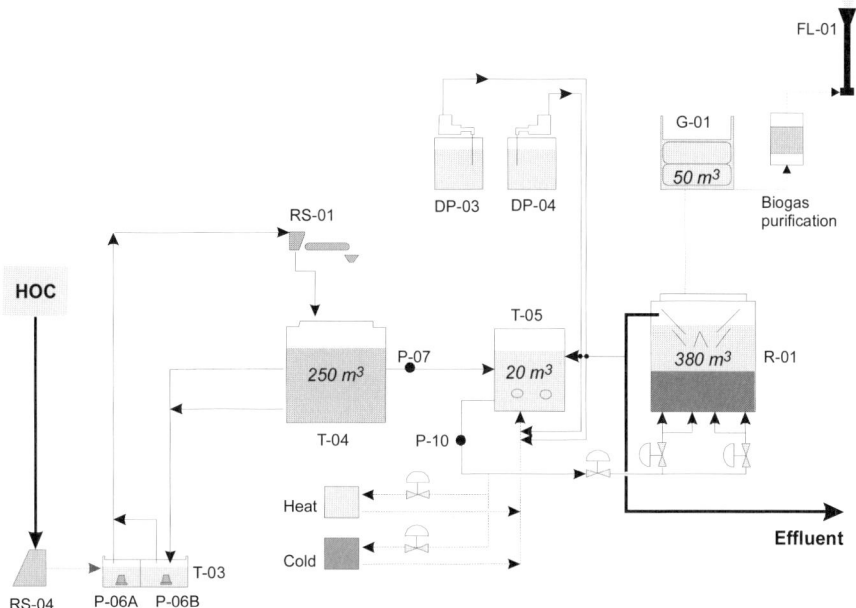

Fig. 2. Anaerobic biological treatment of the HOC stream installed in the factory under study

4.2.3 Treatment of LOC wastewaters

The LOC stream is the largest one according with the strategy followed. Its COD, TSS and VSS content is not high, being in the range of urban wastewaters. However, the contribution of the organic solids is significant (more than 65% of TSS), which indicates that the organic content is mainly related to this fraction of solids, part of them present as colloidal matter. Thus, the treatment considered for this stream comprises basically a high rate flotation device, as those based in Dissolved Air Flotation (DAF) technology. Table 7 shows the main values of the units considered.

4.2.4 Case study

Figure 3 shows the diagram of the process implanted in this factory to treat these effluents. Basically, it comprises three elements: a) tank of receiving waters and homogenisation (T-02), b) automatic bar screen (RS-02), c) DAF unit (D-01).

Apart of the streams M1, M2 and the HOC group, the rest of wastewaters are treated in this plant. Firstly, they are passed through an automatic bar screen device (RS-02) and finally discharged in 150 m^3 tank (T-02), also equipped with stirring devices. In the bar screen a high amount of big solids are separated, such as rests of fishes and different vegetables. Afterwards, this stream is fed into a DAF unit where FOGs are very efficiently removed. In fact, the separation of this floating matter makes possible also an additional separation of solids in the bottom. As a result, the efficiency of this unit is very high (around 40-50% COD, 70% SS and 60% FOG).

Table 7. Approximation of the characteristics of the HOC streams after an specific treatment comprising a flotation unit and an anaerobic reactor

Composition	Flow	pH	TSS	VSS	COD_T
	m^3/d		g/l	g/l	g/l
LOC influent	1699	7.98	0.67	0.44	1.21
Bar screen	1699	7.98	0.34	0.22	0.88
Flotation unit effluent	1699	7.98	0.20	0.13	0.35

Dimensions and efficiencies	Volume	Surface	Red (%)		
	m^3	m^2	TSS	VSS	COD_T
Bar screen	-	6	50	50	27
Flotation unit effluent	150	50	40	40	60

Other streams	Air needed	FOGs
	kg/d	m^3/d
DAF	200	1.6

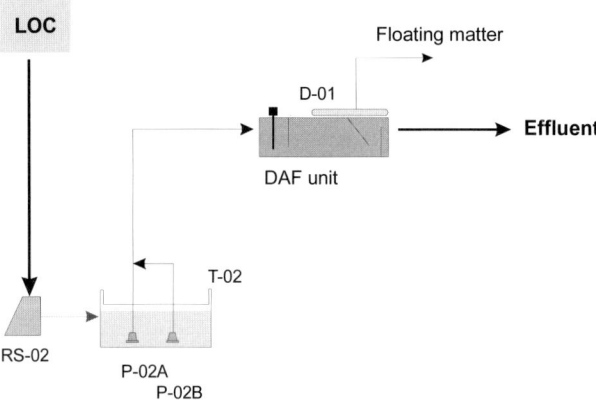

Fig. 3. Treatment of the LOC stream installed in the factory under study

4.2.5 Treatment of HSC wastewaters

In this case, this stream is only M1, which contain all the solids (mainly inorganic) resulting from mussel beard trimming step. In this case, screen bars in combination with rotatory sieves are commonly used to separate efficiently these particles, being obtained an effluent with a very low COD content. The surface required for these units, although important, does not imply a significant area as compared with the other units commented (biological reactors, sedimentation units, DAF devices, etc.).

Assuming a suspended solids removal efficiency around 70%, the TSS, VSS and COD content in the effluent would be 0.32, 0.14 and 0.37 g/l. On the other hand, the sludges production will be important, around 4 Tm/d (50% dried weight).

When the nature of solids is mainly organic, it is better to manage these effluents as described for the LOC wastewaters, since very often FOGs make very difficult to use sieves or sedimentation units.

4.2.6 Treatment of CW wastewaters

The other different streams produced in these factories can be considered as clean waters, since their characteristics would allow their direct discharge. This is one of the preferred options, being also considered their recycling into the factory, but never to join them with the other streams.

Both options should be carefully studied. In the case of the discharge, the final mixture of this stream with the ones obtained in the other treatment lines should allow to achieve the concentration requirements stated by legislation. In the case of recycling, it should be taken into account what conditioning process should be necessary to reuse them inside the factory.

Another point to have into account is that most of these effluents have a high salinity, whereas others, such as those generated after sterilisation step are fresh water, which should be economically more interesting to recover.

4.2.7 Operation of the whole wastewaters treatment plant

Table 8 shows the complete balance for the treatment of these streams taking into account the removal efficiencies considered, that would allow to achieve a similar quality to that considered in the single treatment in a conventional aerobic plant (Table 4). However, in this case, the volume of the units are much smaller, as well as the surface needed. Besides, the amount of excess of sludges has been drastically reduced since the production of the anaerobic reactor is not significant, and the energy consumption is also much lower.

5 Reuse alternatives and new subproducts developed

Apart from the studies focussed on the development of more efficient strategies to treat the wastewaters generated by these factories, another research line is oriented on the study of specific effluents in order to obtain valuable subproducts or high quality water to recycle into the process.

In the following items, three examples of this new approach are shown. All of them have been implemented very recently in this factory.

Table 8. Approximation of the characteristics of the HOC streams after an specific treatment comprising a flotation unit and an anaerobic reactor

Composition	Flow	pH	TSS	VSS	COD_T
	m^3/d		g/l	g/l	g/l
HOC treatment	162	7.50	0.77	0.65	1.53
LOC treatment	1699	7.98	0.20	0.13	0.35
HSC treatment	1600	7.97	0.32	0.14	0.37
CW	1600	8.06	0.34	0.15	0.23
FINAL EFFLUENT	5059	7.99	0.30	0.16	0.36

Dimensions and efficiencies	Volume	Surface	Removal (%)		
	m^3	m^2	TSS	VSS	COD_T
	520	130	70	65	74

Other streams	Air needed	FOGs	Biogas
	kg/d	m^3/d	m^3/d
DAF	200	1.6	1300

5.1 Water reuse from sterilisation effluents

Thermal processing is widely used in the food industry to ensure the sterility of food. Heating is a traditional method of sterilisation, and canning is perhaps the best known packaging technique. Sea food is packaged, then processed at around 120 °C by applying heat to the outside of the can until the whole of its contents have reached the desired temperature.

The wastewaters generated during sea food manufacturing processes as a consequence of the sterilisation of cans comprise the condensation of the vapour used and, mainly, the cooling water used afterwards. The total flow of these effluents can be very important, achieving for large factories around 10% of the total wastewaters. On the other hand, their COD content is extremely low (0.12 g/l) being only consequence of the presence of small quantities of oils washed from the surface of the cans (0.05 g fat/l). These oils are very easily separated from the liquid simply by natural flotation. Thus, a plant based on this unit operation followed by a cooling step and a final disinfection step should provide fresh high quality water to reuse in the manufacturing process.

5.1.1 Case study

In this factory the flow of this stream amounts to 595 m^3/d (11.5% of total wastewaters), with an average composition similar as described previously. A plant to recover and recycle this stream has been constructed, which comprises a 16 m^3 tank with three different areas: reception (20% of the total volume), natural flotation (60% of the total volume) and pumping zone (20% of the total volume); a plate heat exchanger; a chlorination unit and a final storage tank. An economic evaluation of all the costs implied, especially fresh water and taxes for wastewaters discharge, resulted in a favourable balance for the factory, which recovered its capital investment costs in less than 3 years.

5.2 Oil from the treatment of HOC and LOC wastewaters

As previously described, in recent years the application of flotation devices to these HOC and LOC streams allows the separation of the highest fraction of FOGs, resulting in a final liquid stream easier to be further treated because of their low content in fats and colloids. These substances are especially present in wastewaters generated during cooking steps, in which the fats present in the raw material, especially fish, are released into the liquid effluent. Other steps, such as the sauce filling or the final washing of cans, are also important. The main technologies used in flotation devices are dissolved air flotation (DAF) and cavitation air flotation (CAF) systems.

However, the management of the floating matter stream obtained is not easy, since it can amount to several tons per day, being their final management very costly. On the other hand, this stream is rich in products such as oil and polyunsaturated fatty acids (PUFA), which could be of economical interest.

In this sense, different experiments were conducted at lab-scale in order to develop a process to separate the different fractions contained in this floating matter. Finally, four fractions (oil, fine solids, water and thick solids) were clearly obtained after a thermal treatment followed by centrifugation. The volume corresponding to each fraction was 35% oil, 30% fine solids, 20% water and 15% thick solids (García-Sandá et al. 2003).

The overall solid fraction obtained had a high content of organic matter (ash concentration was 3.5% by weight). Moisture was quite high (72.6% by weight). This stream could be easily used to produce fish-meal.

The characterization of the oily stream revealed a high percentage of carbon and hydrogen (89% by weight); absence of nitrogen; a freezing point of 6.7 °C; a density (15°C) of 957 kg/m^3 and a viscosity (23°C) of 80 kg/m·s. Ash (0.28%) and sulphur (0.09%) contents were lower than the values for Fuel-oil nº 1. In addition, the oil has a high calorific value, although slightly lower that of Fuel n° 1 (Table 9).

Table 9. Comparison of the main parameters as a fuel (sulphur and ashes in % by weight)

	Oil obtained	Fuel-oil n° 1
Gross calorific value (cal/g)	9 273	10 100
Net calorific value (cal/g)	8 716	9 600
Sulphur (%)	0.09	2.7
Ashes (%)	0.28	10

5.2.2 Case study

This process was finally implemented at industrial scale. The average composition of the floating stream processed during the evaluation period of the industrial-scale plant was: 38% oil, 6.6% fine solids, 51.6% water and 3.8% thick solids (%v/v). The final oil production obtained under these conditions was 875 l/day, which was burnt mixed with Fuel-oil n°1 for steam generation. The oil/fuel ratio employed was around 10%. Assuming as a minimum economic value for this oil the cost of Fuel-oil n°1, and taking into account the costs needed to manage properly the separated floating matter (transport to specialised treatment plants, processing costs, etc.), the savings achieved with this plant were very important, since capital costs were returned after only one year.

5.3 Protein concentrates from tuna cooking effluents

Among all wastewaters, tuna cooking effluents have the highest protein content. Cooking in these factories is usually carried out either by steam injection or by immersion in brine tanks. These effluents can amount to 1% of the overall flow of the factory and around 10% of the total organic load discharged. Thus, their contribution to the overall organic content to be removed in the WWTP is very important. Besides, ammonia generation can be a problem especially if anaerobic digestion is considered for organic matter removal, because of its toxic effects (Soto et al. 1991). On the other hand, the valuable nature of the protein present makes interesting to study its use as a complement in animal feedstuff.

Therefore, the application of an specific treatment to these waters in order to recover the protein could be not only interesting because of a potential production of an additional subproduct, but also because of the better function of the whole WWTP.

Table 10 shows the detailed characteristics of the tuna cooking effluent obtained during the first discharge of the factory analysed (steam injection process). A high organic matter can be observed, as well as important contents of organic suspended solids. Total protein was 20 g/l and fat content was 2 g/l. The salinity level was not as high as other effluents since fresh water is used both for steam production and for cooling (García-Sandá et al. 2003).

Table 10. Characterization of tuna cooking effluent carried out in 2001 (concentrations in g/l, flow in m^3/day)

Flow	pH	COD$_t$	TSS	VSS	Cl$^-$	Prt$_t$	Fats
56	6.5	59	6.53	6.40	11.5	20	2

5.3.3 Evaporation assays

Evaporation assays were carried out with this single stream in order to obtain a high content protein concentrate. These assays were carried out on laboratory and pilot scale. Laboratory-scale assays were performed in a rotary evaporator, provided with a vacuum controller, a thermostatic water bath, a rotating flask (0.5 l), a condenser and a receiving flask (0.5 l). Pilot-scale assays were carried out in a rising film evaporator, which consisted of a vertical tube inside a shell arrangement, allowing to obtain condensate and concentrate fractions.

The condensate fraction obtained in these experiences had low organic matter content (COD 400-900 mg/l). This stream had a high pH value, because of the volatilization of alkaline compounds during the operation. For example, the condensate corresponding to the concentrate of highest Volume Reduction ratio (VR) had a pH value of 9.7. Therefore, this stream can be directly discharged after minor treatment.

Physical and chemical properties of the concentrate fraction depended directly on VR. Solids, organic matter, protein and salts content increase, as well as the density, whereas the pH value decreases linearly with VR. Other properties, such as viscosity, exhibit a stronger increase, showing an exponential trend.

Scale change involved different conditions of operation and consequently, different characteristics of the concentrates obtained. The main parameters measured, such as solids, organic matter and protein contents, decreased for the same VR, because of the change in operation conditions and fouling. However, the protein content was the parameter least affected. Protein concentrations of around 10% by weight were obtained at laboratory and pilot- scales for the highest VR.

5.3.4 Nutritional value assays

With the concentrate obtained, nutritional assays were carried out with male Sprague-Dawley rats. The addition of a protein concentrate to a conventional foodstuff was evaluated, from the point of view of nutritional value. For that, it is essential that the diets have the same concentration of protein. So, any modification can be attributed to a different composition of the diet.

Table 11 shows protein quality indicators, food and water intake, urine and faeces excretion, body weight gain and relative weight of the liver, which were determined at the end of the study.

The food intake and the body weight gain were significantly higher in the group of animals that were fed with the protein concentrate (treated group). However, the Protein Efficiency Ratio (PER) and the Feed Conversion Rate (FCR), which relate food or protein intake to body weight gain, did not differ in the two groups. The water intake was double in comparison with the reference group because of the high salts content of the concentrate. However, urine excretion was five times higher in the treated group. There was not a significant difference between the relative weight of the liver in the reference and treated groups.

As for protein quality indicators, the digestibility of treated-group protein was about 94%, similar to the values obtained for egg and meat proteins (Santidrián 1989) and higher than the digestibility of the reference-group protein. Because of the high salinity of the concentrate, which caused an osmotic dragging, the Biological Value (BV) and the Net Protein Utilization (NPU) were significantly lower in the treated group.

Table 11. Protein quality indicators, food and water intake, urine and faeces excreted, body weight gain and relative weight of the liver at the end of the study

	Reference group	Treated group
Body weight gain (g)	44 ± 4	55 ± 4
Food intake (g/day)	16.1 ± 1.1	20.6 ± 1.6
Water intake (ml/day)	16.3 ± 1.1	35.8 ± 3.4
Urine excreted (ml/day)	4.2 ± 0.7	21.6 ± 2.1
Faeces excreted (g/day)	4.2 ± 0.3	4.1 ± 0.4
Liver : body weight (%)	3.96 ± 0.17	4.18 ± 0.14
Biological value (%)	94.5 ± 3.9	68.6 ± 2.2
True digestibility (%)	89.6 ± 2.6	94.4 ± 1.1
Net protein utilization (%)	84.7 ± 3.7	64.8 ± 2.3
Protein efficiency ratio (g/g)	3.06 ± 0.07	3.25 ± 0.10
Feed conversion ratio (g/g)	2.64 ± 0.06	2.56 ± 0.08

Further studies carried out with broiler chickens confirmed the feasibility of using this concentrate as a complement for animal foodstuff. The energy metabolised supplied by the concentrate was calculated as 1396 kcal/kg concentrate.

5.3.5 Case study

The application of these results in a full-scale industrial plant is now under development.

6 Conclusions

Fish canning industries generate a large number of different wastewaters that usually are treated together in a huge and complex plant, which implies high installation and maintenance costs (personal, electricity, supplies, etc.). Their characteristics are conditioned by the diversity of the raw material processed and by the fact that they are generated at many points of the factory and with very different properties. While some of them, especially those from cooking, present a high COD, others have a low COD but a high content in organic or inorganic solids. A high salinity as a consequence of the use of seawater is also common.

Data obtained from the characterization of the processing plant effluents allow to define the most suitable strategy to treat these effluents by anaerobic, aerobic or physic-chemical systems depending on the production conditions and the location of the factories, in order to minimize contamination and reach levels established by legislation. Taking as reference the composition of one of the biggest representative factories located in Spain, it is possible to make a preliminar comparison about the technology and key resources needed by a single plant and those required for a more complex plant comprising specific (and smaller) units for each type of effluents. From this analysis it is clear that the first option only should be considered for small factories.

On the other hand, some of the main polluted effluents generated by this industrial sector can be further processed in order to obtain valuable subproducts, with the additional consequence of a substantial reduction of the organic load and volume to be treated in the wastewaters treatment plant.

The wastewaters generated during the sterilisation of cans can be easily recovered with a simple plant focussed on the removal of oils by flotation. This measure would imply the reduction of the overall wastewaters flow around 10% and also a reduction in the amount of fresh water needed.

The FOGs waste stream separated by flotation devices can be a source of interesting subproducts such as oil, which has very good characteristics as a fuel. In this sense, the proposed process for oil recovery can close the whole management of these wastes, since the oily product can be used as a fuel, the solid fractions can be used to produce fishmeal, and the remaining water fraction can be treated together with the other wastewaters.

The treatment of tuna-cooking liquid effluents in order to obtain a protein concentrate implies an important reduction of the complexity and content of organic matter in the final wastewaters effluent. A process based in an evaporation step has been proposed, which gives a condensate with a low organic matter and solids content (that would be easily treated in the WWTP) and a concentrate with high protein content (up to 10% by weight). Nutritional tests carried out with this concentrate showed its feasibility to use as a complement for animal foodstuff elaboration.

Acknowledgements

This work was financed by Xunta de Galicia through the following projects carried out with different companies: PGIDT01MAM03E (Conservas del Noroeste, S.A.), PGIDIT02TAM06E (Conservas del Noroeste, S.A.) and PGIDIT02TAL8E (Conservas Isabel de Galicia, S.L.). The authors acknowledge Dr. Mª Cristina Taboada for her enthusiastic and positive collaboration.

References

Aguilar ALC, Sant´Anna GL (1988) Liquid effluents of the canning industries of Rio de Janeiro state– treatment alternatives. *Environ. Technol. Lett.,* 9, 421-428

Balslev-Olesen P, Lynggaard-Jensen A, Nickelsen C (1990) Pilot-Scale experiments on anaerobic treatment of waste-water from a fish processing plant. *Wat. Sci. Technol.,* 22, 463-474

Fernández JM, Méndez R, Lema JM (1995) Anaerobic treatment of eucalyptus fibreboard manufacturing wastewater by a hybrid USBF lab-scale reactor. *Environmental Technol.,* 15, 677-684

Fraga F (1976) Photosynthesis in the Ria of Vigo. *Inv. Pesquera,* 430, 151 (in Spanish)

García-Sandá E, Omil F, Lema JM (2003) Clean production in fish canning industries: recovery and reuse of selected wastes. *Clean Technologies and Environmental Policy* (in press)

González M, Caride B, Lamas MA, Taboada MC (2000) Effects of sea urchin-based diets on serum lipid composition and on intestinal enzymes in rats. *J. Physiol. Biochem.,* 56, 347-352

Guerrero L, Omil F, Méndez R, Lema JM (1998) "Protein recovery during the overall treatment of wastewaters from fish meal factories". *Bioresource Technol.,* 63, 221-229

Hickey RF, Wu WM, Veiga MC, Jones R (1991) Start-up, operation, monitoring and control of high-rate anaerobic systems. *Wat. Sci. Technol.,* 24 (8), 207-255

Hu HY, Goto N, Fujie K (1999) Concepts and methodologies to minimize pollutant discharge for zero-emission production. *Wat. Sci. Tech.,* 39 (19), 9-16

IHOBE (1999) Libro Blanco para la minimización de residuos y emisiones: Conservas de Pescado. Gobierno Vasco (In Spanish)

Lema JM, Omil F (2001) Anaerobic treatment: A key technology for a sustainable management of wastes in Europe. *Wat. Sci. Technol.,* 44 (8), 133-140

Macarie H (2000) Overview of the application of anaerobic treatment to chemical and petrochemical wastewaters. *Wat. Sci. Technol.,* 42 (5-6), 201-214

Omil F (1993) Tratamiento integral de las aguas residuales del sector conservero de productos marinos. *Ph. D. Thesis.* University of Santiago de Compostela, Spain (In Spanish)

Omil F, Garrido JM, Arrojo B, Méndez R (2003) Anaerobic Filter reactor performance during the treatment of complex dairy wastewaters at industrial scale. *Water Res.,* 37 (17), 4099-4108

Omil F, Méndez R, Lema JM (1995) Anaerobic treatment of saline wastewaters under high sulphide and ammonia content. *Bioresource Technol.,* 54, 269-278

Omil F, Méndez R, Lema JM (1996) Anaerobic treatment of seafood processing waste waters in an industrial anaerobic pilot plant. *Water SA*, 22, 173-181

Puñal A, Lema JM (1999) Anaerobic treatment of wastewater from a fish canning factory in a full-scale upflow anaerobic sludge blanket (UASB) reactor. *Wat. Sci. Technol.,* 40 (8), 57-62

Santidrián S (1989) Digestión de las proteínas y absorción y metabolismo de los aminoácidos. *Offarm*, 59-70

Soto M, Méndez R, Lema JM (1990) Efluentes residuales en la industria de procesado de productos marinos. *Ingeniería Química*, 22, 203-209 (In Spanish)

Soto M, Méndez R, Lema JM (1991) Biodegradability and toxicity in the anaerobic treatment of fish canning waste-waters. *Environ. Technol.*, 12, 669-677

Speece RE (1996) Anaerobic biotechnology for industrial wastewaters. Archae Press, Nashville, Tennessee

Veiga MC, Méndez R, Lema JM (1994) Wastewater treatment for fisheries operations. *In Fisheries Processing: Biotechnological applications.* Martin, A. M. (Edit). Chapman & Hall, London, 344-369

Kinetic Analysis of Aerobic Composting of Tobacco Industry Solid Waste

Felicita Briški[a]*, Nina Horgas[a], Marija Vuković[a], Zoran Gomzi[b]

University of Zagreb, Faculty of Chemical Engineering and Technology, [a]Division of Industrial Ecology, [b]Division of Reaction Engineering and Catalysis, Marulicev trg 19, 10000 Zagreb, Croatia

* Corresponding author. Tel.: ++385 1 4597 269; fax: ++385 1 4597 260. E-mail address: fbriski@pierre.fkit.hr (F. Briški)

Abstract

Solid waste accumulated during the processing of tobacco for cigarette manufacture mostly contains tobacco particles and flavoring agents. Its main characteristics are a high content of nicotine, which is a toxic compound, and high value of total organic carbon of the aqueous extract. Because of this fact tobacco waste cannot be disposed of with urban waste.

The aim of this work was to stabilize tobacco solid waste by aerobic composting. The experiments were carried out in closed thermally insulated column reactors (1.0 l and 25 l) under adiabatic conditions and at an airflow rate of 0.9 l min^{-1} kg^{-1} of volatile solids for 16 days. During the process, temperature changes in the reactor, CO_2 production and the numbers of mesophilic and thermophilic microorganisms in the mixed microbial culture were closely monitored. Nicotine concentration in the samples was analyzed at the start and at the end of process. It was estimated that at the end of the composting the volume and the mass of total solids in the tobacco waste were reduced about 50% and those of nicotine by 90%. A simple empirical model was used to simulate the biodegradation rate of the organic fraction of the solid waste. It was found that the selected model describes aerobic composting fairly well, although only two kinetic parameters (k_o and n) were estimated.

Keywords: tobacco solid waste, nicotine, aerobic composting, adiabatic closed column reactor, modeling the process

List of Symbols

- c_{pS} specific heat capacity of the substrate, kJ kg^{-1} K^{-1}
- c_{pz} specific heat capacity of air, kJ kg^{-1} K^{-1}
- k specific rate, Eqs[5] and [9], h^{-1}
- k_o constant in Eq. 9, h^{-1}
- m_o mass of wet composting material, kg
- m_S mass of dry substrate, kg
- m_{So} initial mass of dry substrate in reactor, kg
- m_{CO2} mass of evolved CO_2, kg
- n order of the reaction in Eq. 5
- Q_v air flow volume, m^3 h^{-1}
- r_S degradation rate, kg kg^{-1}h^{-1}
- SD   mean square deviation
- t time, h
- T temperature in reactor, °C
- T_o temperature of substrate at the beginning of reaction, °C
- T_K temperature of compost at the end of reaction, °C
- T_u temperature of air at the reactor inlet, °C
- w_S mass rate of compost, $m_s \, m_{So}^{-1}$, kg kg^{-1}
- w_{CO2} Mass fraction of CO_2, $m_{CO2} \times m_{CO2k}^{-1}$, kg kg^{-1}
- ΔH_r reaction enthalpy, kJ kg^{-1} of dry substrate
- ρ_z air density, kg m^{-3}
- τ space time, day

1 Introduction

Solid wastes such as food and yard waste, agricultural waste and waste paper are clogging landfills. This leads to problems in finding new replacement sites (White et al. 1994). Disposal of solid waste in the Republic of Croatia is still a major national problem. The annual production of household waste in Croatia is around one million tons and with only a few modern landfills available, the majority of waste is still disposed of in dumps. Accurate information about the quantity of industrial and hazardous wastes is lacking. In a few municipal landfills in Croatia, hazardous waste used to be disposed of together with municipal waste (Ucur and Nikolic 2000). Nowadays industrial solid waste, with a TOC content above 200 mg l^{-1} in aqueous fraction, can no longer be disposed of in the sanitary landfill (Croatian regulation 1997). Therefore, such types of solid wastes have to be either incinerated or transformed into a humus-like product by composting.

A large quantity of non-recyclable powdery waste is accumulated during the tobacco manufacture. Tobacco waste is categorized as agro-industrial waste. Its aqueous fraction has a TOC concentration above 200 mg l^{-1} and therefore it cannot be disposed of in a sanitary landfill. Furthermore, it contains nicotine (Civilini et al.

1997), which makes it hazardous. As agricultural products can be decomposed by aerobic composting, tobacco solid waste can be managed the same way (Haug 1993; Kayhanian and Tchobanoglous 1993; Gries 1994). Composting generates a stable product that may be used as a soil conditioner or a fertilizer. Control over composting conditions and emission of air pollutants and odor is possible in enclosed vessels or in reactor systems (Bari et al. 2000; Korolewicz et al. 2001).

This work was aimed at investigating aerobic composting of tobacco solid waste in a laboratory scale column reactor and proposing a reactor and a kinetic model that represents the experimental results.

2 Material and methods

2.1 Reactor system

Laboratory-scale aerobic composting tests were conducted in closed thermally insulated column reactors with effective volumes of 1.0 l (245 mm high - 72 mm inner diameter) and 25.0 l (850 mm high - 190 mm inner diameter). The solid waste was composed of tobacco particles and flavoring agents (sugars, humidifiers, organic acids and fruit extracts). Particle size distribution in tobacco waste was determined in 100 g of a dry sample using vibrating sieves (Analysette 3, Fritsch). The results are presented in Table 1. The original moisture content in tobacco waste was 51.4%, but to obtain optimal consistency of the waste, it was adjusted to 65.8% by the addition of water. At the start of the experiment the total solids (TS) of wet waste contained 68,5% of volatile solids and 2000 mg kg^{-1} TS of nicotine. Continuous upward aeration was provided by an air blower.

Table 1. Granulometric analysis of tobacco waste

Particle size	Percentage of weight [%]
Ribs of tobacco leaves 5–10 cm	10
Particles > 2 mm	5
600 µm ≤ d ≤ 1 mm	55
400 µm ≤ d ≤ 600 µm	14
200 µm ≤ d ≤ 400 µm	9
100 µm ≤ d ≤ 200 µm	4
50 µm ≤ d ≤ 100 µm	3

2.2 Experimental run

The reactors, 1.0 l and 25 l, were filed with 0.5 kg and 11.5 kg of wet tobacco waste, respectively. The reactors were operated at an airflow rate of 0.9 l min^{-1} kg

$^{-1}$ volatile solids (VS) and the temperature was monitored by thermocouples for 16 days. The temperature of the air at the inlet of the reactor was between 24 and 26°C during the whole period of composting. The spent hot air containing carbon dioxide was cooled and the condensate collected in a graduated cylinder. Before being released to the atmosphere, CO_2 was collected in a tube filled with solid KOH as an adsorbent. A blind test (the reactor without waste) was also carried out to determine the total amount of carbon dioxide that entered the reactor during the 16 days of aeration. All experiments were performed in duplicate.

2.3 Physico-chemical and microbiological analyses

Air temperatures at the reactor's inlet and outlet and at three levels of the composting mass (bottom, middle and top) were monitored continuously by thermocouples connected to the data logger. The airflow rate was measured by an airflow-meter (0.1-0.5 LPM and 0.2-2.0 LPM, Cole-Parmer). The graduated cylinder was used for measuring daily production of condensate. CO_2 content in the spent air was measured gravimetrically (Kolthoff and Sandell 1951) and the results expressed in g CO_2 day^{-1}. At the start and during the composting tests, key parameters of the original tobacco waste and the compost were analyzed (moisture content as a percentage of the wet weight, and total and volatile solids). The analyses were carried out according to the standard methods for compost analyses (Austrian Standard 1986). Viable plate count was determined by the decimal dilution method and the results expressed as colony-forming units (CFU) of mesophilic or thermophilic bacteria and fungi per gram of the composting mass. Petri dishes were kept in the incubator under 80% relative humidity at 28°C for the growth of mesophilic fungi, at 37°C for the growth of mesophilic bacteria, and at 50°C for the growth of thermophilic bacteria and fungi. Nicotine and total organic carbon concentrations in the samples were determined at the start and at the end of the experiment. The HPLC method (Saunders and Blume 1981) with a DAD (diode-array detector) was used to measure the content of nicotine and a Shimadzu TOC-5000 A carbon analyzer was used to measure total organic carbon.

3 Results and discussion

3.1 Composting process

For the efficiency of composting in the controlled reactor system it is necessary to know the reaction rate of biological degradation in the composting mass, and the level of degradation achieved. Therefore, composting of tobacco solid waste in 1.0 l and 25 l reactors was closely monitored for 16 days. It was recorded that temperature variations in both reactors in duplicate sets were similar. The results of temperature variations of the second set in the closed 25 l reactor were selected and are presented in Fig. 1.

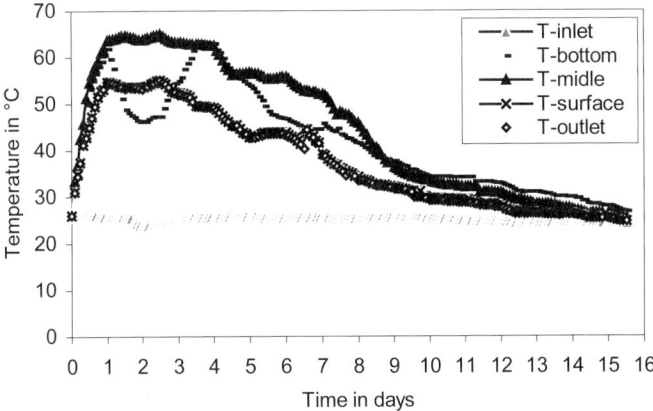

Fig. 1. Temperature over time during 16 days of composting (25 l reactor, second run)

Significant vertical differences in temperatures in the bottom and in the middle layer of the composting mass were recorded during upflow aeration. Maximum difference in temperature between those two layers of the composting mass was about 20 °C. One hour after the start of the experiment the temperature in the composting mass started to rise due to a vigorous biodegradation process. It reached a maximum in the middle layer after day one, remained high to the fourth day and then gradually decreased until day nine. From day ten to day sixteen the decrease became more and more evident until the compost and ambient temperatures leveled. The temperature in the bottom layer also reached a maximum after day one of composting, but between the second and the third day it dropped sharply from 64 to 45 °C. Evidently during these 2 days the waste became deficient in readily biodegradable organic compounds, and microbial activity slowed down, resulting in a lower production of CO_2 and metabolic heat. Under concomitant, continuous aeration the air at the reactor inlet induced cooling of the bottom, layer of the composting mass (Bari and Koenig 2000) but did not affect the middle layer. With the continued biodegradation of tobacco waste the temperature started to rise again.

As shown in Fig. 2, CO_2 generation followed the pattern of the temperature curve. During day one mesophilic microorganisms used initially most readily degradable organic compounds (sugars and organic acids) that resulted in the evolution of carbon dioxide, and a rapid temperature increase. On day two the amount of CO_2 decreased by around 50% because biodegradation of the organic macromolecules was a slower process than biodegradation of the simple molecules. When the microorganisms started hydrolyzing the complex macromolecules to the simple molecules and at the same time degrading them, the production of CO_2 continued. It was estimated that by the end of the composting process the yield of

CO_2 per kg of total solids was 415 g (or 70.9 g for the 1.0 l reactor and 1,630.9 g for the 25 l reactor).

Microbial succession during composting is relatively rapid, since it may take one or two days to reach 55°C. At the beginning the substrate was at the ambient temperature and prevailing organisms were mesophilic bacteria and fungi (Fig. 3). An examination of microflora showed that isolates mainly belong to the genera *Bacillus*, *Pseudomonas*, *Saccharomyces* and *Candida*.

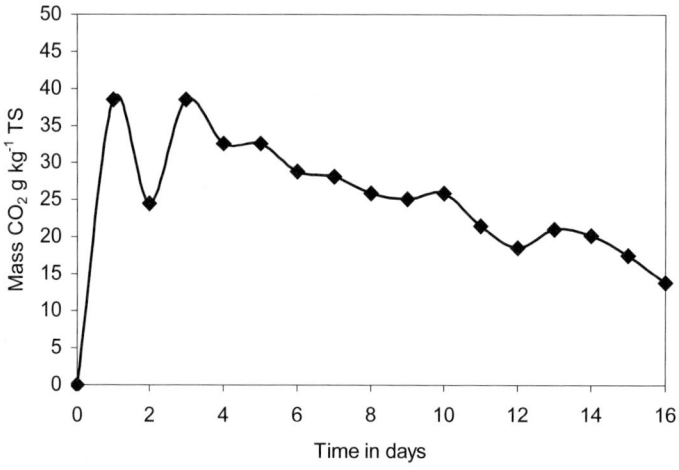

Fig. 2. Mass of CO_2 produced due to substrate biodegradation

Microbial succession during composting is known to be relatively rapid, since it may take 1-2 days to reach 55°C. At the start of the experiment the substrate had the ambient temperature and the prevailing organisms were mesophilic bacteria and fungi (Fig. 3). An examination of the microflora showed that isolates mainly belong to the genera *Bacillus*, *Pseudomonas*, *Saccharomyces* and *Candida*.

Due to the presence of simple organic substrate in tobacco waste the microorganisms started to multiply and the temperature rose sharply to above 60°C, causing the decrease of mesophilic bacteria. It is supposed that sugars from the flavoring agents in tobacco waste were depleted after the day one causing a continuous decrease of fungi until the end of composting. Further biodegradation of ttobacco waste was taken over by thermophilic bacteria. Microbial diversity in the mixed culture during that stage was relatively low. Examination of the microflora showed that the prevailing isolates in composting mass were *Bacillus* and *Thermoactinomyces* genera. Thermophilic spore forming bacteria are known to play a role in the degradation of carbohydrates, whereas thermophilic aerobic bacteria thet form branching filaments (actinomycetes) are engaged in the biodegradation of cellulose and hemicelloleses (Polprasert 1989, Prescot et al. 1996). These processes were relatively slow and decreased the rate of heat generation and, thus, the temperature (Fig. 1). After 7 days at about 47 °C (bottom layer 43 °C and middle layer 53 °C) mesophilic bacteria re-activated and grew again. The bacterial population remained high throughout a further 9 days of gradual temperature decline until the compost reached the ambient temperature.

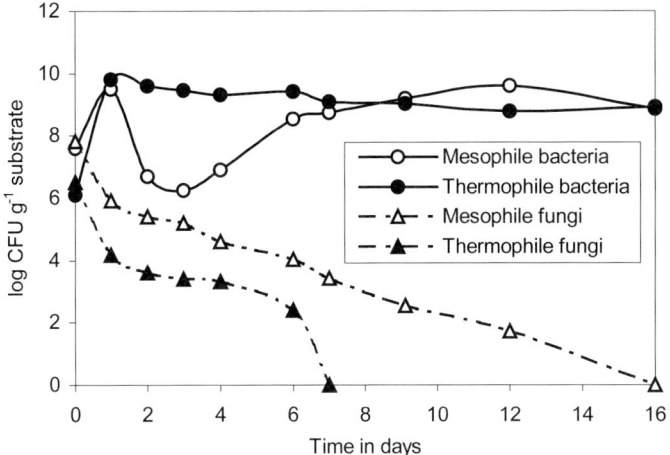

Fig. 3. Growth of bacteria and fungi during composting

Moisture content (MC) and volatile solids (VS) in tobacco solid waste at the start of the experiment were 65.8% and 68.5% respectively. The final MC in the composting ranged from 74.2 to 76.5%. The final total solids in the compost ranged from 55 to 56%, which is around a 45% of reduction. Depending on the amount of the original tobacco waste put in the reactor the range of VS reduction was 39.3-40.2% after composting, which is in agreement with literature data (Bari et al. 2000). Total condensate yields after 16 days were 80.2 ml and 800 ml depending on the reactor volume, as the result of biodegradation of the organic fraction of waste and the production of water. The TOC content in the aqueous fraction of compost dropped to 41.4%. At the end of experiment the nicotine content in compost dropped from 2,000 to 450 mg kg^{-1} of total solids. It is assumed that *Pseudomonas* sp. was responsible for nicotine biodegradation, because of its regular presence in the mixed microbial culture of the composting mass throughout composting (Civilini et al. 1997). Further maturation of raw compost was carried out for another 30 days in open box. During that period the moisture content dropped from 76.5% to 38.4 %.

3.2 Modeling of the composting process

The laboratory scale bioreactor belonged to the group of unsteady state column reactors. Biodegradation of the substrate was slow compared to the airflow rate and oxygen transfer through the boundary gas layer. Consequently, it was supposed that the change in inlet and outlet concentration of oxygen in air was very small. Accordingly, several assumptions were used to develop the model: the process was carried under adiabatic conditions; the air flow rate during composting was constant; the composting rate was formally expressed as a degradation rate of the dry substrate, without taking into account composition of the original moisture content, Eq. 1.

$$r_S = f(w_S, T) \tag{1}$$

Biodegradation (oxidation) of organic matter is known to be a slow process, so it was assumed that only a small difference in concentration of oxygen in the air between inlet and outlet of the column would occur. Consequently, the reactor model could be approximated by two mass balances (for oxygen and for the composting mass or for formation of CO_2) and by one overall heat balance. The equation for the material balance regarding the dry organic substrate in the reactor was described by

$$r_S = -(dw_S/dt) = -\frac{dm_S}{dt}\frac{1}{m_{So}} \tag{2}$$

or

$$r_{CO2} = \frac{dw_{CO2}}{dt} = \frac{dm_{CO2}}{dt}\frac{1}{m_{So}} \tag{3}$$

The heat balance took into account that the process had been carried out under adiabatic conditions and was expressed by

$$m_o c_{pS}\frac{dT}{dt} = (-\Delta H_r)m_{So}r_S - \rho_z Q_V c_{pz}(T - T_u) \tag{4}$$

The real reaction rate dependence as a function of all intrinsic variables was not known, but a simple empirical model could be proposed

$$r_S = k(T) w_S^n \tag{5}$$

Constant k in the Eq. 5 is generally temperature-dependant. Eqs. 2 – 4 and Eq. 5 acted as the basic model for the reactor in the composting process. To apply the model and compare it with experimental data (25 l reactor, second set), several parameters and process variables had to be known. Table 2 shows the necessary experimental data and other known data for application of the proposed model.

Table 2. Dimensions and characteristics of the process

Mass of wet composting material, m_o	11.5 kg
Initial mass of dry substrate in reactor, m_{So}	3.9 kg
Air flow, Q_v	0.9 l min^{-1} kg^{-1} VS
Heat capacity of air, c_{pz}	1.01 kJ kg^{-1} K^{-1}
Heat capacity of mass in the reactor, c_{ps}	3.58 kJ kg^{-1} K^{-1}
Air density, ρ_z	1.3 kg m^{-3}
Inlet air temperature, °C	24 °C – 26 °C

In addition to these variables, available in advance, it was necessary to calculate, on an experimental basis, the reaction enthalpy, to select the kinetic model and to evaluate model parameters. The reaction enthalpy was calculated from experimentally obtained results, by measuring the temperature in the composting mass during the reaction. It was assumed that the process had been carried out under adiabatic conditions and that the released heat was proportional to the progress of biodegradation. Substituting the Eq. 2 in Eq. 4 yields

$$m_o c_{pS} = (-\Delta H_r) dm_{So} - \frac{dw_S}{dt} \cdot - \rho_z Q_v c_{pz}(T - T_u) dt \tag{6}$$

and after integration and rearrangement

$$(-\Delta H_r) = \frac{\rho_z Q_v c_{pz}}{m_{So}} \int_0^t (T - T_0) dt + c_{pS} \frac{m_o}{m_{So}} (T_K - T_0) \tag{7}$$

Thus, at the end of the reaction $T_K = T_0$. The second member on the right side of the Eq. 7 becomes also zero, and the reaction enthalpy was calculated from the Eq. 8, by numerical integration on the basis of the experimental data

$$(-\Delta H_r) = \frac{\rho_z Q_v c_{pz}}{m_{So}} \int_0^t (T - T_u) dt \tag{8}$$

The calculated value of the reaction enthalpy is presented in Table 3. Estimated values for selected substrate was in the range of those from the literature (Haug 1993; Olmsted and Williams 1994) taking into account the percentage of VS in the total solids. Considering that the rate of heat released was proportional to the rate of biochemical reaction, the conversions obtained could be calculated from the proposed model. To estimate the dependence of the reaction rate on temperature, the expression taken from the literature (Haug 1993; Snape et al. 1995) was applied

$$k = k_0 \left[1,066^{T-20} - 1,21^{T-60} \right] \tag{9}$$

Based on the known input variables from Table 2, the calculated reaction enthalpy and the proposed dependence of the rate on temperature, Eq. 9, the kinetic parameters k_o and n were estimated for Eq. 5. Direct nonlinear regression analysis on the basis of the simplex method of Nelder-Mead was performed to determine the composting rate constants. After kinetic parameters were determined, the differential equation describing the system was numerically solved using Runge-Kutta method, and then the theoretical curve was drawn together with the actual data plot in Fig. 4.

Fig. 4 shows fairly good agreement between the measured temperatures and calculated temperatures obtained with the Eq. 5, during the whole period of composting. The heat released was proportional to the progress of biodegradation. In Table 3 are given the values of estimated parameters.

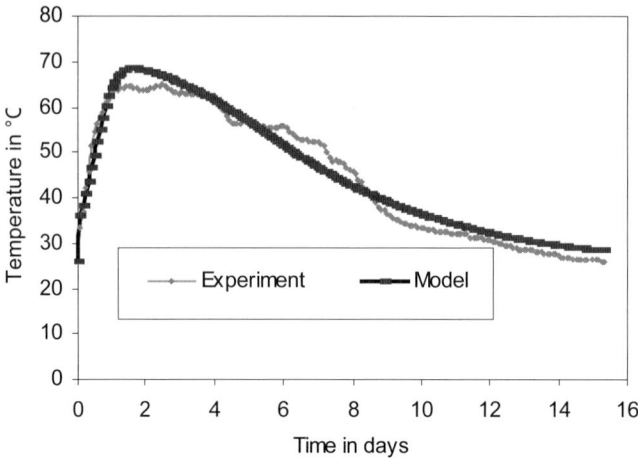

Fig. 4. Comparison of temperature curves obtained in the experiment and by modeling the process (25 l reactor)

Table 3. Results of estimated parameters

Reaction enthalpy, ΔH_i	-7454.2 kJ kg^{-1} TS
Order of reaction, n	1.4
Constant, k_o	0.024 h^{-1}
Mean square deviation, SD	0.2035

Results presented in Table 3 suggested a good selection of the proposed composting model, although only two kinetic parameters (k_0 and n) were estimated (Bari and Koenig 2000; Higins and Walker 2001).

Measured substrate conversion and calculated by the model, Eq. 5, during the 16 days is presented in Fig. 5.

A graphical presentation of substrate conversion versus reaction time during 16 days, calculated using proposed model, shows that during the first 5 days total solids was reduced by 40 %, which was in agreement with the experimental value. It suggested that the empirical model described aerobic composting process fairly well and confirmed the hypothesis that the released heat was proportional to the biodegradation progress.

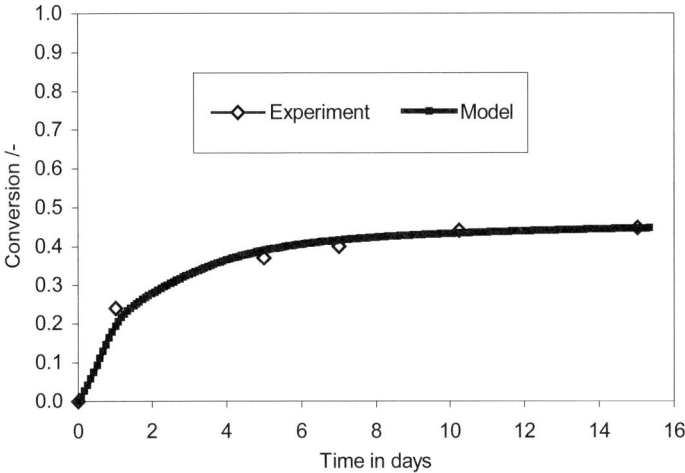

Fig. 5. Comparison between experimentally obtained conversion of tobacco waste and by modeling the composting process (25 l reactor)

4 Conclusion

The results obtained in this contribution to aerobic composting in a column reactor showed good reproducibility. In the laboratory scale reactors a temperature above 50 °C was reached after day one and it was maintained between 50 °C and 65 °C for 7 days. After that it slowly declined until the end of the experiment. After 16 days, the volume of tobacco solid waste was reduced to about 50% and the nicotine content dropped from 2,000 to 450 mg kg^{-1} of total solids.

The temperature in the reactors was between 50 °C and 65°C, which is favorable for microbial activity in the composting mass. Changes in the number of mesophilic and thermophilic microorganisms in the mixed culture were recorded at different stages of composting.

The advantage of the closed reactor over an open one is in that stabilization of organic waste is obtained within a shorter period. Furthermore, CO_2 (air pollutant) and generation of unpleasant odors during composting are much more easily controlled.

The degradation percentage of organic solids could be predicted from the proposed model and might be useful in the pilot plant study. Evaluation of the experimental results suggested a good selection of the proposed composting model, although only two kinetic parameters (k_o and n) were estimated.

References

Austrian standard S 2023 (1986) Analytical Methods and Quality Control for Waste Compost. Vienna, Austrian Standardization Institute

Bari QH, Koenig A, Guihe T (2000) Kinetic analysis of forced aeration composting – I Reaction rates and temperature. Waste Manage. Res. 18 pp 303

Bari QH, Koenig A (2000) Kinetic analysis of forced aeration composting – II Application of multiplayer analysis for the predicition of biological degeadation. Waste Manage. Res. 18 pp 313

Civilini M, Domenis C, Sebastianutto N, de Bertoldi M (1997) Nicotine decontamination of tobacco agro-industrial waste and its degradation by microorganisms. Waste Manage. Res. 15 pp 349

Croatian regulation (in Croatian), Narodne novine No. 123, pp 4112-4116, 1997

Gries G (1994) Composting residential and commercial streams. Biocycle 36 pp 78

Haug RT (1993) The Practical Handbook of Compost Engineering. Boca Raton, Lewis Publishers

Higins CW, Walker LP (2001) Validation of a new model for aerobic organic solids decomposition: simulations with substrate specific kinetics. Process Biochem. 36 pp 875

Kayhanian M, Tchobanoglous G (1993) Innovative two-stage process for the recovery of energy and compost from the organic fraction of municipal solid waste. Water Sci. Technol. 27 pp 133

Kolthoff IM, Sandel EB (1951) Inorganic quantitative analysis (in Croatian). Zagreb, Školska knjiga

Korolewicz T, Turek M, Ciba J, Cebula J (2001) Speciation and removal of zinc from composted municipal solid waste. Environ. Sci. Technol. 35 pp 810

Olmsted J, Williams GM (1994) Chemistry: The Molecular Science, St. Luis, Mosby, St. Luis

Polprasert C (1989) Organic wastes Recycling. Chichester, Wiley

Prescott LM, Harley JP, Klein DA (1996) Microbiology. Dubuque,WCB Publishers

Saunders JA, Blume DE (1981) Quantitation of major tobacco alkaloids by high-performance liquid chromatography. J. Chromatography, 205 pp 147

Snape JB, Dunn IJ, Ingham J, Prenosil JE (1995) Dynamic of Environmental Bioprocesses: Modeling and Simulation, Basel, VCH

Ucur M, Nikolic O (2000) Elements of responsibility in un-sanitary municipal landfills and dangerous waste disposal (in Croatian). Proceedings of the VIth International Symposium on Waste Management, Zagreb, Croatia, pp 237-246

White P, Franke M, Hindle P (1994) Integrated Solid Waste. London, Blackie Academic & Professional

Preventative Measures in Production from the Point of Sustainability

Anna Christianova

Czech Cleaner Production Centre, Dittrichova 6, 120 00 Prague, Czech Republic

Abstract

The strategy of sustainable development faces similar problems as the cleaner production strategy – there is an "evident gap between rhetoric and knowledge needed for implementation." It can be assumed that the conservative behaviour of politicians, companies as well as individuals originates from the same source both for cleaner production and for sustainable development.

Preventive approach and pollution prevention present a continuous challenge for research and development. For example, waste management is not only another source of inspiration for new materials, technologies and construction solutions, it particularly needs a framework definition that would provide sense for these activities. Experience with implementing cleaner production should be an important source of inspiration for defining the requirements for sustainable technologies.

The necessary preconditions for the change of consumption and production patterns include understanding their impacts and managing their aspects. The change in behaviour of both individuals and groups originates from education, networking and exchange of information.

1 Introduction

The strategy of sustainable development faces similar problems as the cleaner production strategy. The seminar in Maribor identified an existing problem of an "evident gap between rhetoric and knowledge needed for implementation." It can be observed that the conservative behaviour of politicians, companies as well as individuals originates from the same source both for cleaner production and for sustainable development. Experience with implementing voluntary activities, e.g. cleaner production, especially at the level of communication among participants, management quality, the role of management in the realisation of preventative measures, approaches to selection of cleaner technologies and, above all, implementation of cleaner production strategy on products, should be one of the important sources of inspiration for defining the requirements and indicators for sustainable development instruments, including technologies. It needs to be said that we have not worked in detail on social indicators of prevention projects.

A joint programme of UNIDO and UNEP targeted at establishing a Network of National cleaner production centres[1] was declared in 1994; these centres were established to:

- disseminate information about preventative strategies not only in a form of seminars, but above all by means of multiple-level education, especially at the company management and university level with the aim of provoking a change in thinking as well as behaviour, i.e. a change of approach towards prevention and source of waste and pollution reduction within the scope of industrial activities
- address both public administration as well as businesses and thus act upon introducing prevention as an important touchstone into decision-making processes; also to create conditions for global support to cleaner production projects (for Cleaner Production Assessment) at the level of both environmental policy and funds
- realise seminars and trainings for cleaner production consultants and company employees to teach them how to individually propose and implement cleaner production programmes
- create new know-how and search tools for implementing cleaner production at the level of „final consumer" – companies, services, public administration as well as individuals
- contribute to reduction of environmental burden, more effective exploitation of raw material sources and energy, and cost reductions achieved through preventative measures within the framework of cleaner production projects

If cleaner production centres want to fulfill this objective, their work needs to be of a creative kind. It has the same character as the work of a research and development institute, which formulates a theory, conducts an experiment to test it and projects its results into refined theory formulation. The objective to achieve is a continual improvement – reducing environmental impacts of human activities through preventative measures implemented at the pollution source and more effective exploitation of raw materials and energy. These objectives are declared by the International Declaration on Cleaner Production adopted in Seoul in September 1998[2].

The Czech Cleaner Production Centre (CPC) was established in 1995 as an independent non- governmental organisation within a framework of the Czech-Norwegian Cleaner Production Project. Starting in 1992, the project was financed by the Norwegian Government. In the years 1994 to 1999 more than 150 cleaner

---

[1] For example, in 1995, the centres in Brazil, China, Czech Republic, India, Mexiko, Slovakia, Tanzania, Zimbabwe were established.

[2] After the Internationale Declaration on Cleaner Production the cleaner production strategy and other preventive strategies such as Eco-efficiency, Green Productivity and Pollution Prevention are preferred options to sustainable development achieving.

production projects were implemented within different industries in a form of long-term courses. Since 1996 the CPC has participated in building cleaner production capacities in other countries (e.g. Croatia, Russia, Uzbekistan, Macedonia) within a framework of UNIDO projects. The CPC today is also competing for projects within public competitions and grants or searching financing to implement prevention projects as an innovation to common procedures.

Cleaner production projects in enterprises have targeted at operational system changes and they also became a tool for continuous improvement of company's environmental profile. Implementation of cleaner production is a voluntary activity; when we implement strategies into a concrete setting, we should repeatedly put a question about the new tools we need. The analysis of weaknesses of cleaner production projects resulted in the fact that CPC started integrating its primary activities with other issues. The first step was the introduction of integrated implementation of cleaner production and environmental management systems (according to ISO 14001 and EMAS) aimed at emphasizing the role of management and increasing its quality (Coll. 2001). Another success story was an extension of both activities mentioned earlier to include the requirements of quality management systems (ISO 9000 standards); this type of project was awarded a prize at the Expo 2000 in Hannover.

Objectives associated with cleaner production are basically identical with the objectives to realise the best production procedures[3]. That is why CPC also concentrated on integrated prevention and pollution control (formulated by the Council Directive 96/61/EC on Integrated Pollution Prevention and Control – IPPC), as well as on ecodesign, LCA and integrated product policy. Generally speaking, CPC deals with environmental, economic and social aspects of activities (e.g. production, services, products) and their management with the objective of elaborating methodologies and tools for preventative measures to improve products towards environmentally friendly products (or to use service instead of a product) and technologies towards best available technologies.

The necessary preconditions for the change of consumption and production patterns include understanding their impacts on the environment in the course of the whole life cycle and managing their aspects. The change in both individual and group behaviour originates from education, networking and exchange of information. From the point of view of educational processes, specific prevention projects oriented towards solution of relevant problems using the win-win strategy take a form of training in establishing and stabilising new habits. Additionally, we can test anticipated impacts of monitored aspects and used tools in these projects.

---

[3] For example, there is a „Best Practice Program" in chemical industry.

2 Preventative strategy and cleaner production methodology as an approach to sustainability

Voluntary activities have different opportunities in a stabilised economy which has gone through an ongoing development, and in an economy undergoing transformation. In our experience, state or department policies and their concrete elaboration into decisions and resolutions help companies to orientate themselves better when defining their own objectives and strategies. Small and medium sized companies, in particular, focus primarily on short-term objectives. It is expected that they will be flexible, e.g. their products and technologies have to be flexible. They do not assess the production process as a whole, i.e. from the moment of getting the input raw materials and energy to the phase of the management of waste or used products. For economic reasons they integrate functioning and responsibility at management and technical levels. They think about voluntary activities only after they reach certain level of economic stability. Briefly, it is very difficult for them to become aware of the fact that voluntary activities offer them new instruments that can help them meet their obligations resulting from legislative requirements in the framework of transformation or reach better position on a market. We have to convince them of this again and again. Large companies, especially exporters to foreign markets, are often better in dealing with this approach.

In economically unstable companies with out-of-date technologies where a cleaner production project can bring the highest benefits, the cleaner production strategy is perceived as a superstructure that „can wait for some time". Such company will not implement cleaner production without support or will not finish the project. At the same time, support can be counterproductive: with respect to the low standard of management or existential problems, it does not result in reducing the environmental load, but it only postpones its breakdown. The knowledge and approaches acquired by employees in the course of cleaner production training are usually surmounted by social instability.

In the course of transition to market economy we need to reckon with certain time period required by companies/enterprises and their management to acquire stability (primarily the economical one). As soon as they achieve it and are able to consider their development at least within the medium-term horizon they start being interested in options for environmental protection solutions (resulting especially from new legislation and control of its compliance), using voluntary activities. A stabilised company can identify the advantages of preventative approach and make the best of it, as well as it is able to estimate the benefits of a cleaner production project and accepts the implementation of environmental management systems.

Cleaner production as a preventative strategy should be logically perceived as a natural approach – an obvious first step – towards viewing the proposed measures. Discussions are held at each cleaner production seminar about experiences with implementing the cleaner production strategy under different conditions. Our goal is to give the preventative approach a status of self-evident criteria for decision-making processes and not a status of a superstructure. For example, it should become a part of other projects and programmes, especially the programmes of regional and territorial economic development or the programmes concentrating on closing material flows (Christianova et al. 1999-2001). However, its application is not yet common place.

A similar approach can be anticipated within the process of enforcing sustainable development; a short-time prospect shall be given a preference against a long-term one.

The concern for preventative approaches and new technologies on the side of big producers in the Czech Republic increased during the preparation and after the approval of the Act on Integrated Prevention, which transposed the Council Directive 96/61/EC, on Integrated Prevention and Pollution Control, into our legal system. Since 1997 many studies on the proposed act on integrated prevention and its impacts on production have been carried out in the Czech Republic. Producers represented by industrial unions had a chance to provide their comments; the Ministry of Environment as well as the Ministry of Industry and Trade opened special information centres. At present most producers/operators of installations are personally interested in comparing their technology to the Best Available Techniques level. This approach could be taken as a challenge for science and research, at least it gives a chance to identify priority problem areas in individual sectors (Christianova et al. 1999).

Preventive approach and pollution prevention present a continuous challenge for research and development.

An important shift can be seen from production towards products („From Cleaner Production towards Cleaner Product"). But it also works respectively: inspired by the complex LCA approach, we aim at monitoring the impact of selected measures of a cleaner production project on a technical life of technology in terms of modernity and efficiency.

3 The impact of preventative measures on consumption and exploitation of raw materials, changes in technology and on the generation and use of wastes

Prevention of waste generation has been given a priority within the system of waste management by the Council Directive 75/447/EU on Waste. That is to say, waste management has been associated with products in this way. Preventative strategy in waste management can be viewed as a reaction to difficulties with waste recycling and closing material flows. Our goal is to accomplish repeated use of components and materials from dismantled products. The analysis of aspects should provide basis for defining conditions for elimination of threats to safety and reliability of installations. We need to specify conditions for reuse of waste that is not disserviceable and does not disproportionally burden the environment; elevation of recycling quota is not just a question of a good will. Waste management is not only another source of inspiration for new materials, technologies and construction solutions, it particularly needs a framework definition that would provide sense for these activities (Christianova et al. 1999).

Reduction of waste generation means approving changes, which can be distributed across the whole product life-cycle and all technologies that the product encounters. An example of this could be end-of-life vehicles and electroscrap – an enormous increase in amounts of waste is in direct connection with an increase of consumption and shortening product life cycles.

It is unreasonable to expect that in the near future man would give up the use of cars, computers, television sets and other home appliances he got accustomed to. We have not yet elaborated any concept of transport and services that would substitute for these products. We can, however, be sure that decommissioned products currently in use shall pile up around us.

Wastes and pollution generated during a product application phase and after the end-of-life-cycle phase attract more attention than wastes and pollution from production itself, as these are reduced and controlled by sectoral legislation in the least. Approximately since 1985 the problems with utilisation of wastes from end-of-life vehicles have been projected into the requirements for:

- new recyclable materials (e.g. micro-alloyed steel)
- unified international codes identifying materials (resulting e.g. in an increased utilisation of plastics)
- new constructional components and combinations of materials enabling disassembly, reuse and recycling of particles (e.g. aluminium sandwich components, especially fasteners, substitution of fixed joints)
- recycling equipment with higher efficiency resulting in decrease of costs for waste treatment (e.g. dismantling lines, separation lines for shredders and mobile equipment).

New materials and constructional solutions development can further be expected after the implementation and compliance with the Directive 2000/53/EC of the European Parliament and of the Council on end-of-life vehicles, especially its detailed requirement for re-use of components, determining the recycling quotas and prohibition to use certain substances. In addition, statistical monitoring, financial flows and control mechanisms have to be implemented (Christianova et al. 2001).

The principle preventative approach in the waste management is a support to the technical development and marketing of products designed in the way that their production, consumption as well as final decomposition do not increase the amount and dangerous properties of wastes and pollution, i.e. by developing products with the help of ecodesign and using the information about the product life cycle.

Prevention in the context of a product must take into account consumer requirements as well as it needs to monitor the opportunities for limiting the impacts of its production and utilisation on the environment, influence the possibilities for the product reuse and increase the proportion of recycled material in the product or the measure of recycling.

It is one of real steps to change the consumption formula that would contribute to sustainable development. A product that retains its environmental friendliness throughout its whole life cycle needs to meet consumer requirements – it must not lose its quality and utility properties as well as it must be safe. Experiences and methodological procedures of cleaner production projects in production can also be used as one of the tools for the change of products[4].

Logically and simply formulated cleaner production strategy (e.g. „It is a pre-vasive value system leading innovatively to new behavior, new organisation and system change of products and processes") opens new opportunities for preventative activities.

4 Conclusion

The way from the political consensus to day-to-day reality can be long and complex. It holds true for the preventative approach towards environmental protection as well.

Economic and environmental impacts of preventative measures in production address business entities thinking pragmatically; however, especially with small and medium sized companies, preventative approach has not yet reached the status of self-evident criteria for decision-making processes. If we really want to achieve sustainable development we need to find out why there exists a "gap between rhetoric and knowledge needed for implementation" and which tools can be used to influence conservative behaviour of politicians, companies as well as individuals.

Prevention projects did not stress out their social aspects even though these are indirectly comprised in them. In terms of achieving sustainable development they offer in the first place a system approach to continuous reduction of the amount of waste and higher utilisation of materials and energy. Although it will be very difficult to formulate social indicators of prevention projects, we should give it a try.

---

[4] **Cleaner production** (CP) is the continuous application of an integrated preventive environmental strategy to processes, products and services so as to increase efficiency and reduce risks to humans and the environment. **For production** processes, CP includes efficient use of raw materials and energy, elimination of toxic or dangerous materials and reduction of emissions and wastes at the source. **For products**, the CP strategy focuses on reducing impacts along their entire life-cycle of the products and services, from design to use and ultimate disposal. *(Definition adopted by UNEP).*

References

Christianova et al. (1999) "Handbook on IPPC for State Officers", for MoE, CCPC
Christianova et al. (1999-2000) "National Cleaner production Programme", for MoE, CCPC
Christianova et al. (1999) "Prevention in Waste Management Plans", Study for draft of Waste Act for the Czech Republic, Parts to Autowracks, Batteries, Oils, CCPC
Christianova (2001) " Implementation of the 2000/53/EU Directive on end-of-life vehicles for the Czech Republic", Study for MoE, CCPC
Coll. (2001) "Cleaner production through EMS", Achieving Cleaner Production using EMS as a tool, developed by the CCPC, Manual

Environmetric strategies to classify, interpret and model risk assessment and quality of environmental systems

Vasil Simeonov

Chair of Analytical Chemistry, Faculty of Chemistry, University of Sofia "St. Kl. Okhridski", 1164 Sofia, J. Bourchier Blvd. 1, Bulgaria, E-mail: VSimeonov@chem.uni-sofia.bg

Abstract

The present paper deals with the application of some environmetric approaches like cluster analysis, projection pursuit, Kohonen maps, neuron gas, principal components analysis, chemical mass balance modeling, multiple regression on principal components, time-series analysis in environmental data mining. Several ecologically important objects including marine sediments, wet and dry atmospheric precipitation are treated in order to obtain relevant information about monitoring data set structure and relationships. Multivariate statistical models are offered, which could help in decision making and problem solution of the local environment. In this way the case studies are contribution to the idea of the increasing role of the environmetrics into the concept of sustainable development.

1 Introduction

In recent years researchers have become increasingly interested in applying sophisticated modeling methods to environmental studies. It helps a lot in understanding of various environmental phenomena and even solving some problems but the complex nature of the dynamic physicochemical models are a more theoretical weapon that an understandable and simple way for risk and quality assessment in a close to the humans environment. Usually, the quality of life in a certain environment is assessed by many parameters but the monitoring of air, water and soil chemistry and biochemistry turns to be the decisive point. The monitoring data are then readily compared to specific patterns of "allowable" values for hazardous chemicals and respective conclusions are offered. Unfortunately, this approach does not correspond to the reality. It is univariate and delivers information only on one possible contaminant or toxic compound. The nature is multivariate and it is much better to collect and interpret information on the relationships and on the effects of many chemicals or pollutants simultaneously. The very idea of sustainable development as defined by many scientists requires taking into account possibly many environmental parameters in order to achieve a reliable pathway: what to do with a system, how to treat it, what involvement of authorities is needed etc.

That is why the application of environmetric strategies for intelligent data analysis (IDA) seems to be an appropriate tool in risk and quality assessment theory and practice. Environmetrcs as part of IDA deals with application of multivariate statistical methods in environmental data mining. The models offered by the environmetric approaches lack the theoretical limitations of the dynamic process modelling and offer a useful way to understand many features of the system in consideration, among them the following:

- similarities and dissimilarities between sampling sites
- detection of site outliers in the monitoring net and their explanation
- identification of data structure features
- detection of latent factors which could be interpreted as anthropogenic or natural sources responsible for the chemical content of the environmental samples
- trends in the behaviour of pollutants
- seasonal effects of the pollutants distribution
- apportioning of the contribution of each identified source to the total species mass
- finding the necessary chemical mass balance and relationships in the environmental systems in consideration

All these opportunities offered by the environmetric approaches turn to be a very important tool for decision making and economic, health and political solutions. I shall try to present some real-time examples about the application of environmetrics to every-day problems on different environmental scales. In all of the case studies a problem arises which needs a quick assessment and even quicker response. Keeping a certain local equilibrium in a local environment should be considered as a global action in the modern world where the relations and the interactions, the atmospheric transfer processes and the mutual dependence on "hot spot" pollution is unavoidable. Therefore, risk or quality assessment at a given location is much more than a local episode. Environmetrics tries to be useful as parameter of the sustainable development.

2 Experimental

2.1 Environmetric Approaches for Exploratory Data Analysis

In the data treatment approaches of the environmetrics both unsupervised and supervised techniques are used. In the first case the data mining is performed spontaneously, in a hierarchical way, from the data set. In the latter case a preliminary step of learning (training) is necessary to derive a treatment (classification) rule based on grouping of objects with known origin or behaviour. This rule allows interpreting new objects with unknown origin or behaviour in the classes offered by the classification rule.

Cluster analysis is a well-known and widely used classification approach for environmetrical purposes with its hierarchical and non-hierarchical algorithms (Massart and Kaufman 1983; Einax et al. 1997).

In order to cluster objects characterized by a set of variables (e.g. sampling sites by chemical concentrations or pollutants), one has to determine their similarity. To avoid influence of the data size, a preliminary step of data scaling is necessary (e.g. autoscaling or z – transform, range scaling, logarithmic transformation) where normalized dimensionless numbers replaces the real data values. Thus, even serious differences in absolute (concentration) values are reduced to close numbers. Then, the similarity (or more strictly, the distance) between the objects in the variable space can be determined. Very often the Euclidean distance (ordinary, weighted, standardized) is used for clustering purposes. Another way of measuring similarity is calculation of the correlation coefficient between two row-vectors x_1 and x_2 characterizing objects 1 and 2. Thus, from the input matrix (raw data) a similarity matrix is calculated. There is a wide variability of hierarchical algorithms but the typical ones include the single linkage, the complete linkage and the average linkage methods. The representation of the results of the cluster analysis is performed either by a tree-like scheme called dendrogram comprising a hierarchical structure (large groups are divided into small ones) or by tables containing different possible clusterings. The hierarchical methods of clustering mentioned above are called agglomerative. Good results are obtained also by the use of hierarchical divisive methods, i.e. methods that first divide the set of all objects in two so that two groups (clusters) are formed. Then each group (cluster) is again divided in two etc., until all objects are separated.

The aim of classification by non-hierarchical clustering is to classify the objects in consideration into certain number of preliminary intended groups, e.g. *K* clusters. For instance, in order to obtain 2 clusters, one selects 2 seed points among the objects and classifies each of the objects with the nearest seed point. Thus, an initial cluster is obtained. For each of these clusters, one determines the centroid (the point of mean values of the variables x_i for each cluster). The whole procedure is repeated; new centroids are calculated for the new clusters. The new centroids have new co-ordinates and it leads to reclassification of the objects. Widely applied are: the Forgy's method, MacQueen's K-means method etc.

All clustering methods mentioned up to now have a general feature in the classification: they consider each object to be part of only one single cluster. A different strategy is typical for the so-called fuzzy clustering, which permits objects to be part of more than one cluster. In the classification fuzzy procedure each object *i* is given a value f_{ik} for a membership function in cluster *k*.

Daszykowski et al. (2001) offer new original clustering algorithms named density-based spatial clustering of application with noise (DBSCAN) and ordering points to identify the clustering structure (OPTICS). In the first case an important advantage is the possibility to detect outliers. A cluster with DBSCAN is defined as a region of the data space where the objects' neighborhood of certain radius ε, contains at least k objects. Only one input parameter is required, namely, the minimal number of objects in the neighborhood, k. Three categories of objects are distinguished – core, border and outlier object, with respect to the neighborhood density. Core and border objects are grouped into clusters.

With OPTICS approach a unique order of objects in the n-dimensional data set is established, which reveals the data set structure. It allows finding zones where many objects are close together (high density zones) using a so-called reachability plot. The OPTICS algorithm calculates the similarity between the objects by reachability distances. In a very first approximation they can be considered as Euclidean distances. Outliers could be easily detected since their reachability distances are higher than those of the other objects. The OPTICS plot can be visualized as a color map. The information of this map can be used to study the inter-variable and inter-object relations and to estimate the contribution of each variable to the data set structure.

Principal components analysis (PCA) is a typical display method, which allows to estimate the internal relations in the data set and to model the ecosystem in consideration. There are different variants of PCA but basically, their common feature is that they produce linear combination of the original columns in the data matrix (data set) responsible for the description of the variables characterizing the objects of observation. These linear combinations represent a type of abstract measurements (factors, principal components) being better descriptors of the data structure (data pattern) than the original (chemical or physical) measurements. Usually, the new abstract variables are called latent factors and they differ from the original ones named manifest variables. It is a common finding that just a few of the latent variables account for a large part of the data set variation. Thus, the data structure in a reduced space can be observed and studied (Massart et al. 1998).

Generally, when analysing a data set consisting of n objects for which m variables have been measured, PCA can extract m principal components PCs (factors or latent variables) where $m < n$. The first PC represents the direction in the data, containing the largest variation. PC 2 is orthogonal to PC 1 and represents the direction of the largest residual variation around PC 1. PC 3 is orthogonal to the first two and represents the direction of the highest, residual variation. Around the plane formed by PC 1 and PC 2. The projections of the data on the plane of PC 1 and PC 2 can be computed and shown as a plot (score plot). In such a plot it is possible to distinguish similarity groups. According to the theory of PCA the scores on the PCs (the new co-ordinates of the data space) are a weighted sum of the original variables (e.g. chemical concentrations):

Score (value of object I along a PC p) = $\gamma_{1p} Y_1 + \gamma_{2p} Y_2 + ... + \gamma_{kp} Y_k$

where Y is indication of the variable value (e.g. concentration) and γ are the weights (called loadings). The information hidden in the loadings can also be displayed in loadings plots. It is important to note that PCA requires very often scaling of the input raw data to eliminate dependence on the scale of the original values.

When time parameters are taken into account quite suitable approaches for three-way data analysis are Tucke3 model and the PARAFAC model (Tucker 1963). The basic principle of the first method is the decomposition of the three-way data set into a three-way core matrix and three two-way loading matrices (one of each mode, e.g. objects, variables, time parameters). To each mode a certain number of factors is assigned chosen in such a way as to be less than the dimensions of the original three-way data set in order to achieve a considerable amount of data reduction. Again, plots for each kind of factors could be constructed showing the relationships of objects, variables or time parameters (seasonality). In PARAFAC again a decomposition of the data set is aimed but the three loading matrices obtained are not necessary orthogonal.

Projection pursuit (Huber 1985) is also a dimension reduction method where the new latent factors are obtained by maximizing projection indices (not variances as in PCA) that describe inhomogeneity of the data. There are different algorithms for projection pursuit. Usually, the first step is to center the data on the mean, then to project all objects onto the possible orthogonal complement of eigenvectors. Once the eigenvector, corresponding to the highest value of the projection index is found, the original matrix with orthogonal complement is adjusted. These steps are repeated until the number of factors is found. The projection indices rely on using an entropy function or using the Yenyukov approach (Yenyukov 1989).

Kohonen network (Kohonen 1990) has the general idea of mapping data from a high dimensional space onto a low dimensional space. The latter consists of a layer of *i* nodes arranged in a 2D plane as neighboring hexagons or rectangles. The mapping preserves the data topology, i.e. objects, which are similar in the original data space are mapped on the same node or onto neighboring nodes. A weight vector represents each node. The training of the net is done in an iterative way by updating the weights of the nodes. At each iteration a randomly selected object is presented to each node and the node whose weight is closest to this object is determined. This is the winner node. The weights of the winner are updated together with the weights of the nodes in its neighborhood, which size is defined by neighborhood function. Once the network is trained, for each object the winning node is found and the object is assigned to this node.

Neural gas network is suggested by Martinez et al. (Martinez et al. 1993). In contrast to Kohonen network, the neural net has no defined topology. Each neuron is associated with vector of weights, which correspond to its co-ordinates in variable space. At each iteration the input vector (object) is presented to the network and the weight vectors are ranked according to their distances to the input vector. Then the nodes are moved according to the rank. After a predetermined number of iterations, the nodes are distributed over the space. During the last step of the procedure a clustering is performed, i.e. for each object from the data set its closest node is found. Similar objects are grouped in the same node but there is no additional information about similarities between the nodes.

Multiple regression on principal components (apportioning models) is a very important environmetric approach (Thurston and Spengler 1985). It makes it possible to apportion the contribution of each identified by PCA latent factor (emission source) to the total mass (concentration) of a certain chemical variable. The first step is performance of PCA, identification of latent factors, then determination of the absolute principal components scores (APCS) and multiple regression of the total mass (dependent variable) on the APCSs (independent variables).

Chemical mass balance (CMB) method for receptor modeling was first applied by Winchester and Nifong (Winchester and Nifong 1972) when the so-called tracer solution was found. The CMB modeling procedure requires: 1) identification of the contributing sources types; 2) selection of chemical species or other properties to be included in the calculation; 3) estimation of the fraction of each of the chemical species which is contained in each source type (source profile); 4) estimation of the uncertainty in both ambient concentrations and source profiles and 5) solution of the chemical mass balance equation consisting of solution to linear equations that express each receptor chemical concentration as a linear sum of products of source profile abundances (i.e., the mass fraction of a chemical or other property in the emissions from each source type) and source contributions. CMB air quality model is one of the several models that have been applied to air resources management.

Time-series analysis (TSA) has, in principle, the following main purposes: 1) display of the series; 2) preprocessing of the data; 3) modeling and describing of the series; 4) forecasting with suitable models and 5) control of predicted values (Einax et al. 1997). Various methods of data smoothing, followed by application of plotting, correlation, regression and modeling techniques, allow to determine time trends and seasonal behavior of environmental data sets. The determination of the time components completes to large extent the environmetric interpretation of data from environmental monitoring.

2.2 Data collection: sampling, sampling sites, chemical analysis

2.2.1 *Black Sea coastal sediments*

Sediment samples were taken from four different sampling sites: Lake Beloslavsko (10 sites, sample number 1 – 10, close to a location of glass production factory), Lake Varnensko (11 sites, sample number 11 – 21, close to a location of steel-work), Varna Gulf, close to Lake Varnensko (7 sites, sample number 22 – 28), Varna Gulf near to coast (7 sites, sample number 29 – 35; close to a location of cement and chemical plant Solvey Soda) and Bourgas Gulf, near to the waste inlets caused by the local oil-refinery (4 sites, sample number 36 – 39). It is worth noting that in the configuration of the coastal line, the two lakes (Beloslavsko and Varnensko) serve as a natural buffer zone between the industrial zone and the gulf of Varna. For the gulf of Bourgas no such zone exists, and there is a direct inlet of contaminated waters into the sea.

The sampling was performed with a standard bottom grab of Smith – McIntyre and the elements measured throughout this study were Cu, Pb, Mn, Zn, Co, Cd, Cr, Fe, Ni and As. Digestion in concentrated hydrofluoric acid and subsequent analysis by atomic absorption were used for quantification. ETAAS (graphite furnace AAS, Perkin Elmer Z/3030) was the analytical method to determine Cu, Pb, Co, Cd, Cr, Ni, As and flame AAS (Perkin Elmer 603) was used for Mn, Fe and Zn. Certified reference materials (MESS-1, BCSS-1 and NBS 1646) were run with each series of samples. Precision for Mn, Fe and Zn was ≤ 5 % (as relative standard deviation); for the other analyses the RSD was ≤ 10 %. The experimental procedure is fully described by Simeonov et al (Simeonov et al. 2000). The aim of the study is to perform a multivariate statistical (environmetric) analysis of metal concentrations in coastal sediments collected at different "hot spots" sites of the Bulgarian Black Sea coast in order to gain information on the marine water quality.

2.2.2 Aerosol monitoring in Kozani region, Greece

Ten sampling sites in Northern Greece were included in a study for chemical mass balance modeling of aerosol monitoring data. The aim of the research work was to determine the contribution of different emission sources presented as source profiles in the formation of the total aerosol mass and in the concentrations of each component of the aerosol. A large data set was available prepared by the analytical team of the Laboratory for Environmental Protection, Aristotle University of Thessaloniki, who was responsible for all sampling and analytical details. Following chemical components were included in the CMB modeling procedure: Mg, Al, Si, S, Cl, K, Ca, Ti, Mn, Fe, Zn, Sr, P, V, Cr, Co, Ni, Cu, As, Se, Br, Cd, Sn, Sb, Te, Pb. The source profiles used were as follows: soil road dust; catalytic vehicles; non-catalytic vehicles; coal fired fly ash; diesel burning and open (agricultural) fire.

The modeling could substantially contribute to the estimation of the air quality of the region and, thus, to the idea of sustainable development.

2.2.3 Austrian wet precipitation monitoring net data

Eleven sampling sites (Reutte, Kufstein, Innervillgraten, Haunsberg, Werfenweng, Sonnblick, Nasswald, Litschau, Lunz, Nassfeld and Lobau) were chosen from the Austrian Wet Only Precipitation Network as they provide almost complete data collections for a long sampling period (between 9 and 16 years within the time interval 1984 – 1999 as follows: 16 years of monitoring for the sites Haunsberg and Werfenweng; 15 years of monitoring for the sites Reutte, Kufstein and Innervillgraten; 12 years of monitoring for the sites Sonnblick and Lobau; 11 years of monitoring for site Nasswald; 9 years of monitoring for the rest of the sites Nassfeld, Lunz and Litschau). Additionally, the trend analysis for each site and each major ion was performed for that particular period for which no missing data were found. Thus, each linear model is calculated for the specific time period assuring a complete data set. The first three sites are located in the Austrian province Tyrol, the next three in Salzburg, the following three in Lower Austria, Nassfeld in Carinthia and the last, Lobau, in Vienna region. The sampling site locations (all of them background stations) are presented in the paper of Puxbaum et at. (Puxbaum et al. 1998).

The Austrian Wet Only Precipitation Network is equipped with the WADOS (wet and dry only precipitation sampler). A slightly heated resistance sensor coupled to a lid ensures the collection of wet only precipitation samples in sampling bottles protected from the light. Local observers carry out the daily control of the samples usually in the morning hours. The collection procedure is performed according to the recommendation of the Austrian Ministry of Health and Environment and details for both sampling and analytical can be found in the study of H. Puxbaum et al., (Puxbaum et al. 2002). Missing data are evaluated by checking the precipitation data (precipitation amount) from the meteorological service to evaluate the rain event and ion concentrations are calculated by regression analysis from the already available measurements. The missing ion concentrations were estimated only for samples with partial analysis. In principle, the samples were analysed for the anions chloride, nitrate and sulfate and the monovalent cations ammonium, sodium and potassium by ion chromatography. The divalent cations, calcium and magnesium, were determined by atomic absorption spectroscopy. The pH was determined electrochemically.

Quality control of the analytical data was performed by comparing calculated *vs.* measured conductivities and by cation *vs.* anion balances of the individual samples. If the differences were less than 10 %, it was assumed that all major ions had been analysed. The majority of the samples presented an anion deficit, probably due to the lack of bicarbonate and the quantity of short – chain organic.

2.3 Results and discussion

2.3.4 Black Sea sediment data

The cluster analysis results (hierarchical clustering, Ward's method) of the sampling sites as objects are shown in Fig. 1.

Altogether four clusters could be interpreted divided into two bigger subgroups: the first contains heavily polluted sites from Varna and Bourgas Gulf (near to the coastal line and waste inlets, sites 29 – 39 located near a big chemical and cement plant – Varna and an oil refinery – Bourgas) and several sites from both coastal lakes located near to industrial sources (sites 2, 4, 6, 7, 8, 9 from the Lake Beloslavsko located near to a glass production factory; sites 14, 17, 19 from the Lake Varnensko located near to a steel-work); the second one indicates a moderately polluted buffer zone consisting of lake and near to the lake Varna gulf sites. In both big clusters two subgroups could be found. In the first one they represent the most severely polluted gulf areas (sites from Varna gulf 29 – 35 and from Bourgas gulf 36 – 39) and the less contaminated lake industrial inlets (sites 2, 4, 6 – 9 from Lake Beloslavsko and sites 14, 17, 19 from Lake Varnensko). In the second one they reflect the separation between one (Varnensko lake and non-affected Varna gulf parts, sites 11 – 13, 15, 16, 18, 20, 21 and sites 22 – 28, respectively) or another part (Lake Beloslavsko, sites 1, 3, 5, 10) of the buffer zone moderately affected by pollutants.

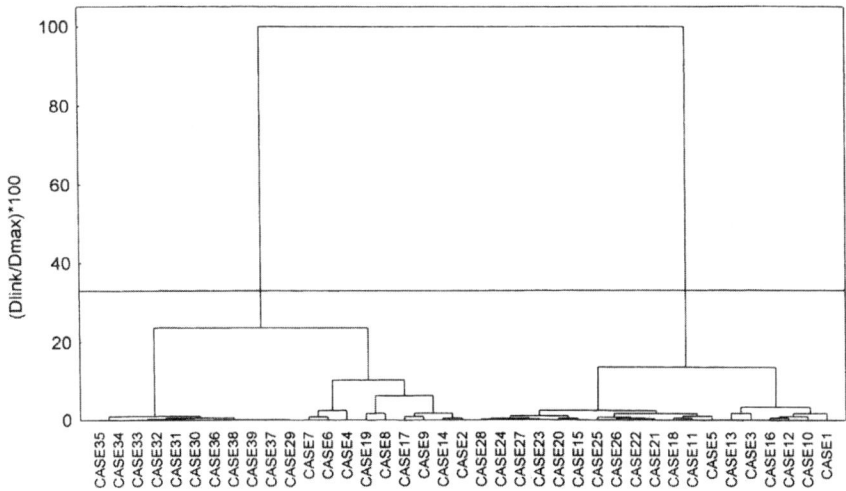

Fig. 1. Hierarchical dendrogram for sediment data clustering of sampling sites

The next step in the multivariate statistical analysis was application of PCA in order to group the chemical components by the loadings plots and the sites by the score plots. It is interesting to note that the site score plot (Fig. 3) reveals a more detailed description of the polluted coastal region.

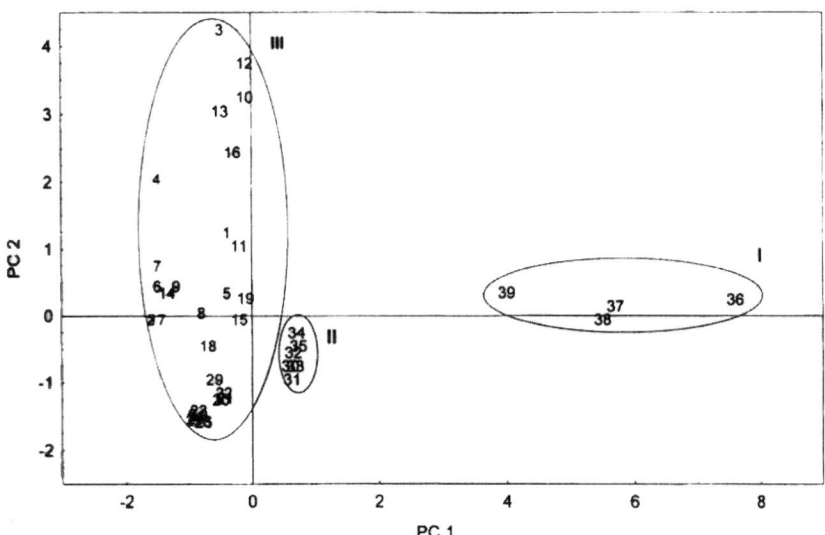

Fig. 2. PC score plot of sampling site for Black Sea sediments

The sites in the Bourgas gulf (36 – 39) represent an independent group (I) of heavily polluted area (oil refinery). They are definitely separated from all other sites and this is due to the enhanced determinant concentrations. The next well – formed group (III) comprises sites from the moderately contaminated lake buffer zones (sites 1 – 28), which indicates sites from the two lakes and Varna gulf sites located near to Lake Varnensko. The third group (II) indicates the intermediate level of pollution (higher than the buffer zone contamination but less than the Bourgas gulf area) of the sites originating mainly from the Varna gulf area (sites 29 – 35). The factor loading matrix is listed in Table 1.

Table 1. Factor loadings (Varimax normalized; marked loadings are higher than 0.7) for four principal components in sediment ambient data

Element	PC1	PC2	PC3	PC4
Cu	0.95	0.08	– 0.09	0.04
Pb	0.04	– 0.04	– 0.88	– 0.04
Mn	0.96	0.04	0.16	0.17
Zn	0.27	0.90	0.04	0.11
Co	0.31	– 0.04	0.17	0.89
Cd	– 0.15	0.88	– 0.26	– 0.05
Cr	– 0.13	0.92	0.05	0.27
Fe	0.95	– 0.14	0.07	0.11
Ni	– 0.01	0.48	– 0.04	0.81
As	– 0.14	0.16	– 0.86	– 0.06
% Expl.var	29.5	27.1	16.5	15.8

Four factors describe almost 90 % of the total variance of the system. The first one contains dominantly copper, manganese and iron and could be conditionally named "natural" since these elements are typical major constituents of Black sea coastal sediments. The second factor includes zinc, cadmium and chromium, the third – lead and arsenic and the fourth – nickel and cobalt. The last three factors reflect typical anthropogenic influences of heavy metals from various sources such as chemical and glass production plants, oil refineries, steel-works and smelting plants. The detected pollution pattern indicates in a semi-quantative way the emission sources.

2.3.5 Aerosol ambient data

The CMB modeling of the ambient aerosol data delivers information on the source apportioning for each day of sampling (in the treated data set the sampling events for each of the 10 observed sites are about 35 within a 2-year period of observation). So, it may be seen what part of the total aerosol mass (or the total element concentration). The models for a certain site could be then summarized for the whole period of monitoring and a more complete impression of the air quality is obtained. This is presented as illustration of the summarized modeling in Table 2.

Table 2. Summarized CMB modeling results for a site in Northern Greece (source contributions in %)

Element	Soil road dust	Catalytic vehicles	Non-catalytic vehicles	Coal fired fly ash	Diesel burning	Open fire
Al	96.95	0.46	0.01	0.36	2.22	0.21
As	1.50	18.88	0.00	32.46	39.87	7.56
Br	0.15	0.39	6.20	67.66	27.07	2.90
Ca	47.71	11.99	0.12	20.51	18.51	1.76
Cl	2.25	0.03	0.05	25.53	0.06	74.62
Co	19.05	12.70	0.13	16.22	48.93	3.18
Cr	78.08	4.12	0.01	6.70	8.81	2.42
Cu	0.65	42.54	0.04	6.66	48.42	1.79
Fe	68.46	1.88	0.01	14.95	14.24	0.59
K	74.88	0.47	0.03	23.72	0.99	0.14
Mg	86.53	3.13	0.01	3.25	6.48	0.89
Mn	47.25	8.32	0.10	18.66	24.29	1.61
Ni	55.95	15.81	0.02	9.76	17.67	0.89
Pb	1.47	22.35	25.21	27.89	34.35	6.32
P	31.66	66.54	0.32	0.86	0.93	0.12
Se	0.00	11.12	0.06	61.09	20.54	7.51
S	1.11	0.54	0.02	8.44	88.97	0.97
Si	98.50	0.24	0.00	0.15	1.24	0.00
Sr	27.59	7.28	0.09	43.07	19.34	2.88
Ti	81.23	1.12	0.00	10.03	5.75	2.08
TSP	63.66	3.03	0.11	7.63	23.69	2.18
V	27.16	15.34	0.05	20.82	32.40	4.45
Zn	0.88	38.60	0.25	10.34	48.52	1.65

It is readily seen what part of the total mass of the aerosols (or the separate element concentrations) is explained by one or another emission source. It is obvious that in the region the most substantial part of the fine aerosol mass is formed by soil road dust emission but other sources also contribute to the formation. The modeling offers a direct way of interpreting the risk assessment options.

2.3.6 Wet precipitation monitoring data

The large data set collected regularly over 15 years of sampling for the most of the sites in consideration offers a rich variety of interpretation and modeling with respect to the air quality over Austria and the level of the sustainable development for this European region.

Following options were considered:

Trend study: The data collection is big enough to give an idea about the time trend for all major ion concentrations or depositions over Austria. The most important conclusion is that a statistically significant decreasing trend is found for sulfate and pH values, which is an important indication for the acid rain reduction at all sampling sites. In Table 3 the trends of sulfate ion concentrations for all sampling sites are shown.

The decreasing trend is parallel to the overall reduction of the SO_x emissions in Austria, which is a sound explanation of the acid rain reduction due to local legislation.

Detailed description of the trends found for the other major rain components (sodium, potassium, ammonium, chloride, nitrate, calcium and magnesium) could be found in various studies on this topic (Puxbaum et al. 1998; Tsakovski et al. 2000; Puxbaum et al. 2002). In principle, the decrease of the acid components is accompanied by slight increase of base metals concentration.

Table 3. Summary for sulfate ion concentration (mg.L^{-1}) changes and trend significance at all sites

Ion/Site	Predicted first year	Predicted final year	Total change in %	Change per year in %	R^2 and trend significance
SO_4^{2-}					
Reutte	0.53	0.32	-40.85	-2.72	0.49
Kufstein	0.71	0.37	-47.66	-3.18	0.41
Innervillgraten	0.63	0.28	-55.86	-3.72	0.53
Haunsberg	1.15	0.53	-53.54	-3.35	0.61
Werfenweng	0.80	0.33	-59.27	-3.70	0.65
Sonnblick	0.55	0.26	-53.40	-4.45	0.72
Nasswald	0.79	0.75	-4.94	-0.45	0.01 (n.s.)
Litschau	1.31	0.76	-41.66	-4.17	0.62
Lunz	0.84	0.52	-37.80	-4.20	0.42 (n.s.)
Lobau	1.74	0.80	-53.94	-4.49	0.80
Nassfeld	0.98	0.72	-26.18	-3.27	0.07 (n.s.)

Seasonality in lead concentration in rainwater: In a data collection from urban and remote sites in the Austrian wet precipitation network the seasonal behavior of lead concentrations and deposits were determined using time-series analysis and ARIMA modeling of the seasonality (Einax et al. 1997). It is shown that for urban sites a typical winter peak in the lead concentration is observed as for rural sites the concentration peaks are in spring and summer. This behavior is illustrated in Fig. 3. Also a decreasing time trend for the period of observation is found.

Sampling site similarity: The application of cluster analysis to a data set consisting of 5 Alpine sampling sites monitored for wet precipitation over 15 years has shown that two stable clusters are obtained after hierarchical, non-hierarchical or fuzzy clustering. The first group includes the sites *Reutte*, *Kufstein* and *Haunsberg* and the second – *Innervillgraten* and *Werfenweng*. These findings confirm some observations for the role of the sampling site in the explanation of the acidity decrease over Central Austria. The site group from the inner Alpine region (second cluster) is subject to less deposition of anthropogenic components that that from the northern Alpine rim (first cluster). This is an important ecometrical conclusion about the role of the geographical factors in atmospheric chemistry.

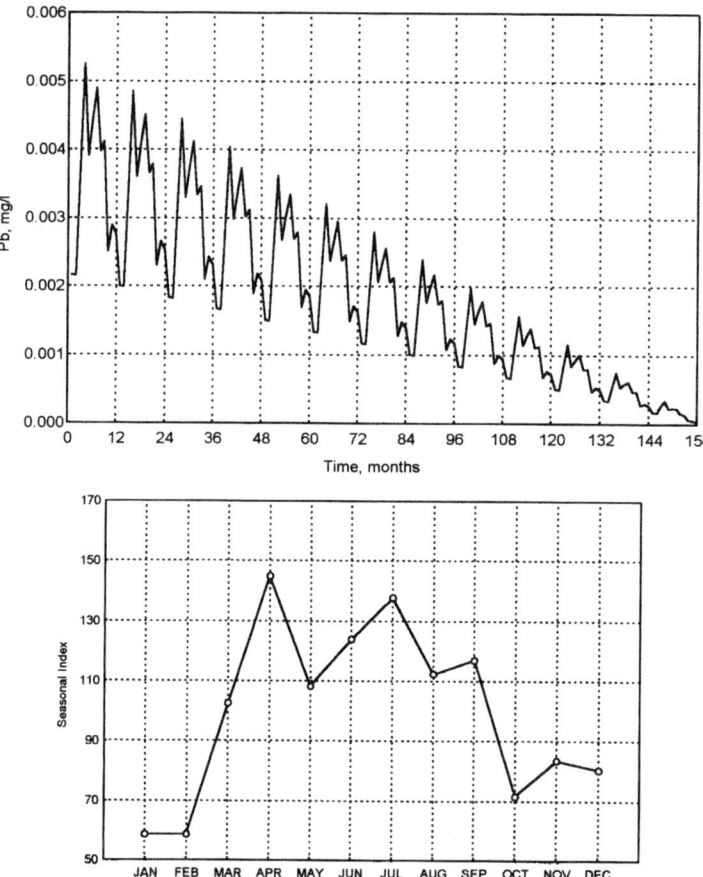

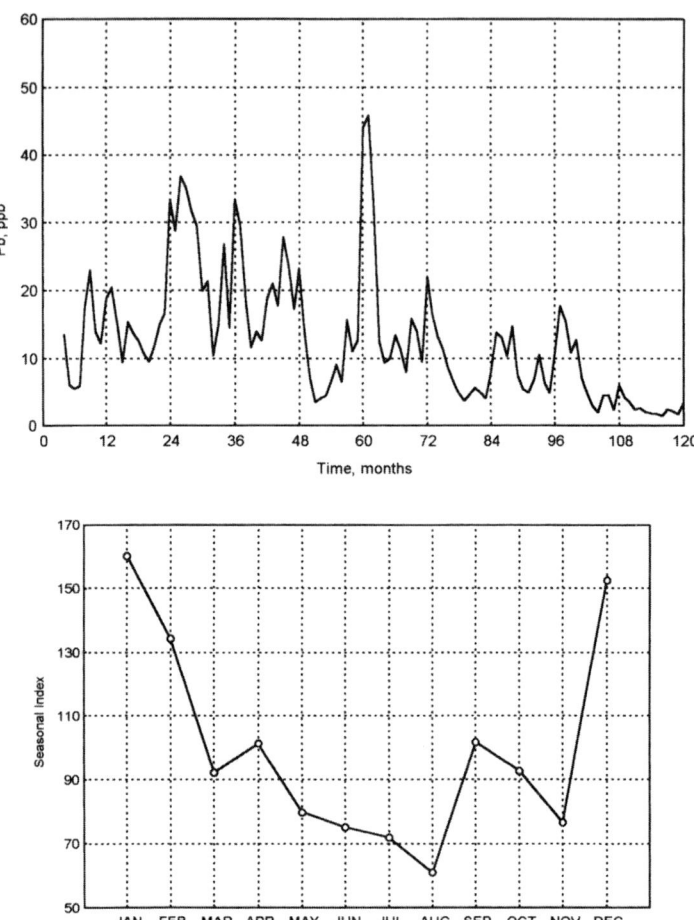

Fig. 3. Time-trend and seasonal models of lead concentrations in rainwater (urban (right side of figure) and rural (left side of figure) sites)

Data set structure: Using PCA (11 sites, over 15 years of observation) it was found that three latent factors explain more than 75 % of the total variance of the system. The factor structure based on the factor loadings values is presented in Table 4.

It is interesting to note that despite differences between site locations, time trends and seasonality, the latent factor structure is relatively similar. This is an indication of the similarity of the emission sources over the region. It is possible to distinguish a typical anthropogenic factor related to ammonium, nitrate and sulfate ion concentrations. The second latent factor comprises mainly sodium and chloride ion concentrations, reflecting the influence of mixed salts in the precipitation. The third latent factor, denoted crustal, is also common to most sites and includes dominantly calcium and magnesium, often accompanied by potassium. Thus, a complete picture of the emission sources could be obtained for the whole region. It could be stated that this idea is more or less qualitative.

Table 4. PCA modeling of the sampling sites from the Austrian network

Site	Factor 1	Factor 2	Factor 3
Reutte	$NH_4^+, NO_3^-, SO_4^{2-}$	Na^+, Cl^-	Ca^{2+}, Mg^{2+}, K^+
Kufstein	$NH_4^+, NO_3^-, SO_4^{2-}$	Na^+, Cl^-	Ca^{2+}, Mg^{2+}, K^+
Innervillgraten	$NH_4^+, NO_3^-, SO_4^{2-}, Mg^{2+}$	Na^+, Ca^{2+}, Cl^-	K^+
Haunsberg	$NH_4^+, NO_3^-, SO_4^{2-}$	Na^+, Ca^{2+}, Cl^-	K^+, Mg^{2+}
Werfenweng	$NH_4^+, NO_3^-, SO_4^{2-}$	Na^+, Cl^-	Ca^{2+}, Mg^{2+}, K^+
Sonnblick	$NH_4^+, NO_3^-, SO_4^{2-}$	Na^+	$Ca^{2+}, Mg^{2+}, Cl^-, K^+$
Nasswald	$NH_4^+, NO_3^-, SO_4^{2-}, Ca^{2+}, Mg^{2+}$	Na^+, K^+	Cl^-
Litschau (7)	NO_3^-, SO_4^{2-}, Na^+	Ca^{2+}, Mg^{2+}, Cl^-	NH_4^+, K^+
Lunz (7)	$NH_4^+, NO_3^-, SO_4^{2-}, K^+$	Ca^{2+}, Cl^-	Na^+, Mg^{2+}
Nassfeld	$NH_4^+, NO_3^-, SO_4^{2-}, Ca^{2+}, Mg^{2+}$	Na^+, Cl^-	K^+
Lobau	$NH_4^+, NO_3^-, SO_4^{2-}$	Na^+, Cl^-	Ca^{2+}, Mg^{2+}, K^+

Source apportioning by multiple regression on absolute principal component scores (APCS): In Table 5 the mean mass contribution of each major ion to each of the three latent factors for the whole region of interest is shown.

Table 5. Component profile of major ions in wet precipitation

Ion	Anthr.	Crustal	Mixed salt	Sum estimated	Sum observed	R
H^+	0.20 (93.27%)	—	0.01 (6.73 %)	0.21 0.102(s.d)	0.22 0.113(s.d)	0.9623
NH_4^+	5.62 (93.65%)	—	0.38 (6.33 %)	6.00 2.272(s.d)	6.03 2.436(s.d)	0.9057
Na^+	—	0.30 (13.85%)	1.84 (86.15%)	2.14 1.750(s.d)	2.23 1.732(s.d)	0.9397
K^+	0.45 (41.21 %)	0.11 (9.58 %)	0.54 (49.21%)	1.10 0.546(s.d)	1.10 0.631(s.d)	0.8518
Ca^{2+}	1.43 (18.71 %)	6.21 (81.29%)	—	7.64 5.587(s.d)	7.50 6.356(s.d)	0.9099
Mg^{2+}	—	0.77 (78.73%)	0.21 (21.27)	0.98 0.721(s.d)	0.99 0.819(s.d)	0.8637
Cl^-	2.51 (45.79 %)	0.71 (12.99%)	2.26 (41.22%)	5.48 2.433(s.d)	5.32 3.471(s.d)	0.8083
NO_3^-	3.93 (92.22 %)	—	0.33 (7.78 %)	4.26 (1.600s.d)	4.25 1.698(s.d)	0.9582
SO_4^{2-}	6.08 (90.83%)	—	0.61 (9.17%)	6.79 2.495(s.d)	6.78 2.451(s.d)	0.9259

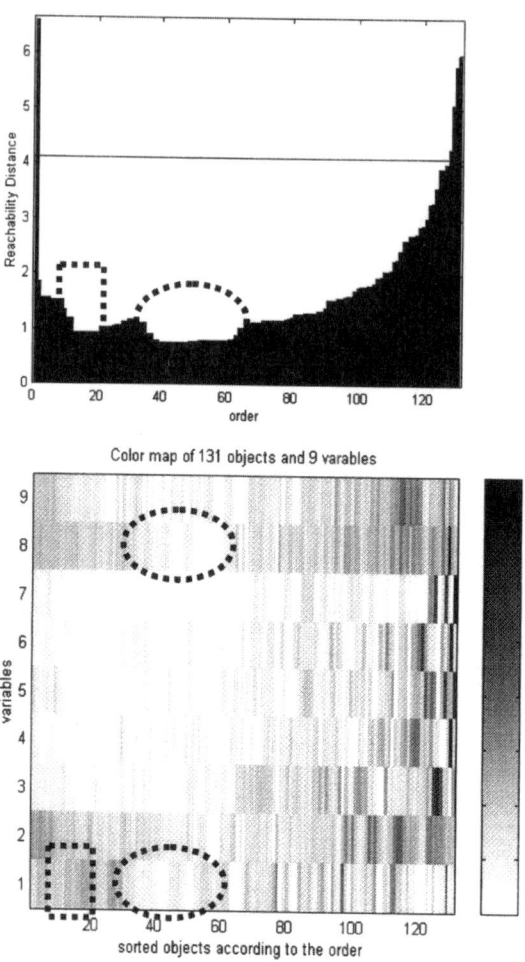

Fig. 4. OPTICS reachability plot with variables map for wet precipitation data

The results show that not one of the components is a unique tracer, occurring in only one of the factors. In particular, the base cations as well as chloride are distributed in different patterns among the three factors, indicating that the base cations are not derived exclusively from one (e.g. a crustal) source and chloride is not derived exclusively from sea salt. The same holds true for the other source contributions, which indicate the variety of possible tracers for this atmospheric event.

Some other options: The introduction of some new environmetrical approaches like DBSCAN, OPTICS, Kohonen maps, neural gas to the data set from the Austrian wet precipitation network gave information about the clustering of the sites and the location of some typical outliers. In a recent study (Stanimirova et al. 2002) several applications of advanced clustering and classification methods for environmental data analysis (Austrian wet precipitation network) are discussed. In Fig. 4 a typical reachability plot of the OPTICS approach is presented.

It is readily seen that only one cluster could be identified. Only five objects have a high reachability distance (RD) and are identified as outliers. If a color map is added to the RD plot the relationship between the outliers and the chemical variables can be found, i.e. the possible reasons for the outlying position, e.g. high calcium and chloride ion concentrations for site Haunsberg in 1994, 1995, 1996 and low nitrate, sulfate and ammonium ion concentrations (strong acidic decrease). Same approach could be applied to the other outliers.

3   Conclusion

The modern world exists in a very dynamic mode. The environmental problems, despite their already long history, are not completely solved but the strategy of the sustainable development is a fact, which helps in solving problems. It is our deep conviction that a serious contribution to the overall solution is the application of environmetrics to the observations, which makes the solution finding and the problem solving a more intelligent and reliable process.

Acknowledgement

The author would like to express his sincere gratitude to NATO for the financial support to present his research work as keynote lecture at the NATO –ARW, Maribor, Slovenia (October 13-17, 2002).

References

Daczykowski M, Walczak B, Massart DL (2001) Looking for natural patterns in data. Part I. Density based approach, *Chemom. Intel. Lab. Syst.*, 56, p 83

Einax J, Zwanziger H, Geiss S (1997) Chemometrics in Environmental Analysis. Weinheim, J. Wiley

Huber P (1985) Projection pursuit. *The Annal. Stat.*, 13, p 435

Kohonen T (1997) Self-Organizing Maps. Berlin, Springer Verlag

Martinez T, Berrkovich G, Schulten K (1993) "Neural-Gas" net work for vector quantization and its application to time-series prediction. *IEEE Trans. Neural Networks*, 4, p 558

Massart DL, Kaufman L (1983) The Interpretation of Analytical Chemical Data by the Use of Cluster Analysis. New York, J. Wiley

Massart DL, Vandeginste BGM, Buydens LMC, De Jong S, Lewi PJ, Smeyers-Verbeke J (1998) Handbook of Chemometrics and Qualimetrics. Amsterdam, Elsevier

Puxbaum H, Simeonov V, Kalina M (1998) Ten years trends (1984-1993) in the precipitation chemistry in Central Austria. *Atmos. Environ.*, 32, p 193

Puxbaum H, Simeonov V, Kalina M, Tsakovski S, Loeffler H, Heimburger G, Biebl P, Weber A, Damm A (2002) Long-term assessment of the wet precipitation chemistry in Austria (1984-1999). *Chemosphere*, 48, p 437

Simeonov V, Massart DL, Andreev G, Tsakovski S (2000) Assessment of metal pollution based on multivariate statistical modeling of "hot spot" sediments from the Black Sea, *Chemosphere,* 41, p 1411

Stanimirova I, Daszykowksi M, Massart DL, Questier F, Simeonov V, Puxbaum H, Multivariate statistical interpretation of the wet precipitation chemistry from the Austrian monitoring network (1984-1999). *J. Envir. Monit.*, (in press)

Thurston G, Spengler D (1985) A quantitative assessment of source contribution of inhalable particulate matter in metropolitan Boston. *Atmos. Environ.*, 19, p 9

Tsakovski S, Puxbaum H, Simeonov, V, Kalima MS, Loeffler H, Heimburger G, Biebl P, Weber A, Damm A (2000) Trend, seasonal and multivariate modelling study of wet precipitation data from the Austrian monitoring network (1990-1997). *J. Envir. Monit.*, 2, p 424

Tucker LR (1963) Problems of Measuring Change. Madison, The UW Press

Winchester JW, Nifong GD (1971) Water pollution in lake Michigan by trace elements from aerosol fallout. *Water Air Soil Pollut.*, 1, p 50

Yenyukov IS (1989) Indices for Projection Pursuit. New York, Nova Science Publishers

Carbon storage: the economic efficiency of storing CO_2 in leaky reservoirs

Minh Ha-Duong[a], David W. Keith[b]

[a] Corresponding author. Engineering and Public Policy department, Carnegie Mellon University, 5000 Forbes Avenue, Pittsburgh, PA 15213 USA, E-mail address: minh.ha.duong@cmu.edu

[b] Engineering and Public Policy, Carnegie Mellon University

Abstract

Fossil fuels can be used with minimal atmospheric emissions of carbon dioxide by capturing and storing the CO_2 away in geologic structures. However, stored CO_2 can leak back to the atmosphere reducing the utility of this technology. To explore the trade-offs between discounting, leakage, the cost of sequestration and the energy penalty (the energy necessary to capture, transport and inject carbon underground), we derive analytic expressions for the value of leaky CO_2 storage compared to perfect storage when storage is a marginal component of the energy system. If the annual leak rate is 1% and the discount rate is 4%, for example, then CO_2 mitigation using leaky storage is worth 80% of mitigation with perfect storage. Using an integrated assessment numerical model (DIAM) to explore the role of leakage when CO_2 storage is non-marginal, we find that a leakage rate of 0.1% is nearly the same as perfect storage while a leakage rate of 0.5% renders storage unattractive. The possibility of capturing CO_2 from the air, not only from flue gases, makes storage with higher leakage rates interesting. Finally, we speculate about the role of imperfect carbon storage in carbon accounting and trading.

1 Carbon storage

1.1 Introduction

Geologic carbon storage is a means of storing carbon dioxide (CO_2) away from the atmosphere by injecting it at depths greater than about 1 km into porous sedimentary formations using technologies derived from the oil and gas industry (Holloway 2001, Herzog2001, Bachu 2001). Natural underground reservoirs have held natural CO_2 in place for thousands of years, but leakage does and will occur. Each year a fraction of the gas stored underground can be expected to return to the atmosphere. The purpose of this work is to discuss economic implications of leakage.

Geologic CO_2 storage might enable the use of fossil fuels without contributing to climate change. Doing so requires a set of technologies for capture, transportation and injection of CO_2. While much is uncertain about future technology and its costs, the multiplicity of technical options and the fact that most if not all of the component technologies have already been demonstrated at commercial scale strongly suggests that capture and storage is a viable near-term option for managing CO_2 emissions. The cost of capture generally dominates the cost of transport and injection. In the electric sector, previous studies suggest that the cost of avoiding CO_2 emissions using these methods is in order of 50 to 150 $/tC (Johnson et al. 2003, Herzog 1999).

The long-range transportation of CO_2 and its injection into deep underground reservoirs is comparatively well understood: The upstream oil and gas industry routinely injects CO_2 underground to enhance oil recovery (CO_2-EOR). A bit more than two thousand kilometers of CO_2 pipelines have been laid in Texas to provide for CO_2-EOR. In these operations the goal is to maximize oil return and minimize the carbon left underground so that it can be re-used, since operators must pay for the CO_2. Yet at the end of an EOR operation a major fraction of CO_2 purchased remains underground.

Industrial experience with CO_2-EOR and with the disposal of CO_2-rich acid gas streams, as well as related experience with natural gas storage and the underground disposal of other wastes, suggests that this technology can be implemented with acceptable local risk and that it could therefore play a significant role as a response to the challenge of global warming. This is why geologic carbon storage has become more and more relevant as a climate policy option during these last five years. Section "Discussion about leakage" extends this introduction with a short review of the literature on leakage.

We assess the economic implications of CO_2 storage in leaky reservoirs from two perspectives. First, in Section "Microeconomics of leakage", we take a microeconomic viewpoint, considering only the cost effectiveness of mitigation options while assuming that storage is a marginal component of the energy system. We assess the relative value of perfect and imperfect storage, or equivalently of imperfect storage and a non-carbon alternative energy source that is adopted to mitigate CO_2 emissions. An efficiency ratio involving the leakage rate, the discount rate, and the energy intensity of storage is derived to compare the two technologies.

Second, in Section "Numerical results in a long run cost–benefit model" we address the economics of leakage when CO_2 storage plays a significant (non-marginal) role in the energy system so that the flux of leaking CO_2 can be large compared to emissions from other sources. This analysis adopts the perspective of optimal climate policy in which trade-offs between costs and benefits play out over time. The problem of finding the efficient mix of two abatement technologies (one being carbon capture and storage) is solved using a numeric optimization model: DIAM. Simulations of optimal long-term global CO_2 trajectories confirm the orders of magnitude previously found: a leak rate of one tenth of a percent per year is roughly equivalent to perfect storage. For higher leak rates, the availability of air capture can make a significant difference.

The last section discusses policy implications for regulating storage activities.

2   Discussions about leakage

Natural analogues show that carbon dioxide can remain trapped underground for very long periods, but they also show that releases can lead to serious local environmental consequences. Excess local concentration of CO_2, for example, can lead to acidification of ground-water, and elevated carbon dioxide concentration in soils can kill plants. While local environmental issues are certainly important they will be ignored here, as this paper is concerned about the global implications of leakage.

Current research can be organized into two categories: descriptive and normative. The descriptive research tries to predict the magnitude of leakage, studying for example rock formations, existing wells, or natural and artificial analogues. The normative approach asks "how small is small enough", framing the problem as a question of resource management over time. Our focus is on the normative problem, but we first review some of the descriptive literature.

2.1   Describing leakage

Both Jimenez (2002) and Celia et al. (2002) explore the mechanisms of leakage. They stress that leakage is possible through or along existing wells, stating for example that in the state of Texas in the United States, more than 1,500,000 oil and gas wells have been drilled. Precisely assessing the status of these wells is difficult since more than one-third have been abandoned, some more than a century ago. The authors conclude that transport models for leakage analysis must include proper representation of existing wells.

Saripalli et al. (2002) presents a risk-assessment pointing out that cap-rock integrity, leading to slow leakage, is a greater cause of concern than the risk of catastrophic failure at the well head during the injection process. This does not contradict the previous point that existing wells are an important factor that compromise cap-rock integrity.

The comparative study by Benson et al. (2002) confirms both these ideas: "Long industrial experience with CO_2 and gases in general shows that the risks from industrial sequestration facilities are manageable using standard engineering controls and procedures. [...] On the other hand, our understanding of and ability to predict CO_2 releases and their characteristics in any given geologic and geographic setting is far more challenging". They also state that in natural gas storage projects, "in the vast majority of cases, leakage is caused by defective wells (poorly constructed or improperly plugged abandoned wells)".

2.2 How small is small enough? Geophysical aspects.

On the normative point of view, Hepple et al. (2002) and Pacala (2002) assessed the maximum leakage (or seepage) rates that would be compatible with stabilization of the atmospheric carbon dioxide concentration. Both studies find that residence times greater than 1000 years (in other words, seepage rates less than 0.1 per cent per year) allow for an effective storage policy. But the later study finds that mean residence time as low as a hundred years could still allow one to meet a stringent environmental target, whereas the former states that with few exceptions, a one percent per year leakage rate is unacceptably high.

This difference can be traced to different assumptions on the long term evolution of the mean leakage rate. Pacala assumes that injection is randomly distributed across a collection of heterogeneous unlimited reservoirs. Consequently in the long run the fraction of carbon remaining in the less leaky reservoirs increases, so the average leak rate decreases. On the contrary, Hepple et al. assume that reservoirs have limited capacity, so as very large quantities of CO_2 are sequestered underground, the probability of selecting less favorable sites with higher leak rates will increase.

2.3 How small is small enough? Economic aspects.

Herzog (2003) calculate the storage effectiveness for injecting CO_2 at various depths in the ocean. Their analysis can be transposed directly to geological storage, since deeper oceanic injection is equivalent to less leaky reservoirs. They use a Hotelling model, in which the critical parameter is the long-term evolution of the marginal damages from climate change, assumed to be equal to the carbon price. If this rises at or near the discount rate, then temporary storage is not interesting. If on the contrary marginal damages are constant, or there is a backstop technology that caps abatement cost, then temporary storage is nearly equivalent to permanent storage.

Hawkins (2002) notes that, considering the world's fossil fuel reserves as underground stored carbon, the present global emissions from energy use represent an annual leak rate of about 0.1% per year, which is unsustainable. He also points out that the leak rate of the current carbon storage sites are unknown: we can not be sure it is less than one per thousandth per year. The conclusion is that while carbon storage should not be ignored, it should not crowd out other mitigation options, and the upper bound on leak rates should be on below a level of concern given the amount stored.

Dooley (2002) used the MiniCAM 2001 integrated assessment model to examine two leak rates: one percent per year and one per thousandth per year. They conclude that the smaller leak rate does not lead to a substantial impact on required net annual emissions reductions, in line with the findings of Hepple et al. (2002) and Pacala (2002). They also find that one percent leakage per year is likely intolerable, as it represents an unacceptably costly financial burden moved to future generations. The implication is that monitoring technology should progress to the point where it can resolve the fate of injected CO_2 with this level of specificity.

Keller et al. (2002) analyze leakage in an optimal economic growth framework using both a simple analytic model and an numerical integrated assessment model. They conclude that CO_2 storage (at a constant marginal cost of a hundred dollars per ton of C, with a reservoir half-life of two hundred years) could reduce mitigation costs and climate damages considerably, and that a subsidy for the initial non-competitive storage is sound economic policy.

They also introduce the notion of an efficiency factor of storage: for example, a hundred tons of sequestered CO_2 would be worth fifty tons of avoided CO_2 emissions at an efficiency factor fifty percent. This factor decreases when the leakage rate increases or when the energy needed for storage increases. On the other hand, increasing the discount rate tend to increase the storage efficiency.

3 Microeconomics of leakage

This short review of the literature shows that the leakage rates over one percent per year tend to be on the high side, while leakage rates less than one tenth of a percent per year tend to be acceptable. This section uses a simple microeconomic model to discuss the relation between leakage, the discount rate and the relative cost of carbon capture and storage. The argument is based on the equality of marginal costs across substitutable technologies, and will also discuss the energy penalty of capture and storage. This leads to an estimation of a maximal acceptable leakage rate that depends of a plausible estimate of the ratio between the cost of perfect storage–or equivalently non-fossil energy–and that of leaky storage.

3.1 Permanent storage by re-capturing leaks

Consider two technological options to deliver energy without CO_2 emissions:
- The first is to use non-fossil primary energy source so that paying some incremental cost a above the conventional energy price results in one ton of carbon being not emitted in the atmosphere.
- The second option achieves the same result by producing energy burning one ton of carbon from fossil fuels, and then–instead of exhausting it in the atmosphere–capturing and injecting it underground. For the sake of simplicity we start by neglecting the energy needed for capture and storage, we will come back on this assumption later.

Alternatively, one may view the first option as being perfect storage where the second is imperfect.

To achieve the same environmental result as the first, the second option has to offset any carbon that leaks out of underground storage, for example by capturing and storing additional CO_2. If c is the marginal cost to capture one ton of carbon and inject it underground, the net present cost of this technological option will be c plus the cost of offsetting future leaks. The standard way to assess net present value NPV of a flow of costs $x(t)$ occurring over time in the future is to use a parameter called the discount rate δ (similar to an interest rate) and sum up the discounted costs over time (see Portney et al. (1999) for a recent discussion of this standard methodology's limitations in the context of climate change):

$$NPV = \int_{t=0}^{\infty} x(T)e^{-\delta t} dt \qquad (1)$$

Assume that leakage is proportional to the amount of carbon stored, and denote λ the annual leakage rate of the underground carbon reservoir. The storage option entails an initial cost of c and a subsequent annual cost of $x(t)=\lambda c$ forever. The total net present cost of the storage option is thus $NPV = c+\lambda c/\delta$, where $\lambda c/\delta$ is the geometric sum of the cost to keep the same total amount of carbon underground by injecting additional CO_2 to make up for leaks.

The question is not to determine which of these two options is cheaper than the other. Both have cost curves with increasing marginal costs. Basic economic reasoning suggests that to minimize the cost of meeting any emission constraint, it is best to spread the effort across the two technologies so that their marginal cost is the same. The economic efficiency condition is thus $NPV=a$, assuming of course the absence of other strategic, environmental or political externalities.

This is why, to compare the two technologies, we determine the ratio $r=c/a$ that corresponds to the economic efficiency condition. This ratio corresponds to the "efficiency factor" recently derived in a similar way by Keller et al. (2002). Economic efficiency implies:

$$r = \frac{\delta}{\lambda + \delta} \qquad (2)$$

Intuitively, the ratio is less than unity because leaks make capture and storage less environmentally efficient than abatement. This is why it has to be cheaper by a factor of r in order to be as interesting.

This result shows that as long as the leak rate λ is an order of magnitude lower than the discount rate, then the penalty for leakage is very small (r is close to one). A public discount rate of a few percent per year is usually recognized as a sensible order of magnitude, in line with observed population and macro-economic growth rates in the long term. This implies that storage with leak rates of a few thousandths per year is economically very close to perfect avoided emissions.

If the leak rate is a few percent per year, then sensitivity to the discount rate becomes important. Consider for example a discount rate of four percent per year and a leakage rate of one percent per year. Then $r=0.8$. Carbon storage should be pushed to the point when its marginal cost is eighty percent of marginal abatement cost. Supposing for example that the value of non-emitted carbon is ten dollars per ton, this lead to a value of temporarily stored carbon of eight dollars per ton of carbon[1]. The penalty is not overwhelming.

Another assumption in Eq. 2 that needs discussion is that the leakage rate λ is constant. Actually, even assuming that carbon capture and storage operates at a small scale in front of the energy system, one can expect the storage conditions to change in the long run as different geologic reservoirs and new technologies are used. Supposing for example that $\lambda(t) = \lambda e^{-\tau t}$, that is a constant decrease at exponential rate τ, then the leakage efficiency ratio becomes:

$$r = \frac{\delta + r}{\lambda + \delta + r} \tag{3}$$

Intuitively the faster the sinks improve (larger τ), the closer is r to unity. We ignore whether λ can be expected to increase or decrease in the long run.

3.2 The energy penalty

Carbon capture and storage has another disadvantage compared to abatement: it needs energy. For example, a coal-fired power plant would take an efficiency penalty when fitted with a system to capture the carbon dioxide from flue gases. Herzog's (1999) studies show an energy penalty in the 14 to 20 percent range using existing technology, and 7 to 17 percent using 2012 assumptions. The numbers depend largely on the existing energy market conditions, since the penalty is relative to the reference technology for electricity production.

Define the energy penalty μ as follows. To produce the same amount of energy services that would have emitted one ton of carbon in the air, one has to capture and store $1/(1-\mu)$ tons of carbon underground. Another way to see μ is to say that the carbon capture and storage process uses fossil energy, and thus emits μ tons of carbon in the air per ton of carbon stored underground, so that the net removal from the atmosphere is therefore $1-\mu$ ton per ton of carbon processed.

The energy penalty makes air capture less interesting than previously. Offsetting leaks by storing more carbon underground would result in the underground stock growing exponentially. But future leaks can also be compensated by abatement instead of storage.

Consider a one-time atmospheric removal of one net ton of carbon, for a storage of $1/(1-\mu)$ ton underground. The initial cost to do this is $c/(1-\mu)$. Assume that this store of carbon in the ground declines at a rate λ without being replenished. Leaks get smaller and smaller with time, since the stored carbon depletes at an exponential rate λ, and at date t leakage is

---

[1] The carbon value decreases but r increases with the discount rate, so the net effect of δ on the dollar value of storage is ambiguous.

$$\frac{\lambda}{1-\mu}e^{-\lambda t} \tag{4}$$

The cost to compensate for this leakage through abatement is

$$x(t) = a\frac{\lambda}{1-\mu}e^{-\lambda t} \tag{5}$$

Assuming a constant discount rate, leakage rate and energy penalty, the efficiency condition $NPV=a$ leads to:

$$r = \frac{\delta}{\lambda-\delta} - \mu \tag{6}$$

This equation makes explicit the trade-off between the leak rate and the energy penalty of carbon capture. This kind of trade off is likely to be important when comparing different storage options, which would differ both in energy requirements and in leak rates. If the energy penalty is too large, then carbon storage does not make economic sense unless $c<0$, that is there is a joint benefit to storage (as in CO_2-EOR).

4 Application

For a given a and c, the ratio r can be used to determine up to what leakage rate λ the storage option is environmentally as efficient as the abatement option. However, it is necessary to remember that for each technology there is a portfolio of actions that can be ordered by increasing marginal costs along a cost curve.

Freund (2001) published explicit estimates of storage costs curves. Some storage has been achieved at negative costs as an ancillary benefit of enhanced oil recovery. Another way to sequester carbon dioxide, less explored but maybe also profitable, is injection into deep, unminable coal seams because this allows the recovery of the natural gas that was adsorbed at the surface of the coal (Reeves 2002; Wong 2000). Injection in depleted gas fields is also possible, with maybe the option of CO_2 enhanced gas recovery. Beyond those, the economics of injection into saline aquifers or into the sea are presently even more uncertain. Of course, the curve depends on how the portfolio of actions is delimited, and for each specific technology costs vary at each particular potential underground reservoir with geometry, geology, location and market forces.

There is also a cost curve for producing CO_2 streams, as discussed for example by Johnson et al. (2003). Opportunities at very low cost come as by-products of hydrogen and natural gas production, but quantities are limited. Higher up on the curve, the majority of today's market production comes from natural resources: CO_2 is mined from underground reservoirs such as the Mac Elmo Dome in Colorado. Presently the average delivered price is 10–20 dollars per ton. Herzog (1999) studied the cost of existing carbon capture systems from power plant flue gas, which uses an amine-based absorption technology. He reports mitigation cost in the 20–60 dollars per ton of CO_2 using present-day technologies (the price per ton of carbon is 44/12=3.7 times higher). Beyond that, Ha-Duong et al. (2002) discussed how carbon could be captured directly from the air at over 150 dollars per ton of carbon, for example as a joint product of bio-ethanol energy.

The cost curves show that the acceptability of a leakage rate depends on values of c that may differ between specific applications, and that our example with plausible and significant numbers is just that (an example). For a discount rate of four percent per year and a leakage rate of one percent per year and an energy penalty of 20 percent, then $r = 0.6$. With a five percent discount rate, a two percent leak rate and a fifteen percent energy penalty, the efficiency ratio is still over fifty percent (0.56).

Since there is evidence that some carbon capture and storage options are substantially less expensive than alternatives, this suggests that one percent leakage may be acceptable in some cases. In the electric sector, for example, when large reduction in emissions are requested (greater than 50%), then mitigation using CO_2 capture and storage may be half the cost of mitigation achieved using non-fossil alternatives (Biggs et al. 2001; Johnson et al. 2003).

5 Numerical results in a long run cost-benefit model

We now turn to the long-term implications of leakage. This section explores the consequences for climate policy of leaks in the artificial carbon store, complementing the previous section by taking a more macro-economic perspective on two key issues: non-marginal storage and cost-benefit analysis.

Concerns about possibly large amounts of carbon stored underground in the long run can be quantified using the following orders of magnitude. At the global scale, if industrial carbon management plays a big role in mitigating emissions, then as much as 500 GtC could be stored by 2100. If the average leak rate is only 0.2 percent annually, there would be a 1 GtC per year source undermining CO_2 stabilization. However, storing that much carbon underground by the end of the century means storing 5GtC per year on average, which is several times larger than the annual leakage in the end.

In order to keep all these numbers and other long-term assumptions consistent, we resort to a numerical model. That model also allows us to explore the implications of a potentially important technology: capturing CO_2 directly from the air. One air capture technology for example could be to use biomass as a fuel for a power plant, capturing and storing the CO_2 in this plant's flue gases.

In this study a simple integrated assessment model, DIAM (Dynamics of Inertia and Adaptability Model), is used to compare optimal global CO_2 strategy with and without air capture, and with or without leaks. The model maximizes the expected discounted inter-temporal sum of inter-temporal utility. DIAM does not represent explicit individual technologies or capital turnover, but does include a representation of the inertia related to induced technical change. The inertia of the worldwide energy system induces adjustment costs, related to the rate of change of abatement.

The DIAM version 2.5 used here[2] is derived from the version described by Ha-Duong and Keith (2002). This numerical experiment is comparable to the previous section's micro-economic model in that carbon stored underground leaks at a constant annual rate, and two reduction technologies are available: a generic abatement technology; and capture with storage. However, the cost curve for carbon capture and storage is flat because we are interested in costs for non-marginal quantities. As Ha-Duong and Keith (2002) discussed, carbon capture and storage can be modeled as a backstop technology, that is available at a constant marginal cost of around $150 per ton of carbon (about half of this is adjustment costs).

This section briefly describes the model, focusing first on the damage function, and then on the mitigation cost functions, before reporting the sensitivity of optimal CO_2 trajectories to variations in the leakage rate and to the possibility of capturing carbon directly from the air.

5.1 The model

The benefits of avoiding climate change, or alternatively the cost of climate impact, is represented using a non-linear damage function (Fig. 1). This frames optimal climate policy as a problem of action facing a known threshold of abrupt climate change. While other versions of DIAM represent uncertainty regarding climate and ecosystems sensitivity, DIAM 2.5 was run here in deterministic mode to better focus on the role of leakage.

The impact is a function of atmospheric carbon dioxide concentration lagged twenty years. While it is measured in monetary units, it represents a global willingness to pay to avoid the given level of climate change, including non-market values. The impact at any date is defined as a fraction of wealth at this date. Therefore it scales over time with the size of the economy. The assumption is that, even though a richer economy is structurally better insulated against climate variations than an poorer economy, the overall desire to limit interference with the biosphere increases linearly with wealth.

The model represents emissions abatement occurring by two activities X and Z, each with its own cost function. Activity X represents emissions abatement through conventional existing energy technologies, its marginal cost increases with mitigation, and X is constrained below 1. Activity Z represents carbon capture and storage at constant marginal cost. The unavailability of air capture is represented as the constraint $X+Z<1$, stating that the total abatement can not exceed the overall demand for energy in the baseline. This constraint can be relaxed to represent the possibility that Z captures the CO_2 directly from the air, as discussed by Ha-Duong and Keith (2002).

---

[2]The GAMS source code is available at
http://www.andrew.cmu.edu/user/mduong

Before leaks, anthropogenic carbon emissions at any given time are $E^{landuse}+(1-X-Z)E^{ref}$. Land use emissions are exogenous and considered irreducible. In addition, the underground reservoir S leaks at annual rate λ into the atmosphere. The energy penalty on carbon capture and storage is $\mu=15\%$, so an activity level Z corresponds to an increase of the underground stock by $[Z/(1-\mu)]E^{ref}$.

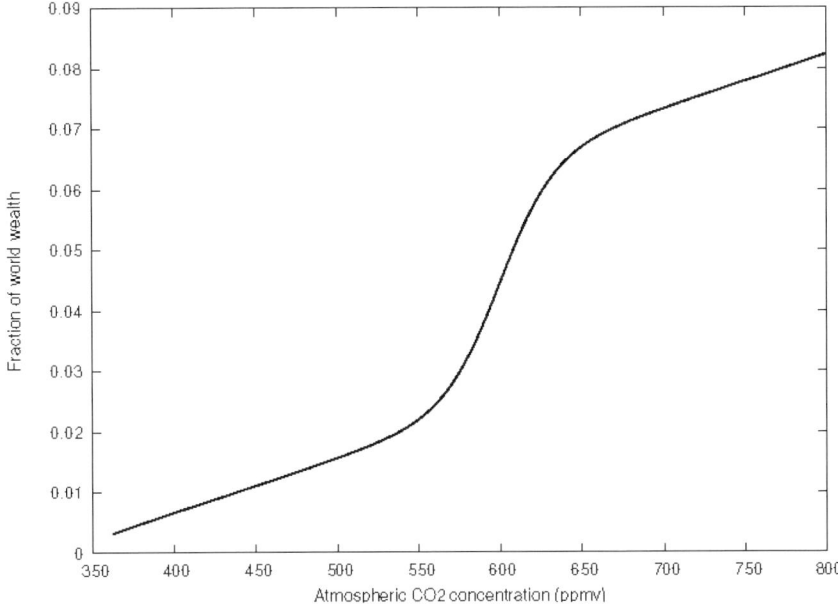

Fig. 1. The impact of climate change. Fraction of global wealth lost each year as a function of carbon dioxide concentration, as used in the cost-benefit model DIAM version 2.5. Damage depends on concentration 20 years before and is assumed to be zero in 2000

Table 1 displays the cost of achieving mitigation activities X and Z. The cost of each activity depends both on its scale X or Z, and on the rate at which it is being increased.

Calibration of cost functions were unchanged from the previous version, and are comparable to the DICE-98 model by Nordhaus et al. (2002). Ignoring adjustment costs, activity X incurs quadratic abatement costs up to full abatement. This leads to a marginal carbon price increasing linearly. In marginal terms, the order of magnitude is a $100 carbon tax for a 20% abatement of world emissions, a common ballpark number.

Table 1. The cost of reducing carbon emissions in DIAM 2.5 for each activity. Gross World Production (GWP) was about 18×10^{12} for the base year. All base costs decline at an autonomous technical progress rate of 1 per cent per year. The $\tau=50\ yr$ inertia parameter in adjustment costs is the characteristic time of the world's energy system

Activity (Unit)	Total cost = U.S.$	Base cost U.S.$/t C	× Scale tC	× Multiplier (dimensionless)
Conventional abatement	$C_X =$	2.45% GWP (t_0)	$\dfrac{E^{\text{ref}}}{E^{\text{ref}}(t_0)}$	$\dot{X}^2 + (r\dot{X})^2$
Backstop Carbon capture + storage	$C_Z =$	75 U.S.$/tC	$\dot{Z}\ E^{\text{ref}}$	$1 + (r\dot{Z})^2$

5.2 Results

Results are displayed numerically Table 2 and graphically Fig. 2. Fig. 2 shows two variables: global anthropogenic carbon emissions (excluding leakage) on the top panel, and carbon dioxide atmospheric concentration on the bottom panel. The top dashed curve corresponds to a business as usual reference scenario. The continuous line corresponds to the Table 2's first row (no leak, no air capture) while the dashed line next to it corresponds to Table 2's next to the last row (leak rate 0.5 percent per year, air capture available).

Table 2. Results and sensitivity analysis. Optimal levels of emissions abatement X are given as a percentage of baseline emissions. The atmospheric CO_2 concentration M is in parts per million. The annual amount of carbon capture and storage Z is given as a percentage of baseline emissions, and is zero in 2050 for all scenarios. The possibility of capturing carbon directly from the air (lower part of the table) is represented by relaxing the constraint $X+Z<100\%$ in the optimization program

		Optimum in 2050		Optimum in 2150		
Air capture?	Leak rate (%/yr)	Abatement X %	[CO_2] ppmv	Abatement X %	Storage Z %	+[CO_2] ppmv
No	0	17	496	52	47	512
No	0.1	18	494	61	38	525
No	0.5	23	491	93	6	533
No	1	23	490	100	0	529
Yes	0	17	496	58	57	494
Yes	0.1	17	495	61	54	507
Yes	0.5	20	492	86	57	521
Yes	1	23	490	100	0	529

The results displayed in Fig. 2 illustrate the model calibration. The overall shape of the optimal trajectories tells the following plausible story. During the next few decades, there will be a slow departure from current trends, because of the considerable inertia in the world's energy system. Late in this century, the atmospheric carbon dioxide concentration stabilizes below what constitutes in this model a soft ceiling at around 550 parts per million. In the next century, the atmospheric carbon dioxide concentration will decline.

Our point is not to discuss the desirability of this storyline in itself. Rather, the model is used to study the sensitivity of optimal trajectories to two parameters: the leak rate and whether the backstop technology can capture carbon from the atmosphere. Results, presented Table 2, were remarkably insensitive to the value of the energy penalty parameter μ so this parameter is kept constant in the simulations.

The table's rows correspond to various leak rates with and without the availability of air capture. The columns show (in 2050 and 2150) three variables: the percentage of abatement using conventional technologies X; the percentage of abatement using the backstop Z and the atmospheric CO_2 concentration M.

First consider the effect of leakage in the absence of air capture. This corresponds to the top half of Table 3, or technically the constraint that $X+Z$ can not go over 100% of reference emissions. We explored leakage rates ranging from zero to one percent per year. In 2050 the backstop technology Z is not used in any scenario. It is because at this date, the marginal cost of X has not risen to the backstop's cost. The fact that X differs across rows for this date reminds us that the model is finding a global optimum and is thus forward looking. Since it optimizes intertemporally, $Z=0$ does not imply that X should be the same across all four rows.

The model sensibly finds that the larger the leakage rate, the smaller the carbon capture Z, and the larger the abatement X should be. This applies at all periods. We find that with perfect storage, the amount of carbon capture is lower than but comparable to the amount of abatement. This remains true with a one tenth of a percent leakage rate. Carbon capture plays a marginal role at 0.5 percent leak rate, and does not enter the optimal technology mix at all at any date with a 1 percent per year leak.

Consider now the atmospheric concentration M. Table 2 suggests that optimal M in 2050 and in 2150 varies in opposite directions when increasing the leak rate. The intuitions behind this result is that a zero leak rate implies the availability of a perfect long term pollution sink. This in turn makes it cheaper to control CO_2 emissions. That has two effects on the optimal trajectories. In the long run the optimal balance of costs and benefits is tilted toward a cleaner environment. At the same time, the optimal burden sharing is also tilted toward future generations: abatement effort in the first periods is comparatively lower.

How does this change when allowing for air capture? The lower half of Table 2 presents results where the constraint $X+Z<100\%$ is relaxed. As the top panel in the Fig. 2 shows, net emissions indeed become negative around 2110 on the optimal path. As the table shows, in 2150 this ($X+Z$ greater than 100%) remains true for the lower three leakage rates. Overall, the qualitative results presented above remain the same: more leakage implies less reliance on carbon capture and storage.

Compared to the no-air capture scenarios, there is more carbon storage everywhere along the way, and ultimately in 2150 the atmospheric concentration is lower. Air capture also pushes up the acceptable leak rate to 0.5 percent per year. This next-to-the-last row illustrates a scenario where CO_2 stored underground contributes significantly to the emissions, about 10GtC per year, but that source is actively offset by capturing carbon from the air even through most (85%) of the energy system is carbon free.

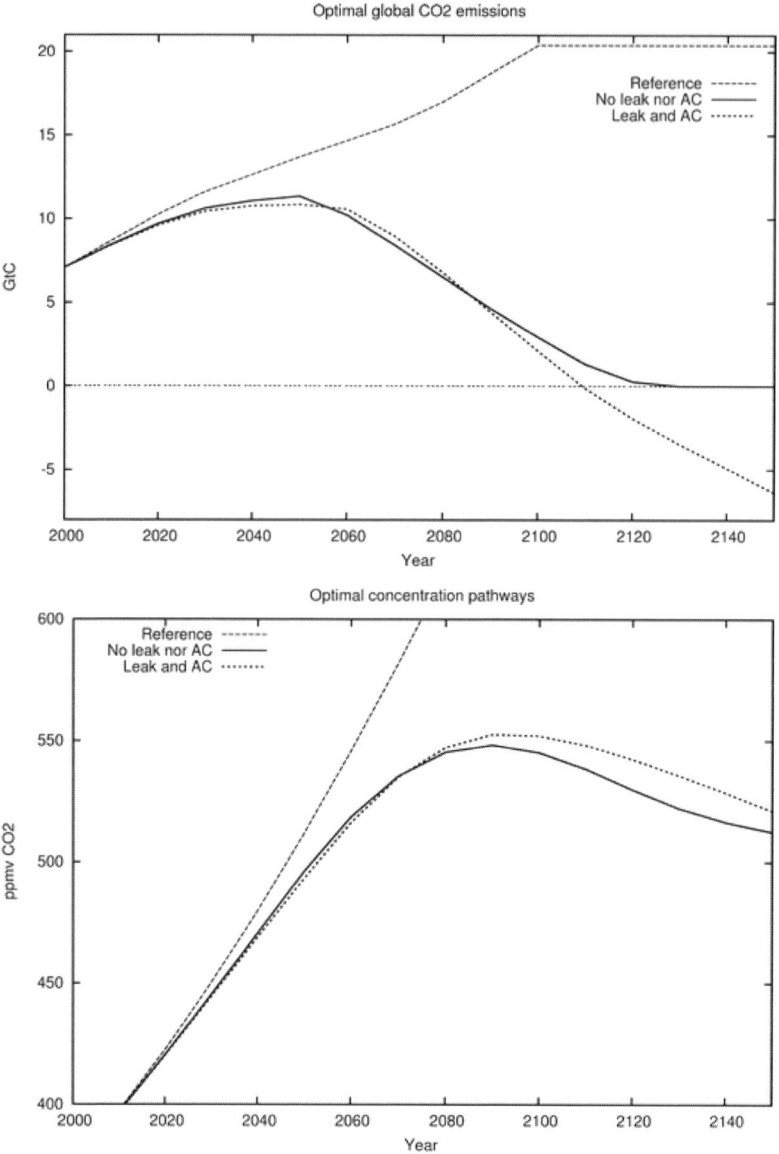

Fig. 2. Optimal CO$_2$ trajectories. Top panel, global carbon emissions, not including emissions from leaks due to underground storage. Bottom panel, atmospheric CO$_2$ concentration in parts per million

6 Concluding remarks

6.1 Policy implications for the value of carbon

Assuming that market-based instruments will be used implementing a carbon emissions reduction policy, how would carbon storage fit within the environmental regulatory framework?

The IPCC defined emission trading as "a market-based approach to achieving environmental objectives that allows those reducing greenhouse gas emissions below what is required to use or trade the excess reductions to offset emissions at another source inside or outside the country". Imagine for example an operator owning two power plants. In the first plant A the operator can reduce emissions easily, but the other plant B uses a different technology and it is more expensive to reduce emissions there. Under a flexible regulation regime, if the operator was ordered to reduce its plants' overall emissions by 10 percent, then he would be allowed to concentrate his efforts on A, go way above 10 percent, and assign the excess reduction to the plant B. Emission trading extends this flexibility to situations where plants A and B are not owned by the same firm.

Firms engaged in carbon capture and storage clearly have a role to play in this market for certificates of emissions reduction or, in less diplomatic language, pollution permits. We assert that because of leakage and the energy penalty, one ton stored underground should correspond to less than one ton of carbon permanently removed from the atmosphere. With the usual caveats about the efficient markets assumption, about the absence of externalities and about discounting, the ratio r derived in this paper can be interpreted as the socially desirable ratio for discounting carbon storage.

The energy penalty should not be left out of the picture, or one risks creating the opportunity to make money by simply moving carbon up and down. In an economy where carbon already has a price reflecting the climate externality, then storage projects already internalize the energy penalty. In this situation they should be regulated using Equation 2. It is conceivable however that a firm involved in carbon storage faces a carbon price not reflecting the climate change externality. For example energy intensive industries have obtained a differential treatment in some countries. In this situation or for a carbon storage project that occurs in a country not controlling emissions at all, for example as a Clean Development Mechanism/Joint Implementation project, Equation 6 should be used instead.

6.2 Conclusion

This paper examined leakage of artificially stored CO_2 from an economic perspective, using first a cost-efficiency microeconomic model, and then a global cost-benefit integrated assessment model.

Leakage of stored carbon is at heart a problem of inter-temporal distribution of abatement costs and benefits. Having decided to mitigate global warming for the benefit of future generations, the present generation should allocate its efforts as efficiently as possible across the various technological options. This is why in a normative economics analysis the discount rate plays the central role, and gives the numeric anchor needed to assess what is an acceptable leakage rate.

The simplest interpretation of our results is that leakage rates one order of magnitude below the discount rate are negligible. In line with previous findings from the literature reviewed, the numerical simulations presented in this paper found that longer than a thousand years is practically as good as infinity. Storage with residence time as short as a few hundred years may still be valuable.

The microeconomic analysis provides a more detailed answer for higher leakage rates in term of storage efficiency ratio r. We use this ratio for projects that remove carbon only temporarily from the atmosphere, to adjust the credit they can claim and be free from further liabilities from leakage. Assuming a public discount rate δ, a leakage rate λ then a project should be credited only the fraction $\delta/(\delta+\lambda)$ of the carbon value initially injected. If the project does not internalize the climate externality in its energy prices, then the energy penalty term μ should be subtracted from this ratio.

These results hold even for one-time storage opportunities, such as enhanced oil recovery. With a one percent annual leak rate and a one to four percent discount rate, the economic efficiency ratio is between fifty to eighty percent. This is not overwhelming.

Acknowledgments

This research was supported by the Centre National de la Recherche Scientifique, France and by the Center for Integrated Assessment of Human Dimensions of Global Change, Pittsburgh PA (created through a cooperative agreement between the National Science Foundation (SBR-9521914) and Carnegie Mellon University).

We thank three anonymous referees, Laurent Gilotte and Patrice Dumas at CIRED, Elizabeth Casman, the organizers and participants of the NATO Advanced Research Workshop on Technological Choices for Sustainability held in Maribor, Slovenia, Oct. 13–17 2002 for useful comments.

References

Bachu S (2001) Geological sequestration of anthropogenic carbon dioxide: Applicability of current issues. In Gerhard LC, Harrison WE, Hanson BM (eds) Geological perspectives of global climate change, number 47 in AAPG Studies in geology. American Association of Petroleum Geologists, pp 285–303

Benson SM, Hepple R, Apps J, Tsang CF, Lippmann M (2002) Comparative evaluation of risk assessment, management and mitigation approaches for deep geologic storage of CO_2. Technical report, Earth Sciences Division, E.O. Lawrence Berkeley National Laboratory

Biggs S, Herzog H, Reilly J, Jacoby H (2001) Economic modeling of CO_2 capture and sequestration. In Willams et al.

Celia MA, Bachu S (2002) Geological sequestration of CO_2: is leakage unavoidable and acceptable? In Kaya et al.

Dooley JJ, Wise MA (2002) Potential leakage from geologic sequestration formations: Allowable levels, economic considerations and the implications for sequestration R&D. In Kaya et al. Pacific Northwest National Laboratory, PNNL-SA-36876 & PNWD-SA-5847

Freund P (2001) Progress in understanding the potential role of CO_2 storage. In Willams et al., pp 272–277

Ha-Duong M, Keith DW (2002) Climate strategy with CO_2 capture from the air. Submitted to Climatic Change

Hawkins DG (2002) Passing gas: policy implications of leakage from geologic carbon storage sites. In Kaya et al.

Hepple RP, Benson SM (2001) Implications of surface seepage on the effectiveness of geologic storage of carbon dioxide as a climate change mitigation strategy. In Kaya et al.

Herzog H (2001) What future for carbon capture and sequestration? Environmental Science and Technology, 35: 148A–153A

Herzog H, Kaldera K, Reilly J (2003) An issue of permanence: assessing the effectiveness of temporary carbon storage. Climatic Change, Accepted

Herzog HJ (1999) The economics of CO_2 separation and capture. Technical report, MIT Energy Laboratory, August, Presented at Second Dixy Lee Ray Memorial Symposium, Washington, D.C.

Holloway S (2001) Storage of fossil fuel-derived carbon dioxide beneath the surface of the earth. Annual Review of Energy and the Environment, 26: 145–166

Jimenez JA, Chalaturnyk RJ (2002) Are disused hydrocarbon reservoirs safe for geological storage of CO_2 ? In Kaya et al.

Johnson TL, Keith DW (2003) Fossil electricity and CO_2 sequestration: How natural gas prices, initial conditions and retrofits determine the cost of controlling CO_2 emissions. Energy Policy, Proof available online (18 march 2003)

Kaya Y, Ohyama K, Gale J, Suzuki Y, editors (2002) *GHGT-6:* Sixth International Conference on Greenhouse gas control technologies, Kyoto, Japan, September 30–October 4 2002, Elsevier

Keller K, Hall M, Bradford DF (2002) Carbon dioxide sequestration: when and how much? Unpublished, cited with permission

Nordhaus W, Boyer J (2002) Warming the World: Economics Models of Global Warming. MIT press

Pacala SW (2002) Global constraints on reservoir leakage. In Kaya et al.
Portney PR, Weyant JP (eds) (1999) Discounting and Intergenerational Equity. Resources for the Future, Washington, D.C.
Reeves S (2002) Coal-seq project update: field studies of ECBM recovery/CO_2 sequestration in coalseams. In Kaya et al.
Saripalli KP, Mahasenan NM, Cook EM (2002) Risk and hazard assessment for projects involving the geological sequestration of CO_2. In Kaya et al.
Willams D, Durie B, McMullan P, Paulson C, SmithA, editors (2001) Greenhouse gas control technologies: Proceedings of the 5th international conference on greenhouse gas control technologies. Collingwood, Australia, CSIRO Publishing
Wong S, Gunter W (2000) Economics of CO_2 sequestration in coalbed methane reservoirs. In SPE/CERI Gas Technology Symposium, 3–5 April, Society of Petroleum Engineers Inc.

An Integrated Computer Aided System for Generation and Evaluation of Sustainable Process Alternatives

Niels Jensen, Nuria Coll, Rafiqul Gani[*]

CAPEC, Department of Chemical Engineering, Technical University of Denmark, DK-2800 Lyngby, Denmark

Abstract

This paper presents an integrated system for generation of sustainable process alternatives with respect to new process design as well as retrofit design. The generated process alternatives are evaluated through sustainability metrics, environmental impact factors as well as inherent safety factors. The process alternatives for new process design as well as retrofit design are generated through a systematic method that is simple yet effective and is based on a recently developed path flow analysis approach. According to this approach, a set of indicators are calculated in order to pinpoint unnecessary energy and material waste costs and to identify potential design (retrofit) targets that may improve the process design (in terms of operation and cost) simultaneously with the sustainability metrics, environmental impact factors and the inherent safety factors. Only steady state design data and a database with properties of compounds, including, environmental impact factor related data and safety factor related data are needed. The integrated computer aided system generates the necessary data if actual plant or experimental data are not available. The application of the integrated system is highlighted through a number of examples including the well-known HDA-process.

List of Abbreviations and Symbols

A	Energy allocation factor
AF	Accumulation Factor
ATP	Aquatic Toxicity Potential
CA	Cost Allocation Factor
CEI	Chemical Exposure Index
E	Parameter denoting the effect of a path component on a reaction
EAF	Energy Accumulation Factor
EC	Energy Cost
EWC	Energy and Waste Cost
FP	Total number of desired products in the process
F_v	Fraction of liquid flashed
GWP	Global Warming Potential
HD	Hazard Distance
$HTPE$	Human Toxicity Potential by Exposure both Dermal and Inhalation
$HTPI$	Human Toxicity Potential by Ingestion
I_{ci}	Chemical inherent safety index
I_{cor}	Corrosivity index
I_{eq}	Equipment index
I_{ex}	Explosiveness index
I_{fl}	Flammability index
I_i	Inventory index
I_{int}	Chemical interaction index
I_p	Process pressure index
I_{pi}	Process inherent safety index
I_{rm}	Heat of the main reaction index
I_{rs}	Heat of the side reactions index
I_{Isbl}	Inside Battery Limit Area
I_{osbl}	Outside Battery Limit Area
ISI	Total Inherent Safety Index
I_{st}	Process structure index
I_t	Process temperature index
I_{tox}	Toxicity Index
LEL	Lower Explosive Limit
m	Mass flow rate
M	Molecular weight
MVA	Material Value Added
ODP	Ozone Depletion Potential
$PCOP$	Photochemical Oxidation Potential
PD	Total number of products in the overall reaction equation
PE	Utility Price
PEI	Potential Environmental Impact
PP	Value outside process boundaries
PR	Purchase price

PS	Value of steam
P_v	Vapour pressure of the liquid at the pool temperature
Q	Energy consumption/duty in a unit
R	Total number of reactive unit operations in the path
RK	Total number of reactions in a unit operation
RM	Total number of raw materials involved in the process
RQ	Reaction Quality
T	Process Temperature
T_b	Normal boiling point of the liquid
TLV	Threshold Limit Value
T_m	Mean temperature
TTP	Terrestrial Toxicity Potential
TVA	Total Value Added
TWA	Time Weighted Average
U	Total number of sub-operations
UEL	Upper Explosive Limit
UK	Total number of component path flows in a sub-operation
w	Mass Distribution Factor
WAR	Waste Reduction Algorithm
WC	Waste Cost
W_p	Total mass entering the pool
W_t	Total liquid release
v	Stoichiometric coefficient
η	Efficiency for fuel credit in furnace
Δh	Height of the liquid above the release point
ΔH_c	Heat of combustion
ΔH_{vap}	Heat of vaporization
ΔH_r	Heat of reaction
ρ	Density
ρ_l	Liquid density
ξ	Extent of reaction (kmol/h)

Subscripts or Superscripts

c	Component index
fp	Index of desired products
k	Index of mass paths
pd	Product index
rk	Index of reactions
rm	Raw material index
u	Index of sub-operations
uk	Index of components path flows in a sub-operation

1 Introduction

The increasing concern and focus on pollution prevention and the never-ending quest for higher profits in order to be competitive on the market, makes the necessity of process optimisation obvious. It is well known that because of conflicting effects of operational costs, operability, sustainability, environmental impact and safety & hazards on the process performance, it is difficult to determine any process (retrofit) design alternative that can simultaneously satisfy all these criteria. If the various process performance criteria are to be considered simultaneously, one solution strategy is to formulate and solve complex multi-objective mathematical programming (optimisation) problems. The resulting optimal solution is usually a trade-off between the various conflicting performance criteria. Lange (2002) argues for sustainable development as way to run businesses by balancing economic, environmental and social responsibilities, where the important aspects of environmental responsibility are the utilisation of natural resources and the disposal of wastes. The case for process efficiency and material reuse (recycle) is advocated. Anastas and Lankey (2002) promote the use of green chemistry and engineering in meeting the challenges of today and tomorrow in terms of global climate change, sustainable energy production, food production, depletion of non-renewable resources and the dissipation of toxic and hazardous materials in the environment.

Alternatively, it is possible to contemplate a method for generating process (retrofit) design alternatives that would either improve the various performance criteria or keep them unchanged. This is feasible if the generated alternative is determined on the basis of largest driving forces employed by each process operation to achieve their designed objectives, on the following postulation – "the cost and operability of process operations such as heating, cooling, separation and reaction are indirectly proportional to their corresponding driving forces". That is, if the driving force is large, the operation is easy and less costly. For heating and/or cooling operations, the temperature difference is the driving force that affects the heat transfer; in reaction operations, it is the reaction kinetics that influences the reaction rate; while in separation operations, it is the composition difference that drives the separation. All these affect the mixing operations (that also include addition of solvents, process fluids, etc.) by requiring fewer quantities to be mixed. Indirectly they promote the reduction of waste and better control of operation (including emission). Koller et al. (1998) developed a flexible assessment framework for consideration of environmental and safety aspects in addition to economic and technical considerations in the early development phase of a new process. Young et al. (2000) combined environmental impacts, energy consumption and engineering economics through their waste reduction algorithm. Hertwig et al. (2000) presented a prototype system for economic, environmental and sustainable optimisation of a chemical process.

Recently, Uerdingen (2002) and Uerdingen et al. (2003) developed an indicator-based method to generate process (retrofit) alternatives that is able to reduce waste and increase process efficiencies by indirectly maximizing the various driving forces, which are based on differences in mass (composition) and energy (temperature difference). Given a simulation report corresponding to an actual process (or base case design), the method identifies the material and energy path flows and calculates a set of indicators. The most important design variables are then identified through a sensitivity analysis in terms of their effect on the indicator values, which also defines the limiting values of these variables. The indicators, by definition, also provide estimates of process improvements (targets) in terms consumption of resources and cost of operation since they involve the same set of variables. The application of this method, however, requires data in the form of process and equipment data (steady state design data) and a database with physical properties of the compounds involved in the process.

The objective of this paper is to present an integrated computer aided system that helps to generate process design (retrofit) alternatives through the use of an extended version of the Uerdingen (2002) indicator-based method. In this extended method, the evaluation of the generated process alternatives in terms of sustainability metrics (Tallis 2002), environmental impact factors as defined by Young et al. (2000) and safety and hazards factors (Hekkila 1999) have been incorporated. A property prediction feature is added to estimate the needed missing properties (Marrero 2002). A simulation-engine has also been added to generate the steady state process flowsheet stream data, if they are not available. Finally, a simple spread-sheet based algorithm, developed by Andersen (2002), for the calculation of the indicators of the Uerdingen method (2002) has been interfaced to the computer aided system so that all the necessary computations may be performed in the same computer environment. This makes the data transfer more efficient and the problem solution less time consuming.

2 Theoretical Background

2.1 Mass and Energy indicators

Introduced by Uerdingen et al. (2001), the method uses information about accumulation, cost and benefit over the total process to compute a set of indicators related to mass and energy being circulated in closed- and open-loops within a process with respect to mass and energy entering and/or leaving the process (or loop). In this way, the indicators provide important information about the process in terms of which operation of a process flowsheet is comparatively more expensive than others. Through a sensitivity analysis, it is then possible to identify the set of operational variables that affect the indicators, thereby helping to define design targets as well as to generate better process (retrofit) design alternatives.

Decomposition of Flows: A process flowsheet consists of unit operations and streams linked to them. Uerdingen (2002) applied the graph theory to convert the flowsheet into a process graph where the process units become the vertices and the process streams become the edges that connect the vertices. This helps to identify the flow paths in terms of mass (of each component present in the system) and energy and classify them as supply flows, demand flows or incident flows. A supply flow is either a flow that enters the process or a flow generated in the process. Demand flows are either outlets or flows that are consumed in a reactor. Incident flows are flows that run between the unit operations. An open path is a flow of a component that enters the process in a supply and leaves in a demand. The path flows are either open, if no recycle occurs, or closed, if recycles are present in the flow sheet. For each recycle, a cycle path is defined for each component in the recycle. The energy demand can exist in two ways - energy leaving with a mass demand flow or as heat transfer.

Material-value added (*MVA*): For a given open path it is desirable to calculate the value generated from start to end point. This is done by calculating the difference between the value of the component path flows outside the process boundaries and the costs in raw material consumption or feed cost. Negative values for *MVA* indicates value losses and show that there are potentials for improving the economic efficiency. *MVA* is calculated in cost units per year.

$$MVA = (mass)(sales\ price - raw\ material\ cost)$$

Energy and waste cost (*EWC*): The *EWC* indicator consists of two parts: *EC* considers the energy costs and *WC* the process waste costs associated with a given path, by allocating the utility consumption and waste treatment costs. The results will indicate the maximum theoretical saving potential for a given path. High *EWC* values indicate high energy consumption and waste costs that could be reduced by decreasing the path flow or the duties. For component c in path flow k the equation is:

$$EWC = EC + WC$$

$$EC = (duty)(cost) \frac{component\ mass \times characteristic\ physical\ property}{sum\ of\ all\ components\ (mass \times characteristic\ physical\ property)}$$

$$WC = (mass) \cdot (waste\ treatment\ cost)$$

$$EWC_k^{(c)} = m_k^{(c)} \cdot \left(\sum_{u=1}^{U} PE_u Q_u \frac{A_{u,k}^{(c)}(T_m, p_m)}{\sum_{uk=1}^{UK} m_{m,uk} A_{u,uk}(T_m, p_m)} + (\rho_k)^{-1} WAx_k^{(c)} \right) \qquad (1)$$

A_u is an allocation, ρ_k is the density and $WAx_k^{(c)}$ is the waste allocation factor. PE_u represent the unit price of the utility Q_u. Subscript u is the sub-operations index, uk is the index of all component path flows in u and k is the path flow index. *EWC* is calculated in cost units per year.

Reaction quality (*RQ*): This indicator measures the effect a component path flow may have on the reactions that occur in its path. If the *RQ* value is positive, the path flow has a positive effect on the overall plant productivity. Negative values indicate an undesirably located component path flow in the process.

$$RQ = \frac{\text{extent of reaction} \times \text{reaction parameter}}{\text{sum of desired products}}$$

Accumulation factor (*AF*): *AF* is a way of measuring the accumulative behaviour of individual components in recycles. Note that the term "accumulation" is not used to mean inventory in this method. It indicates the amount of material being recycle relative to their input to the process and/or output from the process.

$$AF = \frac{\text{mass of component in recycle}}{\text{sum of component mass leaving recycle}}$$

(2)

$$AF = \frac{m_z^{(c)}}{\sum_{i=1}^{I}\left(\sum_{a=1}^{EN} f_{i,a}^{(c)} + \sum_{op=1}^{OP} d_{i,op}^{(c)}\right)}$$

where $m_z^{(c)}$ is the cycle flow rate. High values of this indicator show that there is a build-up of a component within the system, which possibly could be caused by poor separations or low conversion in the reactive unit.

Total value added (*TVA*): This indicator describes the economic influence a component path flow may have on the variable process costs. Negative *TVA* values indicate improvement potentials in the process. Still, if a path flow has an high *EWC* value that is compensated by a high *MVA* value and gives a positive *TVA* value it can still be possible to reduce the energy cost. *TVA* is calculated in cost units per year.

$$TVA_k^{(c)} = MVA_k^{(c)} + EWC_k^{(c)}$$

Energy accumulation factor (*EAF*): The energy accumulation factor (*EAF*), calculates the accumulative behaviour of energy in an energy cycle path flow. Since it is of interest to recycle or recover energy, these factors should be as large as possible in order to save energy. The energy accumulation factor can be calculated as:

$$EAF = \frac{\text{energy recycled}}{\text{energy leaving the recycle}}$$

$$EAF_{ec} = \frac{ebl_{ec}}{\sum_{i=1}^{I}\left(\sum_{a=1}^{EN} f_{i,a}^{h} + \sum_{op=1}^{OP} d_{i,op}^{h}\right)} \quad (3)$$

where I is the total number of vertices encountered in the energy cycle path flow, and i is the index of these. ec is the index of the cycle energy path flows. ebl is the amount of energy recycled in the particular recycle.

Total demand cost (*TDC*): *EAF* calculates the accumulative behaviour of energy in an energy cycle path flow. Since it is of interest to recycle or recover energy, these factors should be as large as possible in order to save energy. However, the penalty of this could be an increased size of heat exchanger equipment. The indicators, demand cost *(DC)* and total demand cost *(TDC)*, are based on energy open paths; they are used to trace energy inputs (and thereby energy costs) through the process. With the *DC* factor it is possible to see exactly how the energy from each supply is distributed to the different demands. By summing up the *DC* values for each demand, it is possible to see exactly the amount of energy leaving the process in each demand. In order to assign costs to the energy open paths, a demand cost is calculated for each path,

$$DC_{s,d} = PE_s\, OP_{s,d} \quad (4)$$

Where PE_s is the utility cost in units of price/energy for each supply s and $OP_{s,d}$ are the open-paths for energy connecting a supply to a demand d. The total cost associated with the demand, *TDC*, is calculated as,

$$TDC_d = \sum_{s=1} DC_{s,d} \quad (5)$$

The idea of the supply cost is to have a method to measure "the quality of the energy". This is illustrated by looking at heat integration. For example, a heat exchanger removing large quantities of low quality energy (low temperature) is not as interesting as a heat exchanger removing a smaller amount of very high quality energy. A high *TDC* value will identify the demands that consume the largest amounts of high quality energy, that is, the demands that hold the greatest potential for heat integration.

2.2 Sensitivity Analysis

The objective of the sensitivity analysis is to identify the most sensitive indicators and the corresponding design variables. Also, the sensitivity analysis should provide an estimate of the effect of changes in the design variables in terms of cost benefits (note that the indicators include the cost data). The sensitivity of an indicator (X_r) due to disturbances in a design variable (Y_r) is calculated from the following equation,

$$\Delta X_r = 100 \, [X_r - X_r \, (Y_r + \Delta Y_r)/Y_r]^2 \tag{6}$$

where, ΔY_r is the change in the specified design variable and $X_r \, (Y_r + \Delta Y_r)$ is the value of the indicator calculated at the perturbed value of the specified design variable. The expression is squared as in least squared methods where small errors become smaller and large errors become larger. By squaring the expression all derivatives become positive. In order to provide a good visual presentation of the sensitivity, the expression in equation (7) should be plotted as a function of Y_r. The vertical axis intersects the horizontal axis at the reference Y_r value. The axes are in units of percent. If it is discovered that some design variables cause large variations in the indicator values, the user should be careful when specifying values for these design parameters. If they are not accurate, the indicator values calculated will not be reliable. This may lead to incomplete or false conclusions.

2.3 Generation of Alternatives

Uerdingen (2002) and Uerdingen et al. (2003) provide a systematic method for analysis of the indicators and based on this, a procedure for generation of retrofit alternatives. In this paper, a brief description of this methodology is given. When the indicators are calculated, all mass flow paths are sorted into four main categories (see Table 1), according to their indicator values. Categories 1 and 2 contain the open paths, while categories 3 and 4 are used for cycle paths. The next sorting is based on the *RQ* values, since it is important to know, whether or not a given path has positive, negative or no influence on the overall plant productivity. Open path reactants and inert will have a value equal to or lower than 0. These are located in the first category. In the second category, only reactant open paths are located since the *RQ* values for these should be positive. The third and fourth categories contain the cycle paths. Unlike the open paths, the cycle paths cannot generally be divided into a reactant and a product/inert category. The product cycle paths from one reaction can affect other reactions in its path and if these reactions are affected in a positive way, the path can score *RQ* values above 0. The third category contains cycle paths with a *RQ* value equal to 0 or lower. This category is divided into two sub-categories, containing *AF* values above and below 1, respectively. A negative *RQ* value with high *AF* value is not desirable in the process and attention should be given to this path. The last category contains cycle paths with *RQ* values above 0. In each of the four categories, the paths are sorted in ascending order according to their *TVA* values. This will ensure that the paths are listed in order of economic impact potential. The second column in Table 1 introduces a list of heuristic rules used to find generic design (retrofit) alternatives that can be applied for the paths in the different categories. The more detailed specifications of the alternatives should be defined by investigation of the different parameters that can be manipulated in the process. The analysis of the energy indicator values is carried out a little different from the mass indicators, and does not require distribution into a series of categories.

Table 1. Feasible design alternatives based on the calculated indicator values

Component path flows	Design (retrofit) action	Indicators that may be affected
All categories	Reduce duty by changing temperatures, by reducing flows or through heat integration	*EWC, MVA*
Category 1: Open path flows, $RQ \leq 0$	Reduce/remove flows	*EWC, MVA*
	Improve separation	*EWC, MVA*
	Increase value of the open path	*MVA*
	Introduce a separation method	*EWC, MVA*
	Change conversion	*EWC, MVA, RQ*
Category 2: Open path flows, $RQ > 0$	Reduce flow if no consumed in reactor	*EWC, MVA*
	Recycle the open path	*EWC, MVA*
	Improve separation	*EWC, MVA*
	Increase value of the open path	*MVA*
Category 3: Cycle path flows, $RQ \leq 0$		
a) $AF > 1$	Reduce flow	*EWC, AF*
	Change catalyst	*EWC, RQ, AF*
	Reduce AF by increased purge	*EWC, AF*
	Remove path at source	*EWC, AF*
	Reroute path (change to open path)	*EWC, AF*
b) $AF \leq 1$	Remove path at source	*EWC, AF*
	Reduce flow	*EWC, AF*
	Reroute path (change to open path)	*EWC, AF*
Category 4: Cycle paths, $RQ > 0$	Optimise the flow rate	*EWC, AF*
	Increase conversion	*EWC, AF, RQ*

The energy open paths are not listed in the same way as the mass paths, since the *EOP* (energy open paths) values are summed up when calculating the *TDC* values and not used separately. As discussed above, the *TDC* values can be used to trace high quality energy through the process. When high *TDC* values are detected, an effort should be made to investigate possible alternatives for energy integration, by coupling them with suitable supplies. Low *EAF* values for the energy cycle paths indicate lacking or poor energy integration, which will also display as high *TDC* values for the demands in the cycle. When analysing the retrofit alternatives, the *EAF* values will clearly indicate any improvement in heat conservation.

A summary of the meaning of the indicator values in terms of which direction they should move is given in Table 2. Note that the recommendations given in Table 1 and the meanings given in Table 2 are based on observations and analysis of results from a number of process flowsheets that have been studied. It is possible, however, that exceptions exist to these observations.

Table 2. Meaning of the indicator values

Indicator	Negative value	Positive value
MVA	Value lost in path	Value gained in path
RQ	Negative impact on plant productivity	Positive impact on plant productivity
TVA	High potential for improvement	Low potential for improvement
	Low value	High value
EWC	Low energy & waste reduction potential	High energy & waste reduction potential
AF	Low accumulation of component	High accumulation of component
EAF	Low energy utilization	High energy utilization
TDC	Low energy loss	High energy loss

New design (retrofit) alternatives are generated by moving the indicators in the correct direction through changes in the identified most sensitive variables. Since the indicator values include cost estimates, a separate cost analysis is not necessary and reaching a targeted indicator value automatically provides the cost reduction (or profit improvement). Feasibility of change of a design variable limits the possible alternatives that can be generated. For example, a design variable may be very sensitive (such as conversion in the reactor) but it may not be feasible or allowed to change it. Note also that since the costs (related to operation) are included, once the potential savings for a feasible alternative is evaluated, it is possible to verify if investment with respect to cost of equipment will be justified. At least, this method provides a simple and easy way to define the targets for generating alternatives. Note also that the optimisations related calculations involve simply moving a design variable in the direction that improves the sensitive indicator(s), where the sensitivity analysis provided a good estimate of the solution. It is possible that more than one indicator may be sensitive to the same design variable. In this case, a numerical optimisation technique may be used. Based on the problems solved, if more than one indicator were found to be sensitive, they moved in the same direction and therefore, did not cause any difficulties in solving the optimisation problem.

2.4 Sustainability Metrics

Defined by the American Institute of Chemical Engineers-USA and the Institution of Chemical Engineers-UK for chemical process industries, the sustainability metrics help engineers to address the issue of sustainable development. Engineering for sustainable development means providing for human needs without compromising the ability of future generations to meet their needs. They also enable manufacturing companies to set targets and to monitor progress on a yearly basis. Use of the metrics indicators of sustainability follows the simple rule that the lower the metric the more effective the process. A lower metric indicates that either the impact of the process is less or the output of the process is more. In this paper, the sustainability metrics as proposed by the IChemE (Tallis 2002) have been employed. These include the following,

Sustainability Metrics: Energy
    Total Net Primary Energy Usage Rate = Imports – Exports (GJ/y)
    Percentage Total Net Primary Sourced from Renewals (%)
    Total Net Primary Energy Usage Rate per kg Product (kJ/kg)
    Total Net Primary Energy Usage per Unit Value Added (kJ/$)

Sustainability Metrics: Material
    Total raw materials used per kg product (kg/kg)
    Total raw materials used per unit value added (kg/$)
    Fraction of raw materials recycled within company (kg/kg)
    Fraction of raw materials recycled from consumers (kg/kg)
    Hazardous raw material per kg product (kg/kg)

Sustainability Metrics: Water
    Net water consumed per unit mass of product (kg/kg)
    Net water consumed per unit value added (kg/$)

2.5 Environmental Impact Factors

There is an increasing trend in chemical process design to consider minimizing of cost (equipment as well as operation) simultaneously with satisfying environmental constraints defined through a set of environmental impact factors. This essentially leads to a reduction of waste. One way to achieve this has been proposed by Cabezas et al. (1999) based on the theory proposed by Hilaly and Sikdar (1994). Cabezas et al. (1999) proposed a waste reduction (WAR) algorithm, which describes the flow and the generation of potential environmental impact through a chemical process and this algorithm has been employed in this work to evaluate the generated process design (retrofit) alternatives.

The theory in the WAR algorithm defines indexes that characterize the generation and output of potential environmental impact (PEI) from a process. The PEI is a relative measure of the potential for a chemical to have an adverse affect on human health and the environment. The result of the PEI balance is an impact (pollution) index, which provides a quantitative measure of the impact of the waste generated in the process. The goal of this methodology is to minimize the PEI for a process instead of minimizing the amount of waste (pollutants) generated by a process. The potential environmental impacts are calculated from stream mass flow rates, stream compositions and a relative potential environmental impact score for each chemical present obtained.

There are several categories of impact. This can be subdivided into physical potential impacts (acidification, greenhouse enhancement, ozone depletion and photochemical oxidant depletion), human toxicity effects (air, water and soil) and ecotoxicity effects (aquatic and terrestrial). The important parameters are: HTPI (Human Toxicity Potential by Ingestion), HTPE (Human Toxicity Potential by Exposure both Dermal and Inhalation), TTP (Terrestrial Toxicity Potential), ATP (Aquatic Toxicity Potential), GWP (Global Warming Potential), ODP (Ozone Depletion Potential), PCOP (Photochemical Oxidation Potential), AP (Acidification Potential). All of these are gathered in one total factor, the Total PEI (Total Potential Environmental Impact), which indicates the unrealised effect or impact that the emission of mass and energy would have on the environment on average. The US-EPA has developed a computer-aided tool to compute these factors (Young et al. 2000). This tool contains a database of chemicals with their corresponding property values and a calculation algorithm to compute the impact factors, given the stream properties (composition, temperature and pressure) of all input and output streams belonging to the process.

2.6 Safety Indicators

For reasons of general legal requirements, company image, economic reasons, *etc.*, it is required that the safety of a process plant fulfils a certain required level. Also, an unsafe plant cannot be profitable over a period of time due to losses of production and capital. According to Kharbanda and Stallworthy (1988) safety is a concept covering hazard identification, risk assessment and accident prevention. In practice the main purpose of process plant design is to minimize the total process risk for the limitation of effects.

The best-known measure for safety is risk, which is the defined potential for loss or in mathematical terms the probability of a specified undesired event occurring, in concrete circumstances or within a particular period of time and the consequence of the undesired event:

Risk = f (probability of undesired event, consequences of undesired event)

In a process plant the loss may be damage to equipment, damage to environment, loss of production or even human injury or death. Risk involves two measurable parameters: consequence (what happens?) and probability (how often does it happen?). A hazard is a condition with the potential of causing a loss. A chemical process has a number of potential hazards, for example toxicity and reactivity of raw materials and intermediates, energy release from chemical reactions, high temperatures, high pressures, quantity of material used, etc. Each of these hazards impacts the overall process risk. A pursuit of safety is a matter of identifying hazards, eliminating them where possible or otherwise protecting against their consequences. In practice risk is often viewed as the product of the probability of an incident times the consequences of the incident. A hazard is simply the potential for an incident.

An inherently safe chemical process is one that avoids hazards instead of controlling them, particularly by removing or reducing the amount of hazardous material in the plant or the number of hazardous operations. The possibility of affecting the inherent safety of a process decreases as the design proceeds and more engineering and financial decisions have been made, so it is much easier to affect the process configuration and inherent safety in the conceptual design phase than in the later phases of the process design. In this paper, a method (Heikkilä 1999) for measuring the intrinsic safety of a process, which is affected by both the process equipment and the properties of the chemical substances present in the process, has been used. The method requires the evaluation of a number of interrelated factors and sub-indices, as listed in Table 3 (Heikkilä 1999).

Table 3. List of Inherent Safety Sub-indices

Total Inherent Safety Index (ISI)			
Chemical Inherent Safety Index, I_{ci}	Score	Process Inherent Safety Index, I_{pi}	Score
Sub-indices for reactions Hazards		*Sub-indices for process conditions*	
Heat of the main reaction, I_{rm}	0-4	Inventory, I_i	0-5
Heat of the side reactions, I_{rs}	0-4	Temperature, I_t	0-4
Chemical interactions, I_{int}	0-4	Pressure, I_p	0-4
Sub-indices for hazardous substances		*Sub-indices for process conditions*	
Flammability, I_{fl}	0-4	Equipment, I_{eq}	
Explosiveness, I_{ex}	0-4	I_{Isbl}	0-4
Toxicity, I_{tox}	0-6	I_{Osbl}	0-3
Corrosivity, I_{cor}	0-2	Process structure, I_{st}	0-5
Maximum I_{ci} score	28	Maximum I_{pi} score	25
Maximum I_{si} score 53			

The indices listed in Table 3 are calculated for each process alternative separately and the results are compared with each other. They are calculated through the following equations:

$$I_{ti} = I_{ci} + I_{pi} \qquad (7)$$

$$I_{ci} = I_{rm,max} + I_{rs,max} + I_{int,max} + (I_{fl} + I_{ex} + I_{tox})_{max} + I_{cor,max} \qquad (8)$$

$$I_{eq} = I_{Isbl} + I_{Osbl} \qquad (9)$$

$$I_{pi} = I_i + I_{t,max} + I_{p,max} + I_{eq,max} + I_{st,max} \qquad (10)$$

The calculations of the Inherent Safety Index are made on the basis of the worst situation. The approach of the worst case describes the most risky situation that can appear. A low index value represents an inherently safe process. The application of the index is flexible. Comparisons can be made at the level of process, sub process, subsystem, or considering only part of the factors. Sometimes a comparison based on only one or two criteria may be interesting. The total inherent safety index is quite easily integrated to simulation and optimisation tools by adding a calculation option for the indices after a simulation has been made. As it is possible to see in Table 3, scores are different for each index. A wider range means greater impact to the plant safety. Scores can be changed depending on the aspect the engineer wants to emphasize, but this modification must be kept for comparison with another processes or alternatives. The guidelines provided by Heikkilä (1999) have been employed in this paper to compute appropriate scores for each sub-index.

3 The Indicator-Based Methodology & Computer-Aided Toolbox

The indicator-based methodology consists of 6 main steps, as outlined below.
1. Obtain process (plant) steady state data.
2. Transform the process flowsheet to a process graph, and decompose this graph to identify the path flow diagrams – algorithm for decomposition of flowsheet into path flows.
3. Calculate the set of indicators (five for mass and two for energy) – See Eqs. 1-6.
4. Perform sensitivity analysis on the indicators with respect to changes in design variables (to identify the important design variables) – Algorithm outlined in section 2.2.
5. Generate retrofit (design) alternatives by changing values of the identified design variables – based on the algorithm outlined in section 2.3.
6. Evaluate the generated alternatives in terms of improvements with respect to resources (mass & energy), sustainability metrics, environmental impact factors and inherent safety factors – use the definitions and algorithms outlined in section 2.4, 2.5 and 2.6, respectively.

The above methodology has been implemented as an integrated computer aided process analysis tool-box by incorporating the indicator-based method of Uerdingen (2002), the sustainability metrics of IChemE (Tallis 1999), the WAR algorithm Young et al. (2002), the inherent safety index (Heikkilä 1999) together with the simulation-engine, the properties database and properties package of ICAS (ICAS Documents 2002). The flow diagram highlighting the main steps of the integrated methodology is given in Figure 1.

Step 1 is simply a collection of plant operational data or generation of the steady state (design) data through process simulation. In principle, any process simulation program may be used for this purpose. In this paper, the simulation-engine of ICAS (ICAS Documentation 2002) and PRO-II (PRO-II User's Guide 2002) has been used.

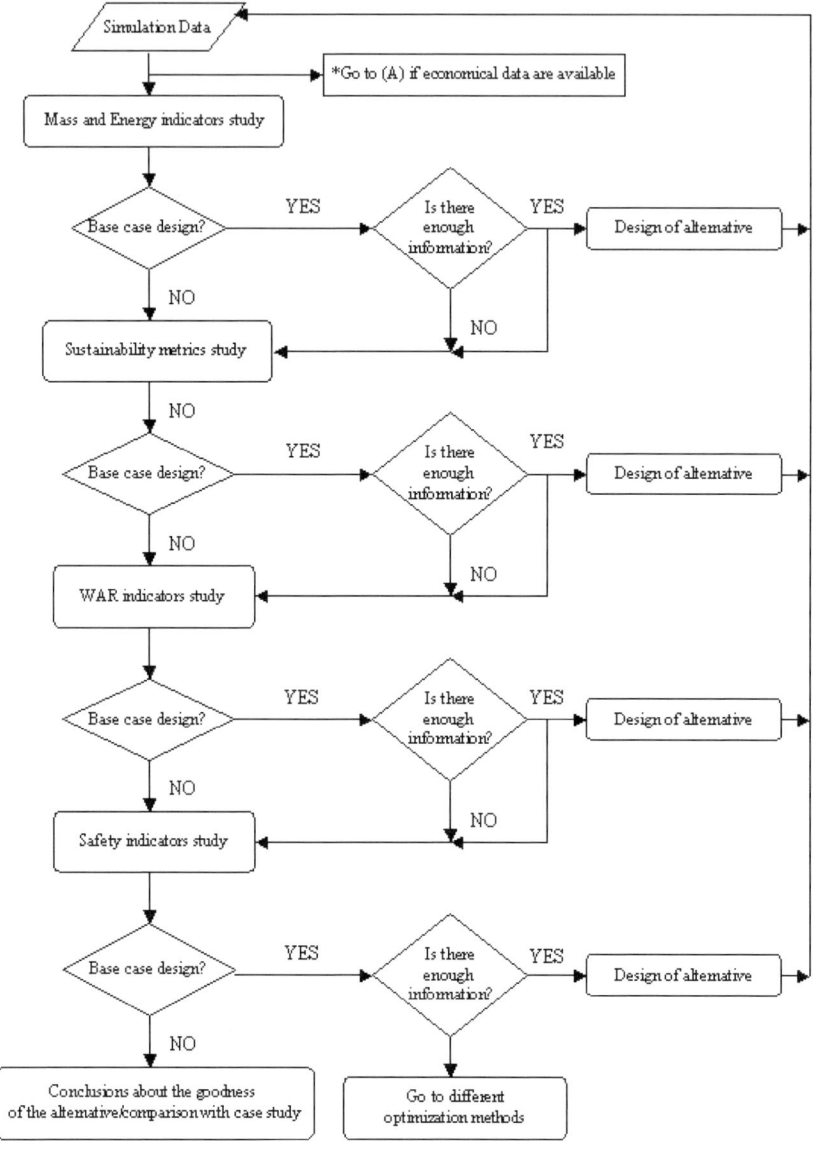

Fig. 1. Flow diagram of the indicator-based methodology

In **Step 2**, the process flowsheet is decomposed in terms of material and energy path-flows as specified in the Uerdingen algorithm. Open-paths as well as closed-paths (recycle-loops) are identified here with respect to material flow of each chemical present in the system and energy. The data generated in step 1 and the path-flow information generated in step 2 are used in **step 3** as input to a specially developed software (RetroFit-Pro), which calculates all the indicators.

Steps 4-6: Once the indicators have been calculated, they are included in a sensitivity analysis in order to pinpoint areas of the process with good improvement potentials and to identify the corresponding set of design variables that may be changed in order to achieve the improvement (design target). The generated retrofit alternatives are evaluated in terms of various performance indices (mass/energy consumption, sustainability metrics, environmental impact factors and safety factors). A process simulation needs to be performed for each generated alternative. Since the generated retrofit alternatives are usually relatively small changes on the original design, and, a reference flowsheet simulation already exists, these additional simulations are not difficult to make. Note that since the indicators by definition include cost data, estimates of the cost of operation are directly included in their calculations. Therefore, by achieving a targeted value of a selected indicator, a corresponding reduction in the cost of operation is automatically obtained. This means that a separate cost-based optimisation calculation is not necessary.

4 Case Study

The application of the indicator-based methodology and the integrated computer-aided toolbox is illustrated through 3 case studies, which have been used by others in studies involving process design, synthesis and environmental impact. The first case study involves the hydrodealkylation of toluene (as described in Seader et al. 1999), the second process involves a solvent-based separation of an azeotropic mixture of acetone-chloroform (Hostrup et al. 1999) and the third example involves a polymeric resin adsorption process (Jimenez-Gonzales et al. 2002). The detailed simulation data (starting information for the analysis in this paper) for the three case studies are not given here. Copies of the detailed simulation results can be obtained from the authors. The objectives of these case studies are not only to determine better design alternatives but also to verify if an existing design needs any improvement (or has potentials for improvements).

In all the case studies, a simulation engine (ICAS and PRO-II) has been used to generate the simulation data, which have been validated with known values from the corresponding references (given above). ICAS has also provided the methods and tools for generation of alternative solvent candidates, for evaluation of physical properties (whenever necessary) as well as evaluation of design and operability analysis for any generated alternative.

4.1 Case Study 1

The hydrodealkylation of toluene involves a process where toluene is converted into benzene by reaction with hydrogen, forming Diphenyl as a byproduct. This is a well-known process, which has been studied in numerous publications. The reference design and flowsheet considered in this paper are taken from Seider et al. (1999), where further details can be found. The necessary steady state simulations have been performed with a commercial process simulator, from where the results have been transferred to RetroFit-Pro to calculate the indicator values. The steady state simulation results are also transferred to ICAS (ICAS Documentation 2002) to determine the environmental impact factors, the sustainability metrics and the safety factors. Table 4 shows the most important indicator-values from the base case design. A detailed calculation results document can be obtained from the corresponding author. The process flowsheet is shown in Fig. 2.

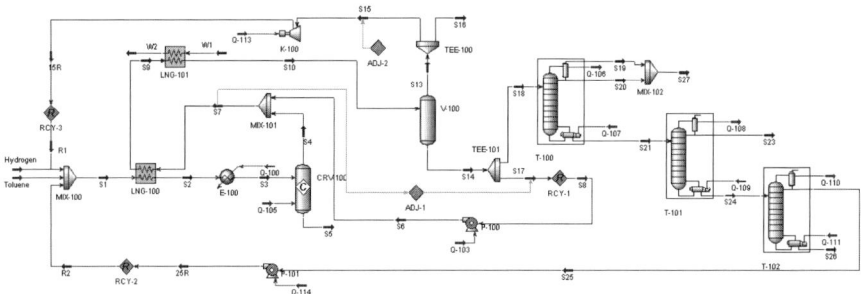

Fig. 2. Flowsheet for the HAD process (reference design

The open path flows O5 and O7 score a high negative MVA value because the production of methane from the raw materials is very high compared to the fuel credit given for these paths. Open path flow O9 shows a high negative TVA value. This is because the hydrogen in this path is lost in the purge and the purchase price of hydrogen is high compared to the fuel credit given by incineration. The methane gas cycle C1, shows an AF value above one, which indicates an unfavourable build-up of methane in the system. The low TVA value is caused by the high EWC value, which is again a result of the high flow rate. This means that the process can be improved in terms of efficiency of operation, cost and waste (environmental impact) by investigating the flows associated with hydrogen and methane.

A sensitivity analysis was performed in order to identify the retrofit alternatives with the largest impact towards an improvement of the process. The sensitivity analysis identified the design variables that affect the indicators most (see Table 5 for a list of design variables considered). Results of the sensitivity analysis are shown in Figures 3 and 4 for two of the most sensitive design variables (flowrate of hydrogen in feed and conversion).

Table 4. Indicator values for base case design

Name	Description	AF	RQ	EWC 10^3\$/y	MVA 10^3\$/y	TVA 10^3\$/y
C1	CH_4 in gas recycle	4.88	0.00	484.5	-	-484.5
O2	Diphenyl in stream 26	-	0.00	266.8	-13.4	-280.2
O5	CH_4 in stream 27	-	0.00	0.43	-481.9	-482.3
O7	CH_4 in purge	-	0.00	7.78	-8748.0	-8755.0
O9	H_2 in purge	-	1.06	33.98	-2948.0	-2982.0

Table 5. List of design variables considered in the sensitivity analysis

Design variable	-10%	-5%	Reference value	+5%	+10%
Flowrate in H_2 feed (kmol/h)	231.78	244.65	257.5	270.41	283.83
Purge fraction	0.144	0.152	0.16	0.168	0.176
Temperature in S10 (K)	306.15	309.04	310.93	311.82	413.71
Pressure in flash unit (atm)	29.72	31.37	33.02	34.67	36.32
Temperature in S2 (K)	733.15	783.04	809.95	836.82	863.71
Temperature in S3 (K)	856.15	888.60	921.05	953.48	985.93
Outlet pressure in compressor (atm)	35.31	37.27	39.23	41.19	43.15
Reflux ratio in T-102	0.90	0.95	1.00	1.05	1.10
Conversion	0.675	0.713	0.760	0.788	0.825

The sensitivity analysis revealed that the most sensitive design variables were the flowrate of hydrogen in the feed, the conversion and the temperature in the inlet stream to the reactor while the most sensitive indicators were the *EWC* and the *TVA*. This matched with the analysis of the indicator values (given above) that the flows of hydrogen and methane needed to be investigated further. One option to reduce the amount of hydrogen feed but keeping the production constraint is to recover more hydrogen for recycle and thereby purge less. As hydrogen is fed in excess compared to toluene, reducing the hydrogen feed rate can reduce the amount of hydrogen lost in purge in O9 (see *TVA* value) as well as the raw material cost, which affects the benefit of the process. In addition, as valuable hydrogen is now not lost in the purge, a decrease of the negative *TVA* value of O9 is also achieved.

By choosing these design variables and the changes (separation of hydrogen from methane) needed to achieve the targeted indicator value, the design alternatives were generated. In order to study the effect of the separation of hydrogen from methane before purge, a component splitter was introduced after the flash operation. Even though a component splitter is not "real" equipment, it serves the purpose since the objective is to evaluate the effect of hydrogen flow and not the cost of equipment or how the separation can be achieved. This confirmed that removal of hydrogen is indeed a good idea since it led to a reduction of the *EWC* values for the gas cycle paths (see C1 value in Table 6).

To achieve the separation (assumed for component splitter calculations), a membrane-based separator is recommended. The reduction in the cost of operation would not be affected as long as the membrane-based separation unit is able to achieve the desired separation. Note that the slight increase in the *MVA* value for O7 is compensated by the decrease in the *MVA* values for the other path flows. Finally, the temperature in the inlet stream to the reactor is optimised (increased) in order to reduce the *EWC* values of the large streams passing unit E-100 (preheater for the reactor). These suggested changes are implemented and Table 6 shows the improvements in the process and thus in the indicator values for the new process design.

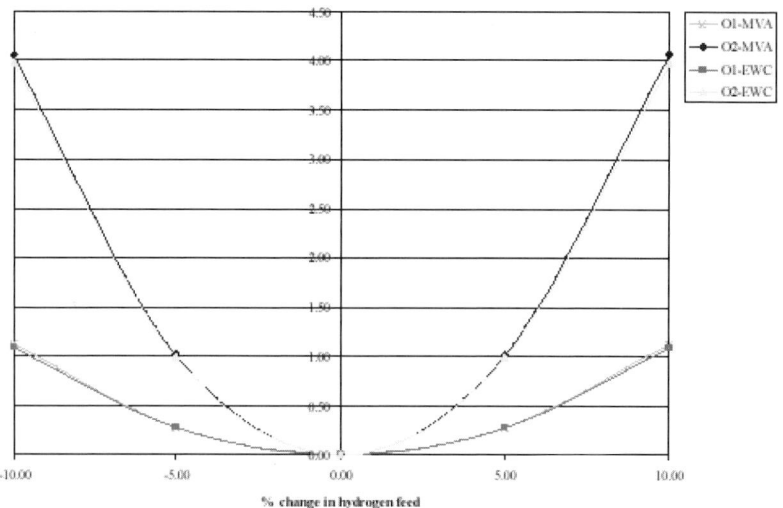

Fig. 3. Sensitivity of *EWC* and *MVA* with respect to hydrogen flowrate in feed

Impact on introduction of hydrogen/methane separation unit: By installing a hydrogen/methane separation unit, the gas recycle and the costs associated with recycling methane have been reduced (as shown by the indicator values). In addition, no hydrogen or other valuable components are lost in the purge as long as a sharp split can be achieved in the separation unit. The total difference in *TVA* values before and after installation of such separation unit is about $3.5 \cdot 10^6$ \$/yr, thereby justifying the additional capital investment.

The sustainability metrics that show any change (compared to the original design) are summarized in Table 7. As it can be seen, the new design gives an improvement, especially for the energy consumption, Finally, results from the WAR algorithm are listed in Table 8, showing a slight improvement in the *GWP* and *PEI* values (all other factors are unchanged).

The total inherent safety index (*ISI*) for the base case process was calculated to be 31. This did not change (as it is closely related to the production rate and identity of the product, in this case, benzene) with the new process design, so no safety or environmental aggravation was encountered by changing the design to a more energy efficient process, using the retrofit alternatives. Besides this HDA case study, the method has been successfully applied to a range of other processes, including cyclohexane, styrene and monochlorobenzene plants (Andersen 2002).

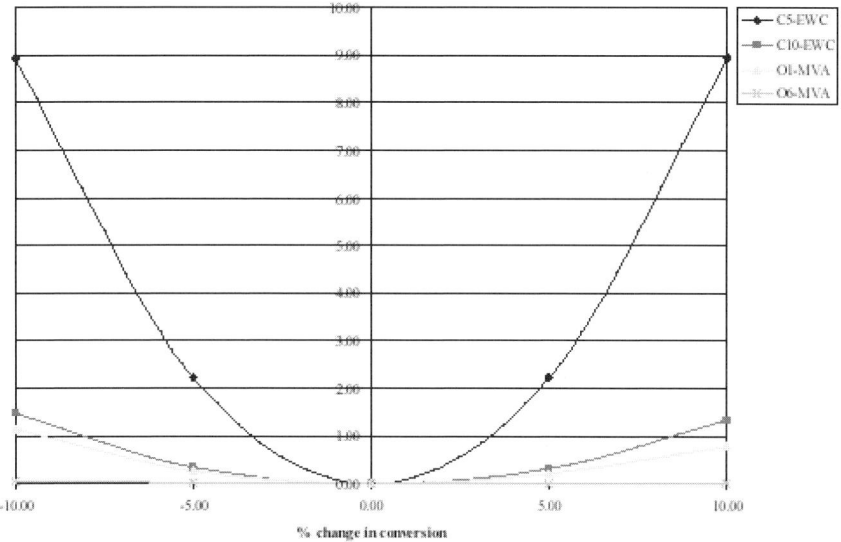

Fig. 4. Sensitivity of *EWC* and *MVA* with respect to conversion

Table 6. Indicator values for improved process design

Name	Description	AF	RQ	EWC 10^3\$/y	MVA 10^3\$/y	TVA 10^3\$/y
C1	CH_4 in gas recycle	0.00	0.00	0.00	-	-484.5
O2	Diphenyl in stream 26	-	0.00	267.2	-12.9	-280.2
O5	CH_4 in stream 27	-	0.00	0.11	-159.0	-482.3
O7	CH_4 in purge	-	0.00	6.45	-9085.0	-8755.0
O9	H_2 in purge	-	1.04	0.06	-7.5	-2982.0

Table 7. Results from sustainability metrics

	Base case design	Improved design
Total net primary energy usage per kg product	78.58E+06 kJ/kg	55.24E+06 kJ/kg
Hazardous raw material per kg product	1.22 kg/kg product	1.19 kg/kg product
Net water consumed per unit mass of product	184.6 kg/kg	171.3 kg/kg

Table 8. Indicators for the WAR algorithm

	Base case design	Improved design
Global Warming Potential (GWP)	123.1	117.6
Potential Environmental Impact (PEI)	573E+03	570E+03

4.2 Case Study 2

In this case study, the separation of acetone from chloroform by solvent-based extractive distillation is considered. The objective of this case study is to show the effect of solvent selection on design and on different sets of indicators. In this case, if the environmental and safety indicators are high, they can be reduced by substitution of the solvent. Alternative solvents have been identified through the CAMD method (Hostrup et al. 1999) and the corresponding tool-box in ICAS.

4.2.1 Process Flowsheet and Design Data

As the base case design, benzene is selected as the extractive agent. A fresh benzene stream and an equimolar mixture of acetone and chloroform are mixed with recycled liquid benzene and fed to the first distillation column at 350 K and 1atm; acetone (C_3H_6O) goes out from the top of the extractive distillation column and benzene and chloroform ($CHCl_3$) from the bottom. This bottom product is sent to a second distillation (solvent recovery) column where benzene and chloroform are separated and the benzene is recycled back to the first column. The feed and solvent stream specifications of the process are given in Table 9. Two base case designs, one with solvent loss and solvent make-up and another with negligible solvent loss and zero solvent make-up have been considered. The flowsheet for the case of solvent make-up is shown in Figure 5.

Table 9. Feed and solvent stream details for case study 2

	Solvent	Feed
Temperature (K)	300	300
Pressure (atm)	1	1
Molar flows (kmol/h)		
Benzene	37	0
Acetone	0	100
Chloroform	0	100

Process simulations have been performed for the two base case designs and the corresponding mass and energy indicators have been computed (see Tables 10 and 11). Since there are no reactions, the *RQ* (Reaction Quality) indicator has not been calculated. Open path flows O3 and O8 (corresponding to benzene and chloroform respectively) show a very high *EWC* value. The main reason is that the mass flows in these paths for these particular components are very high, indicating that the cost of energy in the first distillation column reboiler is high (note that the base case design has a high reflux ratio). The indicators for the benzene cycle path C1 is exactly the same as for paths O3 and O8, because it is favourable to recycle as much as benzene as possible in order to reduce waste and environmental impact factors. Accumulation factor AF is not especially high for any of the paths studied.

For both base case designs, benzene has been replaced with another solvent that improved the target indicators (EWC, the energy and water sustainability metrics, the safety factors, and, at least improved and/or negligible effect on the environmental impact factors).

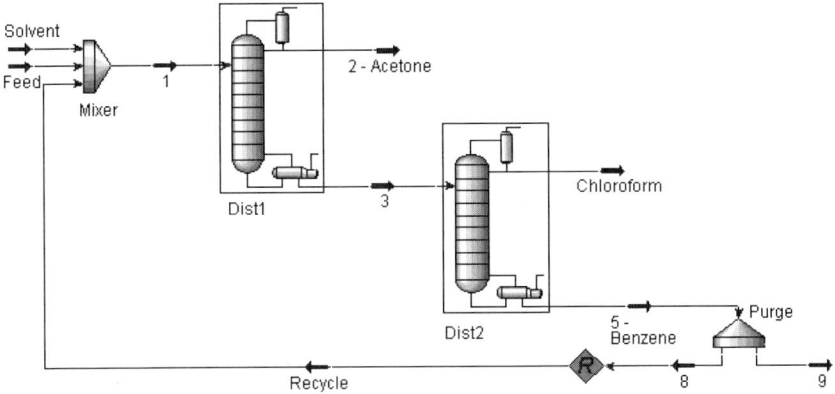

Fig. 5. Process flowsheet for case study 2 (acetone-chloroform separation)

Table 10. Mass and energy indicators for case study 2 (with solvent make-up)

Nr	Component	Out stream	CASE STUDY Mass Flow (kg/h)	AF	EWC	ALTERNATIVE Mass Flow (kg/h)	AF	EWC
O1	Bz/Ether	2	57.049	-	3.028	1.011	-	9.874E-04
O2	Bz/Ether	4	37.740	-	15.799	5.169E-08	-	2.233E-09
O3	Bz/Ether	9	5582.421	-	2577.551	80.731	-	6.850
O4	Ac	2	5763.249	-	358.483	5630.223	-	7.705
O5	Ac	4	44.751	-	21.863	177.779	-	3.662
O6	Ac	9	1.650E-03	-	8.88E-04	2.345E-07	-	8.946E-09
O7	Chlorof	2	4.672	-	1.51E-01	346.223	-	0.260
O8	Chlorof	4	11787.953	-	3006.745	11572.299	-	124.464
O9	Chlorof	9	290.139	-	81.534	1.181	-	2.348E-02
C1	Bz/Ether		53033.058	9.34	24486.766	40284.611	492.82	3418.083
C2	Ac		1.567E-02	2.00E-06	8.44E-03	1.170E-04	2.01E-08	4.464E-06
C3	Chlorof		2756.307	2.28E-01	774.568	589.307	4.94E-02	11.718

Substitute solvents have been found by means of ProCamd (a toolbox within ICAS for finding solvent substitutes through computer aided molecular design).

Two new solvents found as suitable alternatives for benzene are: methyl-n-pentyl ether ($C_6H_{14}O$), with normal boiling point of 372 K and 2-methylheptane, with normal boiling point of 390.8 K. Both of them fulfil the desirable solvent properties of benzene but do not have the undesired safety and hazards related properties of benzene. Also, as they have higher solubility of chloroform than benzene, they require less amount of solvent, indicating that the reflux ratios in the distillation columns can also be reduced. New simulations have been made with respect to the two flowsheet alternatives together with their replacement solvents, respectively. For the case of methyl-n-pentyl ether, the needed solvent feed rate has been found to be only 81.74 kg/h (compared to 2886 kg/h for benzene). The mass and energy indicator values for the two design alternatives are also given in Tables 10 and 11. It is clear that the new solvents matched the required reduction of the target EWC indicators.

Table 11. Mass and energy indicators for case study 2 (without solvent make-up)

Nr	Component	Out stream	CASE STUDY Mass Flow (kg/h)	AF	EWC	ALTERNATIVE Mass Flow (kg/h)	AF	EWC
O1	Bz/Methyl	2	2.246E-03	-	6.010E-05	3.289E-03	-	5.460E-06
O2	Bz/Methyl	4	3.114E-04	-	1.055E-04	4.561E-04	-	6.106E-06
O4	Ac	2	20.903	-	6.574E-01	20.903	-	5.012E-02
O5	Ac	4	4.648E-03	-	1.835E-03	4.648E-03	-	8.938E-05
O7	Chlorof	2	1.113E-10	-	1.824E-12	3.060E-13	-	3.821E-16
O8	Chlorof	4	28.651	-	5.913	28.651	-	2.881E-01
C1	Bz/Methyl		265.456	103808.440	106.488	388.369	103702.080	7.353
C2	Ac		3.437E-14	2.000E-06	1.606E-14	9.986E-16	4.770E-17	2.715E-17
C3	Chlorof		3.814E-02	0.228	9.317E-03	4.574E-02	1.596E-03	6.502E-04

4.2.2 Sustainability metrics

The energy, material and water sustainability metrics are summarized in Tables 12-13. The calculations are based on a 8000 h/year rate of work and a conversion factor of 0.8. The total net primary energy usage rate per kg product has been calculated on the basis of the recovered acetone (stream 2) and chloroform (stream 4) as the product. As this amount is quite small, although, the total energy required is not especially significant, the amount required per kg of product is quite high. For material sustainability indicator, streams Feed and Solvent are the raw material (used) streams, which correspond to the acetone-chloroform mixture and fresh (make-up) solvent. The recovered solvent is recycled. Tables 12 and 13 list and compare the sustainability metrics for the two base case designs and their generated alternatives, respectively. It can be noted that in both cases, the metrics values have been improved.

Table 12. Calculated values of sustainability metrics for case study 2 (with solvent make-up)

Indicator Type	Indicator	Base Case Design	Alternative Design
Energy	Total Energy Primary Energy Usage Rate = Imports – Exports, GJ/y	789672.0	338050.5
	Total Net Primary Energy Usage Rate per kg Product, kJ/kg	5264.1	2374.9
Material	Total raw materials used per kg product, kg/kg	1.17	1.00
	Hazardous raw materials per kg product, kg/kg	1.17	1.00
Water	Net water consumed per unit mass of product kg/kg	243.97	89.6

Table 13. Calculated values of sustainability metrics for case study 2 (zero solvent make-up)

Indicator Type	Indicator	Base Case Design	Alternative Design
Energy	Total Energy Primary Energy Usage Rate = Imports – Exports, GJ/y	3355.8	1080.7
	Total Net Primary Energy Usage Rate per kg Product, kJ/kg	8464.9	2776.1
Material	Total raw materials used per kg product, kg/kg	1.00	1.00
	Hazardous raw materials per kg product, kg/kg	1.00	1.00
Water	Net water consumed per unit mass of product kg/kg	385.7	111.2

4.2.3 Environmental Impact Factors

The calculated stream composition, temperature and pressure of all input and output streams are available, and therefore, they are used by the WAR algorithm option in ICAS to compute the corresponding environmental impact factors (see Table 14 and 15). Only the results for the case of zero solvent make-up are given, as the comparison is similar.

Table 14. WAR algorithm results for the case study (benzene) with zero solvent make-up

Stream No	Total PEI	HTPI	HTPE	ATP	TTP	GWP	ODP	PCOP	AP
Stream 1	32.85	12.37	3.79E-02	2.03E-01	12.37	2.27E-01	0	7.65	0
Input sum	32.85	12.37	3.79E-02	2.03E-01	12.37	2.27E-01	0	7.65	0
Stream 2	10.19	1.27	2.78E-03	2.41E-03	1.27	0.00	0	7.65	0
Stream 4	22.67	11.10	3.53E-02	2.00E-01	11.10	2.27E-01	0	1.82E-03	0
Output sum	32.86	12.37	3.80E-02	2.03E-01	12.37	2.27E-01	0	7.65	0
Impact generated	8.00E-04	1.40E-04	1.64E-04	6.50E-05	1.40E-04	0.00	0.00	2.61E-04	0.00

Table 15. WAR algorithm results for the alternative (2-methyl heptane) with zero solvent make-up

Stream No	Total PEI	HTPI	HTPE	ATP	TTP	GWP	ODP	PCOP	AP
Stream 1	32.86	12.37	3.78E-02	2.03E-01	12.37	2.27E-01	0	7.65	0
Input sum	32.86	12.37	3.78E-02	2.03E-01	12.37	2.27E-01	0	7.65	0
Stream 2	10.19	1.27	2.57E-03	2.89E-03	1.27	0.00	0	7.65	0
Stream 4	22.69	11.10	3.52E-02	2.01E-01	11.10	2.27E-01	0	7.67	0
Output sum	32.88	12.37	3.78E-02	2.03E-01	12.37	2.27E-01	0	15.32	0
Impact generated	2.35E-02	2.88E-03	5.08E-06	4.52E-04	2.88E-03	0.00	0	7.67	0.00

4.2.4 Safety Indicators

The safety indicators calculated through the integrated tool-box are compared for the two process designs with their generated alternatives in Tables 16 and 17. It can be noted that a reduction in the risk with respect to the Chemical Inherent Safety has been achieved in both cases.

Table 16. Calculated safety factors for case study 3 (with solvent make-up)

Chemical Inherent Safety Index, I_{ci}	Score		Process Inherent Safety Index, I_{pi}	Score	
	Ref	Alt		Ref	Alt
Sub-indices for reactions Hazards			Sub-indices for process conditions		
Heat of the main reaction, I_{rm}	0	0	Inventory, I_i	4	4
Heat of the side reactions, I_{rs}	0	0	Temperature, I_t	1	1
Chemical interactions, I_{int}	4	4	Pressure, I_p	0	0
Sub-indices for hazardous substances			Sub-indices for process conditions		
Flammability, I_{fl}	3	3	Equipment, I_{eq}		
Explosiveness, I_{ex}	1	1	I_{Isbl}	1	1
Toxicity, I_{tox}	4	2	I_{Osbl}	2	2
Corrosivity, I_{cor}	1	1	Process structure, I_{st}	2	2
Maximum I_{ci} score	13	11	Maximum I_{pi} score	10	10

Note: Ref - Base Case Design; Alt - Generated Alternative Design

This case study clearly confirms the postulation that improvement of the mass and energy indicators through targeted design changes also (at least) improves the sustainability metrics, the environmental impact and the inherent safety of the process.

Table 17. Calculated safety factors for case study 3 (without solvent make-up)

Chemical Inherent Safety Index, I_{ci}	Score Ref	Score Alt	Process Inherent Safety Index, I_{pi}	Score Ref	Score Alt
Sub-indices for reactions Hazards			Sub-indices for process conditions		
Heat of the main reaction, I_{rm}	0	0	Inventory, I_i	0	0
Heat of the side reactions, I_{rs}	0	0	Temperature, I_t	1	1
Chemical interactions, I_{int}	4	4	Pressure, I_p	0	0
Sub-indices for hazardous substances			Sub-indices for process conditions		
Flammability, I_{fl}	3	3	Equipment, I_{eq}		
Explosiveness, I_{ex}	1	1	I_{Isbl}	1	1
Toxicity, I_{tox}	4	2	I_{Osbl}	2	2
Corrosivity, I_{cor}	1	1	Process structure, I_{st}	2	2
Maximum I_{ci} score	13	11	Maximum I_{pi} score	6	6

Note: Ref - Base Case Design; Alt - Generated Alternative Design

4.3 Case Study 3

The objective of this case study is to highlight the use of the integrated computer aided tools for analysis of a specified process flowsheet with respect to the various indicators. A polymeric resin adsorption process (also studied by Jimenez-Gonzalez et al. 2002), for which important process data are given by (Crook et al. 1975; Fox 1978; Juang and Shiau 1999), has been selected for this purpose.

4.3.5 Process Description

The process involves the removal of phenolic compounds with polymeric resins by physical adsorption, which is carried out in two resin columns, where one is in a loading step while the other is being regenerated. Normally, a 6-h loading cycle is employed, with an average of two cycles per day per resin bed. To regenerate an adsorbent resin loaded with phenols, a solvent system is used.

Traditionally, for the regeneration of the resin and recovery of the solvent, a train of three distillation towers is used. For this specific case study, a 'superloading' process is taken into consideration for the calculations. The resulting solution from the regeneration cycle is distilled at 337 K and a reflux ratio of 1.3 in order to recover the solvent (methanol); and the bottom product is cooled and allowed to separate into two phases. The organic phase with the phenol is recovered (about 72 % - 75 % of phenol in water) and the aqueous phase (with 8 % - 10 % phenol) is recycled to the end of the loading cycle of the resin. All the streams leaving the system are cooled to 298 K. The basis for the steady state mass balance calculations is 1000 kg of wastewater treated with a concentration of phenol of 10 g/l. The process flowsheet is given in Figure 6. The simulated stream compositions are given in Table 18 and a selected set of design and stream variables are given in Table 19 (only those needed to calculate the sustainability metrics).

Table 18. Stream compositions for case study 3

	S1	S2	S3	S4	S5	S6	S11	S12
Water	990 g	992 g	2 g	5.1 g	0 g	0 g	3.1 g	0 g
Phenol	10 g	0.7 g	0.18 g	0 g	0 g	0 g	9.3 g	0 g
Methanol	0 g	3.8 g	3.8 g	0 g	4.6 g	120 g	0 g	0.8 g

4.3.6 Indicators

The calculated stream mass flowrates are given in Table 20 and based on these, the mass indicators have been calculated (see Table 20) for the open and cycle (closed) paths. All the mass indicators are relatively quite small, indicating very little scope for improvements. The only cycle-path with a somewhat higher value is the one related to methanol cycle-path C1, and the reason for this is that methanol is the solvent, and recycling it as much as possible, reduces waste.

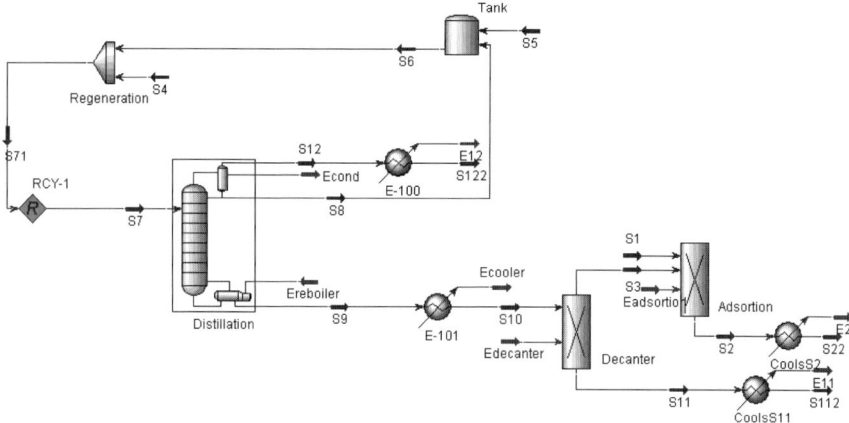

Fig. 6. Process flowsheet for resin based adsorption process

Table 19. Values for selected set of stream and design variables

Process Stream/Design variable	Value
Steam Energy value	
Total raw material used, including packaging, t/y	8077.6
Raw material used which poses health, safety or environmental hazard (described hazard), t/y	116.2
Water used in cooling, t/y	124408.77
Water used in process, t/y	7960.80
Water recycled internally, t/y	3.44
Net water consumed = Total used – recycled, t/y	132366.13

4.3.7 Sustainability metrics

The energy, material and water sustainability metrics are summarized in Table 14. The calculations are based on a 8000 h/year rate of work and a conversion factor of 0.8. The total net primary energy usage rate per kg product has been calculated on the basis of the recovered phenol as the product. As this amount is quite small, although, the total energy required is not especially important, the amount required per kg of product is quite high. For material sustainability indicator, streams S1, S4 and S5 are the raw material (used) streams, which correspond to the main feed through which phenol comes into the process, the water needed for the regeneration and the solvent required (methanol) streams, respectively.

Table 20. Mass and energy indicators for case study 3

Path	Component	Streams	Flow (kg/h)	EWC	AF
O1	Water	S4-S112	2.838	2.975	-
O2	Water	S4-S22	1.830	1.916	-
O3	Water	S1-S22	990.000	0.027	-
O4	Methanol	S5-S122	7.986E-01	0.069	-
O5	Methanol	S5-S22	4.230	1.966	-
O6	Phenol	S1-S22	6.880E-01	0	-
O7	Phenol	Phenol regen-S112	9.300	2.155	-
O8	Phenol	Phenol regen-S22	1.24E-02	0.003	-
09	Water	S4-S122	1.38E-03	0	-
O10	Phenol	S1-Phenol adsorp	9.312	2.149	-
O11	Phenol	Phenol regen- Phenol adsorp	1.677E-01	0.039	-
C1	Methanol		114.970	8.771	22.86
C2	Water		0.430	0.065	9.21E-02

Table 21. Calculated values of sustainability metrics for case study 3

Indicator Type	Indicator	Value
Energy	Total Energy Primary Energy Usage Rate = Imports – Exports, GJ/y	2596.6
	Total Net Primary Energy Usage Rate per kg Product, kJ/kg	34537.63
Material	Total raw materials used per kg product, kg/kg	108.57
	Hazardous raw materials per kg product, kg/kg	1.57
Water	Net water consumed per unit mass of product kg/kg	1779.12

4.3.8 Environmental Impact Factors

The calculated stream composition, temperature and pressure of all input and output streams are available, and therefore, they are used by the WAR algorithm option in ICAS to compute the corresponding environmental impact factors (see Table 22). It can be noted that since all the phenol is recovered and the solvent loss is minimal, this process does not show any appreciable environmental impact.

4.3.9 Safety Indicators

According to the numbers in Table 23, the total inherent safety index is 19. Since the percentage of this value compared to maximum is 36%, it means that for this process, the chemical inherent safety factor (42.8%) is more important. That is, there is a higher risk with respect to the chemical present in the system than the process operational variables.

The Dow Chemical Exposure Index (CEI), which is a measure of the potential for harm to people around the facility, has also been calculated and it is found to be 10.72, while the three Hazard Distance (HD) values are calculated to be 47m, 107 m and 240 m, respectively. The Hazards Distance indicate how far it is generally believed one should be away from the release to avoid permanent injury, disabling injury or minor discomfort respectively.

Table 22. Calculated Environmental factors for case study 3

Stream	Total PEI	HTPI	HTPE	ATP	TTP	GWP	ODP	PCOP	AP
S1	22.63	11.09	0.0153	0.28	11.09	0	0	0	0
S4	0	0	0	0	0	0	0	0	0
S5	1.74	0.028	0.0052	0	0.028	0	0	1.163	0
Input	24.38	11.39	0.161	0.28	11.39	0	0	1.163	0
S22	3.19	1.04	0.016	0.019	1.04	0	0	1.07	0
S122	0.30	0.049	0.0009	0	0.049	0	0	0.02	0
S112	21.05	10.32	0.144	0.26	10.32	0	0	0	0
Output	24.54	11.41	0.161	0.28	11.43	0	0	1.271	0
Impact	0.167	0.027	0.0005	0	0.0267	0	0	0.108	0

Table 23. List of inherent safety sub-index factors for case study 3

Chemical Inherent Safety Index, I_{ci}	Score	Process Inherent Safety Index, I_{pi}	Score
Sub-indices for reactions Hazards		Sub-indices for process conditions	
Heat of the main reaction, I_{rm}	0	Inventory, I_i	1
Heat of the side reactions, I_{rs}	0	Temperature, I_t	1
Chemical interactions, I_{int}	4	Pressure, I_p	0
Sub-indices for hazardous substances		Sub-indices for process conditions	
Flammability, I_{fl}	3	Equipment, I_{eq}	
Explosiveness, I_{ex}	2	I_{Isbl}	1
Toxicity, I_{tox}	2	I_{Osbl}	2
Corrosivity, I_{cor}	1	Process structure, I_{st}	2
Maximum I_{ci} score	12	Maximum I_{pi} score	7

4.3.10 Retrofit Alternative

As stated earlier, the indicator values for this case study are quite small and therefore, the scope for process improvements through changes in design variables are quite small. Two possibilities (not studied in this paper), however, could be to change the solvent and/or the separation technique. This case study illustrates that the indicator-based computer aided tool-box may also be used only for analysis to confirm whether a current design is satisfactory and that changes are not necessary.

5   Conclusions

A systematic and easy to apply computer aided technique for generation and evaluation of sustainable process (retrofit) design alternatives has been developed and validated through three different case studies. The mass and energy based indicators help to identify process alternatives for which sustainability metrics, environmental impact factors as well as inherent safety factors can be made to move in the same direction (towards improvement with respect to the base case design). The application of the methodology requires a combination of tools ranging from databases, to process simulation software, to computational routines for various types of indicators, process synthesis/design tools and many more. An important feature of this technique is that it does not perform any complex calculations but provides feasible retrofit (design) alternatives that show improvements when compared to the base case design. This has been possible because only design variables that are sensitive to the indicators targeted for change in order to generate the retrofit alternatives. At the same time, since the indicators, by definition, are also functions of the same set of variables as the sustainability metrics, environmental impact factors and inherent safety factors, any generated retrofit alternative that improves the indicator values also improves these factors. For the highlighted case studies, it appears that rather than a trade-off between competing factors, the generated retrofit alternatives either improve them or are neutral to them. This means that the process optimisation becomes easier and multiple objectives can be satisfied without resorting to tradeoffs between them. Current work involves developing more case studies so that the question of multiple objectives can be more thoroughly investigated.

References

Anastas PT, Lankey RL (2002) Sustainability through green chemistry and engineering. ACS Symposium Series, 823: 1-11

Andersen NK (2002) An Indicator Based Retrofit Design Method. MSc-Thesis, department of Chemical Engineering, Technical University of Denmark, Lyngby, Denmark

Cabezas H, Bare J, Mallick S, (1999) Pollution Prevention with Chemical process Simulators: The Generalized Waste Reduction (WAR). Algorithm, *Computers Chemical Engineering*, 23 (4-5):623-634

Crook EH, McDonnel RP, McNulty JT, (1975) Removal and recovery of phenols from industrial waste effluents with Amberlite XAD polymeric adsorbents. *Ind. Eng. Chem. Prod.Res.Dev,* 14:113-118

Fox CR (1978) Plant uses prove phenol recovery with resins. *Hydrocarbon Process,* 11: 269-273

Hartwig TA, Xu A, Nagy AB, Pike RW, Hopper JR, Yaws CL (2000) A prototype system for economic, environmental, and sustainable optimisation of a chemical complex. Session 18008 Tools for Sustainability, AIChE Annual Meeting, Los Angeles, Nov 12-17, 2000

Heikkilä A-M (1999) "Inherent Safety in Process Plant Design – An Index-based Approach". Ph.D-Thesis, VTT Automation, Espoo, Finland

Hilaly AK, Sikdar SK (1994) Pollution Balance: New Methodology for Minimizing Waste production in Manufacturing Processes. *J Air & Waste Management, Association*, 44:1303-1310

Hostrup M, Harper PM, Gani R (1999) Design of Environmentally Benign Processes: Integration of Solvent Design and Process Synthesis. *Computers & Chem Eng.*, 23:1394-1405

ICAS Documentations (2002) CAPEC Internal Report. PEC02-23, Technical University of Denmark, Denmark

Jimenez-Gonzalez C, Constable DJC, Curzons AD, Cunningham VL (2002) Developing GSK´s Green Technology Guidance: Methodology for Case-Scenario Comparison of Technologies. *Clean Technologies and Environmental Policy,* 4(1): 44-53

Juang RS, Shiau JY (1999) Adsorption Isotherms of Phenols from Water onto Macroreticular Resins. *Hazardous Mater,* 70: 171-183

Kharbanda OP, Stallworthy EA (1988) Safety in the Chemical Industry: Lessons form Major Disasters. Heinemann, London, UK

Koller G, Weirich D, Brogli F, Heinzle E, Hoffmann VH, Verduyn MA, Hungerbuhler K (1998) Ecological and economic objective functions for screening in integrated development of fine chemical processes.2. Stream allocation and case studies, Industrial & Engineering Chemistry Research, 37(8): 3408-3413

Lange JP (2002) Sustainable development: Efficiency and recycling in chemical manufacturing. Green Chemistry, 4(6): 546-550

Marrero J, (2002) ProPred User's Manual. PEC02-15, CAPEC, Technical University Denmark

PRO-II Users Guide (2002) SimSci an Invensys Company, Brea California

Seider WD, Seader JD, Lewin DR (1999) Process Design Principles, Chapter 3, John Wiley & Sons, Inc., USA

Tallis B (2002) Sustainable Development Progress Metrics. IChemE, Sustainable Development Working Group, IChemE, Rugby, UK

Uerdingen, E., 2002, Retrofit design of continuous chemical processes for the improvement of production cost-efficiency. PhD-Thesis, ETH-Zürich, Switzerland

Uerdingen E, Gani R, Fisher U, Hungerbuhler K (2003) A New Screening Methodology for the Identification of Economically Beneficial Retrofit Options for Chemical Processes. *AIChE Journal* (in press)

Young D, Scharp R, Cabezas H (2000) The Waste Reduction (WAR) Algorithm: Environmental Impacts, Energy Consumption and Engineering Economics. *Waste management*, 20: 605-615

Pollution prevention and environmental management systems – tools to obtain a sustainable development

Victor Teodor Petcu Nitica, Vladimir Gheorghievici

Pollution Prevention Center, Theodor Sperantia Str. 98, Bl. S28, Sc. 1, Ap. 10, Bucharest 3, RO-74316, Romania, Tel: +40 (21) 327 47 95, Fax: +40 (21) 327 47 96, E-mail: cpp@pcnet.ro

1 Introduction

The adverse environmental impacts of rapid economic development issued a warning that the developing countries face grave environmental dangers if they emulated the current economic model of the developed world. This model, which has been consuming natural resources at an ever-increasing rate, is not sustainable.

From this point of view the reduction of the natural resources consumption rate can transform this not sustainable economic development pattern into a sustainable one. For the reduction of the adverse environmental impacts and the consumption of natural resources a very efficient tool is the pollution prevention. The pollution prevention programs can reduce the pollution at source, reduce the specific natural resource consumption and increase the economic benefits of the production processes. Environmental management systems, which include as management programs the pollution prevention programs, can improve the results of these programs and assure, if used properly, a continual improvement of the environmental performance of the companies.

Therefore the pollution prevention programs are an efficient tool of sustainability. The efficiency of this tool can be enhanced by a suitable environmental management system that includes these pollution prevention programs.

2 Clarifying the terms

It is important to define very precisely the concepts of pollution prevention, cleaner production, eco-efficiency and sustainable development to avoid possible misunderstandings. Therefore we will present subsequently the definition of these concepts.

Pollution prevention means "source reduction" and other practices that reduce or eliminate the creation of pollutants through:
- increased efficiency in the use of raw materials, energy, water, or other resources, or
- protection of natural resources by conservation.
- The Pollution Prevention Act defines *source reduction* to mean any practice that:
- reduces the amount of any hazardous substance, pollutant, or contaminant entering any waste stream or otherwise released into the environment (including fugitive emissions) prior to recycling, treatment, or disposal
- reduces the hazards to public health and the environment associated with the release of such substances, pollutants, or contaminants.

Under the Pollution Prevention Act, recycling, energy recovery, treatment, and disposal are not included within the definition of pollution prevention. Some practices commonly described as "in-process recycling" may qualify as pollution prevention.

The international community has adopted the term *cleaner production*. As you can see from the definition of cleaner production, it has a broader meaning than the one we give to the term P2. The final term eco-efficiency is used extensively in the sustainable development arena and it is also defined subsequently.

Cleaner production is the continuous application of an integrated preventative environmental strategy applied to processes, products, and services. It embodies the more efficient use of natural resources and thereby minimizes waste and pollution as well as risks to human health and safety.

It tackles these problems at their source rather than at the end of the production process; in other words, it avoids the "end-of-pipe" approach.

For processes, cleaner production includes conserving raw materials and energy, eliminating the use of toxic raw materials, and reducing the quantity and toxicity of all emissions and wastes.

For products, it involves reducing the negative effects of the product throughout its life-cycle, from the extraction of the raw materials through to the product's ultimate disposal.

For services, the strategy focuses on incorporating environmental concerns into designing and delivering services.

Eco-efficiency is the efficiency with which ecological resources are used to meet human needs. It is expressed as the ratio of an output – the value of products and services produced by a firm, a sector, or the economy as a whole – to the "input" – the sum of environmental pressures generated by the firm, sector, or economy. Measuring eco-efficiency depends on identifying indicators of both input and output.

The World Business Council for Sustainable Development (WBCSD) considers that eco-efficiency places seven demands on a firm:
1. Reducing material intensity of goods and services.
2. Reducing energy intensity of goods and services.

3. Reducing toxic emissions.
4. Enhancing material recyclability.
5. Maximizing sustainable use of renewable resources.
6. Extending product durability.
7. Increasing the service intensity of goods and services.

All three of these terms afore defined - pollution prevention, cleaner production, and eco-efficiency - address:
1. Elimination of process losses at the source without resorting to end-of-pipe pollution control devices.
2. Conservation of resources (including energy, materials, and water) that are used in the process or operation.

There are also some differences between these terms. For example, **eco-efficiency** looks at maximizing the sustainable use of renewable resources while **cleaner production** focuses on the more efficient use of natural resources. **P2** looks at the protection of natural resources by conservation. All of the definitions address hazards to public health and the environment and seek to reduce toxic emissions and the use of toxic raw materials. However, only cleaner production addresses the need to consider whether there is a shift in risk from the environment to worker safety as a result of changes made in the process.

Eco-efficiency and cleaner production address processes, products, services, and life cycle issues. P2 considers "in-process recycling" while eco-efficiency considers "enhancing material recyclability".

P2 plays an important role where the goal is ***sustainable development***. There are many definitions of sustainable development. The following definitions provide broad and operational perspectives to cover the range of components that are commonly included under the sustainability umbrella. According to the World Commission on Environment and Development, *"sustainable development is a process of change in which the exploitation of resources, the direction of investments, the orientation of technological development, and institutional change are all in harmony and enhance both current and future potential to meet human needs and aspirations"*. An operational definition of sustainable development is *"good stewardship of natural resources such that long-term productivity may be maintained or improved with minimal, if any, adverse impacts on the environment and worker health and safety"*.

3 Integrating the pollution prevention programs into core business practices

Seldom there are organizations that don't have already in place other types of *"prevention"* programs. Such programs are: environmental management systems, quality management systems, preventive maintenance programs, health and safety programs, insurance and risk management. Therefore it is often easier to tie P2 programs to any of these in place programs. The integration of the P2 program into existing core business practices can help small organizations find resources to start a new P2 program and large organizations consolidate existing programs, allowing each to remain competitive in the global marketplace as they implement P2.

The implementation of P2 projects can yield some modest, immediate benefits. However, the big payoff from P2 often requires a program that is integrated into the operations of the organization and supported for a minimum of two to three years. Like quality, P2 is a mindset that needs to permeate into the culture of the organization. One of the greatest P2 myths is that a P2 program is a "quick fix" used to turn around organizations. Many P2 programs do not offer instant financial success. P2 is a long-term effort with *both* long- and potential short-term bottom-line benefits.

P2 success requires full financial support as well as management commitment. Resources that will be needed include funds, people, training, facilities, support structure, and, in some cases, the adoption of new technology. Often projects that are already funded can be turned into P2 projects by emphasizing different aspects.

3.1 Environmental management systems

One popular EMS format, known as ISO 14001, has been issued by the International Organization for Standardization (Geneva, Switzerland). ISO 14001 is a management system standard, not a performance standard, providing a general framework for organizing the tasks necessary for effective environmental management. This approach may prove effective in encouraging the organization to take an active, preventive, and systematic approach to managing its environmental impacts.

An EMS protocol requires the organization to consider the prevention of pollution, compliance with all legal requirements, and continual improvement. Like P2, an EMS seeks to integrate environmental concerns into core business practices.

3.2 Quality Initiatives

Quality initiatives focus on preventing defects in processes, products, and services. These initiatives often declare a "war on waste". However, too few also consider air emissions, water discharges, solid and hazardous wastes, and spills and leaks to be a waste. Organizations develop ISO 9000 programs to deal with quality. ISO 9000 programs are prepared in the same format as the ISO 14001 program.

Quality initiatives have evolved just as P2 has been defined and refined. Many people have less than fond memories of certain management fads like "Total Quality Management (TQM)". Despite the approaches and fads that cycle in and out, most organizations would agree that quality refers to everything an organization does to provide goods and services that meet customer requirements, the way that organizations employees interact together, and the organization's expectations of its suppliers and other interested parties. Developers of P2 programs should become familiar with the quality improvement initiatives in the organization.

Some organizations use the Baldrige criteria to judge their overall operating performance. The Malcolm Baldrige National Quality Program is the Presidential Award program in the United States. These performance-based criteria are currently used in approximately 50 countries and 44 of the 50 US states to help improve competitiveness in both manufacturing and service businesses. An environmental excellence program has been developed in New Mexico using the Baldrige model. This Green Zia Program is used to rate organizational environmental programs that "go beyond mere compliance". This program helps an organization establish core values for its program and demonstrates how quality and P2 can be effectively integrated. A set of criteria and a rigorous scoring system allow any organization to track and search for trends in its continual improvement using a unitless score. This eliminates the need to "normalize" for production.

3.3 Preventive and predictive maintenance

Preventive and/or predictive maintenance is designed to keep machinery from breaking down. Unscheduled equipment downtime often leads to the generation of wastes in organizations. The principles from this field are applicable to P2 programs.

3.4 Health and safety

Many environmental managers are gaining some oversight of the safety function in their organizations. Organizations track safety closely because it impacts worker compensation rates and related insurance costs. P2 training and safety training are often combined in organizations to stress the prevention message. Safety has always had its focus on *preventing* incidents and exposures.

3.5 Insurance and risk management

Insurance companies and organization risk management professionals frequently audit organization processes and facilities to prevent property loss and other forms of insurable risk. P2 programs should collaborate with risk management personnel, whether in the company or sent by the insurance company.

4 Pollution prevention tools

The implementation teams can use a broad variety of problem solving and decision making tools available. These tools have been used in quality programs for over half a century. The most used tools system approach tools and checklists.

4.1 System approach tools

An organization acts as a *system* that functions as a whole through the interaction of its parts. The Systems Approach looks at the whole organization, and the parts, and the connections between the parts. The functionality of the parts depend on how they are connected, rather than what they are. The parts of a system are all connected directly or indirectly. Therefore, a change in one part affects all the other parts. Given this interdependence, tools that address the complexity of organizations are important. These tools include:
- Process mapping;
- Determining the cost of loss;
- Pareto analysis;
- Root causes analysis;
- Brainwriting;
- Criteria matrix (selection grid);
- Forced pair analysis (bubble down, bubble up analysis);
- Action planning.

The process mapping tools helps both the implementation team members and the company management understand the processes that take place in the company and identify P2 opportunities. The determination cost of loss provides important information to the Pareto analysis that helps ranking the P2 opportunities identified in the process mapping phase.

The "root cause" analysis of the resource use or of the loss helps identifying the basic reason that a resource is used or a loss occur. If the cause can be eliminated the resource use or the loss can be prevented. This approach is the very basis of the P2. This tool helps identifying the action that would best solve the problem and avoid to take most obvious action.

To solve a P2 problem is important to have many alternative solutions. Thus the generation of these alternatives it is important for the success of the P2 program. A very useful tool of generating these alternative solutions is the brainwriting technique, similar to the brainstorming but less restrictive. This technique use the writing of the alternative solutions of the P2 problem on the paper and then summing all up to have a final list of possible solutions. To be successful the team members should have the appropriate information, especially the one obtained during a root cause analysis.

After the list of possible alternatives is compiled these alternatives have to be ranked. The ranking can be obtained using either a criteria matrix or a force pair analysis. The second tool compares only two alternatives at a time and tends to be more easy to use. If an alternative is more likely to solve the problem it will be ranked above the one it is compared. The comparing ends when the new alternative encounters an old alternative (ranked previously) that is more likely to solve the problem.

After the alternatives to be implemented are selected can be used the action plan tool that provides information about how and who will implement this alternative, providing deadlines and responsibilities.

5 EMS approach to P2 implementation

The new international voluntary standard for environmental management systems (EMSs) known as ISO 14001 is proving to be an effective tool for improving organizational environmental performance and implementing P2 opportunities. The intent of the standard is to establish and maintain a systematic management plan designed to continually identify and reduce the environmental impacts resulting from an organization's activities, products, and services. Currently, no government mandate requires organizations to have a comprehensive EMS, but several states are exploring the effectiveness of having organizations use an EMS in implementing and complying with P2 planning requirements.

The principles of an effective EMS are as follows:
- EMSs should improve compliance with environmental laws, enable organizations to achieve performance "beyond compliance" with legal requirements, and reduce environmental impacts from both regulated and unregulated activities.
- An EMS can serve as a supplementary tool that enables regulatory agencies and others to jointly achieve greater environmental protection.
- The quality of an EMS is linked to environmental performance achieved.
- EMS metrics can document improved environmental performance, which may enable regulatory agencies to achieve policy objectives more efficiently and improve communications with the public.

A growing number of organizations have pioneered new strategies for integrating environmental management into their overall business strategy. Although regulatory compliance remains an important driver of environmental performance and of the adoption of advanced practices, business factors such as cost savings and improved business performance are just as important.

EMSs are motivating organizations all over the world to reconsider their environmental performance and effectiveness and determine how P2 strategies can help them reduce wastes, risks, and costs. These organizations should establish and maintain a systematic management plan that promotes P2 and is designed to continually identify and reduce the environmental harm (impacts) created by the organization's activities, products, and services.

The EMS fosters innovative strategies and a framework for improving environmental performance by encouraging all the employees of the organization to look for ways to reduce environmental impacts by first using P2 techniques.

The goal of the ISO 14001 standard is to establish a common approach to EMSs that is internationally recognized, leads to improved environmental performance, and provides an opportunity for gaining international recognition and market share. ISO 14001 is a management system standard, not a performance standard. Given that ISO 14001 is a system built for industry by industry, it uses a language that management understands, and it will keep top management's attention through involvement. The EMS provides a systematic approach for integrating environmental protection into all business functions and management strategies.

The EMS is designed to continually improve system and environmental performance through creation of an environmental management program (EMP). The EMP is the last element of the EMS planning phase. It sets up action items, assigns responsibilities at all levels of the organization for plan execution, sets specific time lines, and determines the resources needed for implementation to achieve the objectives and targets. With the goals established, the subset of activities defined, and the accountabilities in place, each person with specific responsibilities must now develop EMPs for implementation. One person or several people are assigned the accountability for meeting the goals and objectives in the planned time frame for each task in the action plan and for maintaining the current level of performance on each of these items.

Although setting objectives and targets is treated as a separate function from EMPs in the planning phase, they are related. You have to have an idea of how you will accomplish an objective and target before you set it up as a program in your system. This is the process many organizations now use in their P2 planning effort to accomplish specific projects. After P2 assessment and planning, projects are initiated to implement technically and economically feasible P2 opportunities. Without the continual improvement component of the EMS, however, P2 planning and implementation may be an end point instead of the ongoing process of setting new objectives and targets for other aspects that impact the environment.

The EMS is based on a documented and clearly communicated policy that includes three distinct guiding principles: compliance with applicable environmental requirements, prevention of pollution, and a commitment to continual improvement in environmental performance. In some cases, organizations' environmental policies, especially corporate policies, may have become too long and broad to be understood easily by employees and the public. An organization's EMS policy needs only to focus on the three guiding principles and to drive the accomplishment of the EMS's objectives and targets through training and involvement.

An EMS identifies, translates, and communicates applicable environmental and voluntary requirements to affected employees, suppliers, and contractors. Voluntary requirements may include those addressing P2, company or corporate initiatives, health, process safety management (PSM), and sustainable development.

The EMS specifies procedures for how compliance will be achieved and maintained organizationally. For example, it defines the compliance roles and responsibilities of environmental managers, establishes how they and management will be held accountable for achieving and maintaining compliance, and describes how environmental performance and compliance information will be communicated to relevant employees, suppliers, and contractors. The EMS establishes a mechanism for receiving and addressing environmental and compliance concerns raised by individuals, organizations, or other interested parties.

The EMS includes procedures for identifying changes to applicable environmental requirements – including new ones that may apply as a result of process or material changes – and addressing these changes through the EMS process. For those organizations that are already performing environmentally, the EMS should establish objectives and targets that promote leadership and ensure continued achievement of compliance.

Identifying all aspects and determining their significance is usually the largest gap in most organizations' current environmental systems. The EMS establishes and maintains a procedure to identify all of the environmental aspects of the organization's activities, products, and services that it controls and influences. Current procedures to identify existing process waste streams and review new customer work requests can be used as starting points for identifying all aspects. Also, a procedure to identify which of these aspects have significant impact on the environment is needed, and significant impacts must be considered in setting objectives.

Many organizations focus almost exclusively on negative environmental impacts. Positive environmental impacts are also important. These might include company-sponsored community recycling programs and household hazardous waste collection days. An EMS can develop approaches to procurement, processing, and delivery that reduce or minimize significant environmental impacts for organizations, customers, and interested parties.

An EMS establishes specific objectives, targets, and time frames for implementing P2 initiatives, improving environmental performance, and maintaining compliance. These should be documented and updated. An EMS ensures that the organization has skilled employees and financial and technical resources to achieve its objectives and targets and maintain compliance. In setting objectives and targets for each relevant job within the organization, it is important to consider pollution prevention goals; any additional significant impacts; legal and other requirements; technological options; financial, operational, and business requirements; and views of interested parties. These considerations are important in EMS planning and are used for capital improvement decisions, product and process design, training programs, and maintenance activities.

The organization establishes environmental management programs (EMPs) to achieve its EMS objectives and targets. EMP requirements specifically include designation of responsibility for actions and the means and time frame by which the objectives are to be achieved. The EMP must review new activities, products, equipment, or services and address environmental changes through the EMS. For measuring performance-based improvement, targets must be quantifiable and use metrics that are related to the organization's overall goals. Most organizations have set some quantitative goals for various process waste streams, for example, reducing sludge production 10% by 2002 based on amount of wastewater treated. The EMP establishes the frequency at which the objectives and targets will be reviewed.

Organizations are discovering that their investments in EMSs are leading to improved environmental performance and compliance with benefits for the environment and community. An EMS provides a good method for establishing and implementing a P2 program. To achieve maximum environmental benefits, the EMS should embody the "plan, do, check, and act" model for continual improvement. This model ensures that environmental impacts are systematically identified, controlled, and monitored. The EMS helps ensure more consistency by organizations in achieving and maintaining compliance, promoting results-oriented efforts, and attaining more reliable data on environmental performance. Effective use of an EMS can be viewed as a demonstration of environmental responsibility and leadership by organizations. An EMS provides the basis for collaborating with regulatory agencies to enhance suitability and effectiveness and promote a leadership, performance-based system.

6 Conclusion

As aforementioned the pollution prevention is very important when the goal is a sustainable development. Pollution prevention programs used as core business practices fits the definition of a sustainable development being the tool to implement the change in business practices. That change from the expansive mindset to the intensive mindset in the production growth can provide, by means of the continual improvement required by the P2, a sustainable increase of the production. Most developing countries have a specific consumption (the ratio between the consumption of resources to obtain the final product and the final product output) of natural resources much higher than the developed countries, thus there is a wide margin of growth without increasing the natural resource consumption.

The developed countries have the lion's part from the natural resource consumption and the environment pollution. In this case the goals of "near zero" waste or "near zero" emissions can be obtained by means of pollution prevention, cleaner production or eco-efficiency initiatives.

As was presented before there are several tools to use in the P2 initiatives but them, as a whole, there are also tools for sustainable development. As these initiatives are integrated in the core business management practices, mainly in the environmental management systems, transforms them into tools of sustainable development.

Along with eco-accounting, eco-efficiency, eco-ranking, pollution prevention and environmental management are effective tools to reach the goal of sustainable development both in the developing and developed countries.

Acknowledgments

This paper is inspired from the lessons learnt by means of direct participation at the implementation of P2 programs and EMS in different Romanian industrial companies.

We want to mention the important role of:

- WEC and USAID in supporting technically and financially promotion of Waste Minimization/Energy Conservation/EMS training programs, for industrial manager and experts, and ministries staff, including the technical and financial non reimbursable assistance program for the establishment of PPC as a follower for these type of activities
- UNEP/UNIDO for its ongoing assistance for professional and expertise improvement/increasing for the PPC's staff, trough training and delivering of specific technical materials and advising on important matters and involving it in international programs
- Robert Ferrone, member of the US 207 Technical Committee for ISO 14001, for his excellent contribution to PPCs staff training on project management and ISO 14001 implementation integrated into overall business management of the industrial companies and USAID for its substantial financial support for an one year long distance training program with six demonstration workshops, each one week
- CEMC and Czech Government for its support and delivering of substantial training and financial support to the PPC for the implementation of POEMS EE Program (pollution prevention and EMS implementation in industrial companies with focus on SMEs)
- Prof. dr. Robert B. Pojasek in the implementation of a cleaner production program in an industrial company. Based on his experience he helped us to focus on the issues of the implementation and use a very systematic work procedure and learn important tools that make more straightforward the implementation and the maintenance of such programs.

We also want to thank to the members of the implementation teams that worked very hard to implement these programs based on our consulting and supervision.

References

Pojasek RB (1996) Identifying P2 alternatives with brainstorming and brainwriting. Pollution Prevention Review, 6(4), pp 93-97

Pojasek RB (1996) Using cause and effect diagrams in your P2 program. Pollution Prevention Review, 6(3), pp 99-105

Pojasek RB (1997) Materials accounting and P2. Pollution Prevention Review, 7(4), pp 95-103

Pojasek RB (1997) Understanding a process by using process maps. Pollution Prevention Review, 7(3), pp 91-101

Pojasek RB, Metcalf C (2001) An Organizational Guide to Pollution Prevention. EPA/625/R-01/003, USEPA

Sustainability Metrics

Technology sensitive indicators of sustainability

Andrzej Wasiak

Bialystok Technical University, Wiejska 45 A, Bialystok, Poland and Supreme School of Economics, Choroszczańska 31, Bialystok, Poland. E-mail address: awasiak@pb.bialystok.pl

Abstract

The idea of sustainable development was introduced as a remedy that should improve the situation of environment, should balance living conditions for human beings in various countries and continents, and finally should lead to rational exploitation of natural resources.

The contradiction between needs and ambitions of human societies and the need for conservation of environmental resources can only be, at least partially, removed by change of technologies being used in industrial processes. Only use of modern technology can assure growth of production with simultaneous reduction of consumption of energy, raw materials and reduction of environmental damage.

A number of sustainability indicators have been already proposed and recommended by several international organizations. It should be considered, however, that even those quantitative ones have to be considered as variables consisting s. c. affine space. It means that having two or more states, described by appropriate values of the indicators, we cannot determine distances between those states. From this point of view it is also clear that changing some of the variables we are unable to determine distance between the old and a new position of particular state.

Consequently an attempt should be made to constitute s. c. metric space of variables that would describe, at least most important aspects of sustainable development.

In the second step a trajectory that could be considered as at least approximate realization of sustainability, should be defined. Preliminary attempts and discussion of possible formulations of the model will be presented in the paper.

1 Introduction

Analyzing the idea of sustainable development, the definition of which is usually given in verbal form, one might attempt to specify it in more quantitative manner. It is not an easy task because many qualitative factors are usually mentioned as attributes of sustainable development.

At first, it should be development, i.e. a time dependent process that is associated with economic progress, and with an increase of wealth of individuals and organizations, it is also connected to an increase of industrial and agricultural production in order to fulfill increasing needs of the whole community.

It also should sustain for a long time within unspecified limits of time, presumably always, what can be interpreted as "to the infinity".

Moreover, this growth or development should occur under requirement to reduce consumption of natural resources, in order to protect them for the next generations. This process should also occur in the universal, global scale, in a manner balanced between nations, regions and individual people.

This development should also comply with market driven economy, smoothly shifting towards knowledge driven one.

Further requirement is that the development should not affect quality of environment, which may be exploited only within the limits enabling its self-renovation, conservation of bio-diversity, landscape, etc.

All the requirements, mentioned earlier, make an impression that the exact realization of the idea is nearly impossible. Therefore, it can be considered as an idealized model that can be built only within some approximate limits. Such assumption brings strong need to specify those limits as well as to measure the distance between really achieved course of the development of particular community and the sustainable model. Consequently, it leads to postulation of various indicators and measures of sustainability. Strong needs for sustainability indicators appears also in problems of technology transfer, especially in situations when new technology is being implemented to manufacturing systems located within ecologically sensitive areas, and impact on environment has to be minimized and balanced with economic and technological gains.

Although technology and its progress are important factors determining economic growth, their importance was not recognized in economic theory till quite recent years (Solow 1956, 1957, 1960; Uzawa 1961). Relations, however discussed in the theory above mentioned, consider only the effects of technological progress on overall economic growth, completely neglecting any environmental impacts. It is clear that diverse impacts of particular technology on economic, social or environmental aspects can be expected. These impacts might also depend upon the local situation (population, character of environment, geographic position, etc.). Consequently a set of characteristics enabling impact assessment is needed in order to support decisions in the field of environmental and industrial policy.

2 Sustainability indicators

This paper is not aimed towards reviewing and discussion of existing indicators of sustainability, however it should be briefly mentioned that a number of such indicators were already proposed, and are discussed in various papers, books and international documents. Initially they were based on social wealth indicators, like System of National Accounts (Stone 1956), or on System of Material Product Balances (Borys 1999). Both approaches already considered as controversial when applied to the estimation of sustainability. Later the systems of indicators, taking into account various aspects of social life including ecology, became subject of investigations carried out under support of international organizations like UNDSC (United Nations Commission on Sustainable Development) or OECD (Organization for Economic Cooperation and Development). As a result of those activities, more than a hundred of indicators were proposed (EC 1996; UNEP 1994; UNSCD 1996a, 1996b, 1996c) to characterize sustainability of social and economic development. Interesting and important issues were also presented by several industries defining individual routes for improving eco-efficiency (Jimenes-Gonzales et al. 2002; Saling 2002)

3 Rate processes

The time dependent processes frequently occur in nature, and are discussed and described in various branches of science e.g. physics, chemistry, biology, etc. Those processes may proceed along various schemes, depending upon the nature of the process, boundary conditions and driving forces that cause particular process. Usually some form of mathematical description can be offered for each process, enabling quantitative analysis of the behavior of the system. Several examples seem to be worth of consideration. The simplest case is linear growth that can be represented by mathematical formula:

$$x(t) = \alpha + \beta\, t \qquad (1)$$

where α and β are constants, and $\beta = dx(t)/dt$ is called the rate of the process. Graphically such linear process of growth is shown in Fig. 1.

Ideally, such process may begin from zero ($\alpha = 0$), and proceed to infinity, what in some aspects could be close to the idea of the sustainable development. In the case, however, when such equation is used to describe any real process, in which some material or financial objects are converted into others, such growth is possible only when infinite amount of substrate (raw material) from which the product is formed is available, or when spatial limits causing hindrances at some stage of progress do not exist.

Other imaginable situation can be described by slightly more complex equation of the form:

$$x(t) = \alpha + \beta\, t^n \qquad (2)$$

Graphical representation of such process, which sometimes is called as self-accelerating one, is shown in Fig 2.

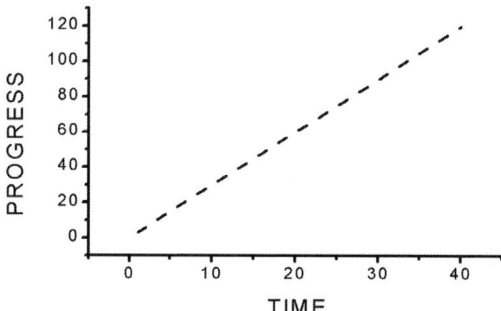

Fig. 1. Linear (constant rate) time dependent process

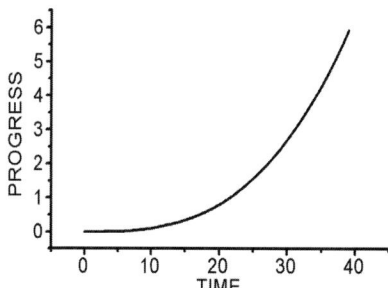

Fig. 2. Progress of self-accelerating process as a function of time

The rate of the process in this case is an increasing function of time, and is represented as:

$$\dot{x}(t) = \frac{dx(t)}{dt} = n\,\beta\,t^{n-1} \qquad (3)$$

while graphical representation of the time dependent rate of the process is shown in Fig. 3.

It is clearly visible that both the progress of the process as well as its rate is increasing functions of time. Such process, when realized in nature must stop at some instant of time or finish with some kind of catastrophe, e.g. explosion, and consequently cannot be considered as sustainable one.

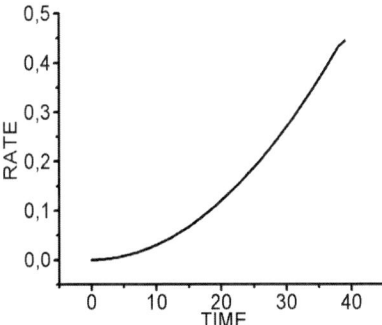

Fig. 3. Time dependence of the rate of self-accelerating process

The other type of time dependent process, that occurs in nature when either resources or space for the product, or both are limited. Such a process is represented by an equation:

$$x(t) = 1 - \exp(-k\,t^n) \tag{4}$$

The plot of above function is presented in Fig. 4.

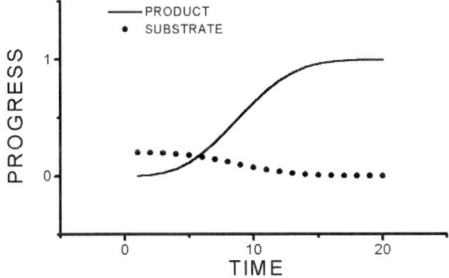

Fig. 4. Plot of the process accelerating at the beginning and decelerating close to the end

The plotted curves represent progress (an increase) of the product and a decrease in the amount of substrate (resource) available for transformation into product.

A rate of such a process is expressed by the equation:

$$\frac{dx(t)}{dt} = [1 - x(t)] \times nkt^{n-1} \tag{5}$$

showing that the rate depends not only on time, but also on the available amount (concentration) of substrate.

Time dependence of the rate is plotted in Fig. 5. The plot shows an initial increase of the rate followed by a decrease after reaching a maximum.

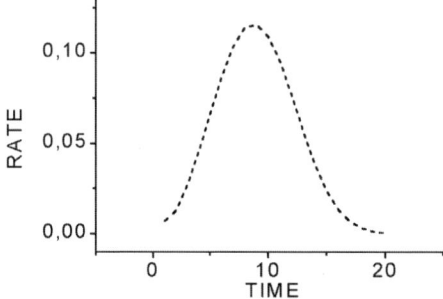

Fig. 5. Time dependence of the rate of the process occurring in conditions of limited resources or limited space

The later process is of special interest for the present discussion. It is because in economic, social or technological processes occurring in reality in the most frequent situations meet constrains of spatial nature or limited resources. Since the rate of the process diminishes to zero at sufficiently long time the requirement of sustainability is not fulfilled for such processes. It can be, therefore, raised a question what could be done to improve the situation and convert such processes into more sustainable.

Direct consequences of above reasoning bring to two imaginable solutions:
- removing constrains
- modifying the process or product

Neither the first nor the second task seems to be simple. However both can be considered.

Focusing our interest on technical aspects of the processes the meaning of removal of constrains might concern both aspects of the process.

Constraints caused by a decrease of available resource might be removed or lowered by diminishing consumption of resources through application of the more efficient technology that use less raw materials, less energy, etc. in relation to the same amount of a product.

Constraints resulting from the "spatial" limits might be reduced by removal of the excess of the product. Products already worn or obsolete may be considered as a new resource, new raw material. This clearly means collecting, and recycling of used products, that not only cleans the "space" for new products (at this moment we focus ourselves only to replacement of exactly the same product), but also replaces virgin raw material by the new (recycled) one. It should be mentioned here, that the purchase, and especially disposal of the products is connected with "scattering" of the material. In consequence it increases entropy associated with this material. Collecting this material back, and reuse or recycling always requires additional energy. Consequently the life cycle of the material cannot support itself without input of energy. In fortunate cases, the amount of energy needed for recycling might be lower than that needed for production of the same material from natural resources, but it is not always true.

The second mechanism, mentioned above, takes into account all kind of eventual innovations that modify the product or the process of its production. In such a case, production of new innovative product can be considered as continuation of the same process of economic development. The hypothetical course of such process is shown in Fig. 6, where several sigmoidal processes are superimposed leading to much longer lasting process of continuous growth can be represented by an equation:

$$x(t) = \sum_{i=1}^{m} A_i \exp\{k_i(t-t_i)\} \quad (6)$$

where A_i – represent intensity, k_i – rate coefficient, and t_i – starting time of individual process.

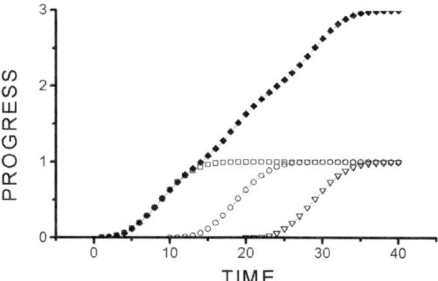

Fig. 6. The progress of several consecutive sigmoidal processes

(Similar curves as presented above have been discussed by Foster (Foster 1986) who referred them to the rate rather than to the progress of the process.)

When dealing with technological processes one should keep in mind that any innovation leading to replacement of technology is never free of negative environmental or social aspects. The assessment of all kinds of impacts caused by a new technological process requires various indicators or measures that we would like to express whether or not given innovation bring us closer to sustainability. In this place we meet another problem. Namely, various indicators based on financial, social. technical, etc data and expressed in very different forms cannot be incorporated into one metric space, constituting only the s.c. affine space. The difference between both types of spaces is schematically presented in Figures 7 and 8. The most important feature of the metric space is that the distance between two arbitrarily chosen points is always defined and measurable (as we used to have in our physical space). In contrast, his feature is lacking in the affine space.

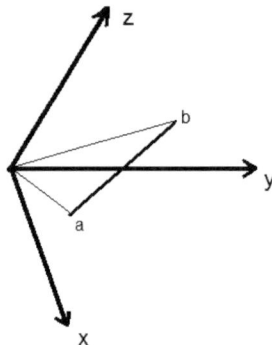

Fig. 7. Schematic representation of a metric space

It means that even having quantitative measures along some vectors defined by particular measures we are not able to determine whether or not those measures really represent any progress that could be considered as an improvement of sustainability.

Examples of the use of affine spaces are well known in many technical applications, and have proved their usefulness as tools enabling to draw various conclusions. It should be kept in mind, however, that distances between points specified on various axes are not defined, and cannot be determined.

Concerning measures of sustainability, an application of measures and indicators constituting an affine space give only possibility for qualitative reasoning and subjective judgments

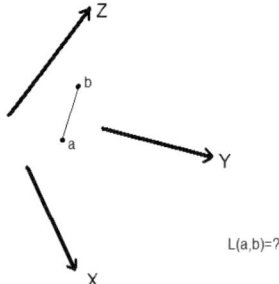

Fig. 8. Schematic representation of an affine space

In spite of the difficulty mentioned above in several cases well chosen indicators have been shown to characterize a local situation leading to useful conclusions. Several examples of such attempts, have been described in the literature, and are also given in other chapters of this book e.g. (Fet 2003; Saling 2003). It should be kept in mind that in any case a chosen set of indicators may be not complete enough, and some of the impacts might be overlooked.

In addition to the approach that consists in defining a set of indicators, each one characterizing particular aspect or impact caused by the activity studied, the other route can be offered basing upon analysis of fundamental physical, biological, etc. driving forces that govern particular process and may provide appropriate characteristics. An attempt at such approach is presented by Lems (Lems et al. 2002) through thermodynamic characterization of sustainability of an industrial process.

Acknowledgment

NATO CCMS support enabling Author's participation in Maribor conference is acknowledged. The work was supported in part by the European Community grant: EVG3-CT-2002-80005 CSDEM and the BTU grant W/WZ/2/02

References

Borys T (ed) (1999) Wskaźniki Ekorozwoju (Eco-development indicators). Bialystok, Poland
EC (1996) Environmental Indicators and Green Accounting. Brussels
Fet AM (2003) Eco-efficiency reporting exemplified by case studies. This book.
Foster R (1986) Innovation: The Attacker's Advantage. Summit Books, New York (cited after Affuah A (1998) Innovation Management. Oxford Univ Press, New York, Oxford)
Jimenez-Gonzales C, Constable DJC, Curzons AD, Cunningham VL (2002) Developing GSK's green technology guidance: methodology for case-scenario comparison of technologies. Clean Techn. Environ Policy 4, pp. 44–53
Lems S, Van der Kooi HJ, de Swan Arons J (2002) The sustainability of resource utilization. Green Chemistry 4, pp. 308–313
Saling P (2003) Realizing More Sustainable Products and Processes in Different Fields of Business by application of the Eco-efficiency Analysis. This Book.
Saling P, Kicherer A, Dittrich-Kramer B, Wittlinger R, Zombik W, Schmidt I, Schrott W, Schmidt S (2002) Eco-efficiency Analysis by BASF: The Method. Int. J. of Life Cycle Assessment 7, No. 4, pp. 203 – 218
Solow RM (1956) A contribution to the Theory of Economic Growth. The Quarterly Journal of Economics MIT
Solow RM (1957) Technical Change and the aggregate production function. Harvard University Press
Solow RM (1960) Investment and Technical Progress. Massachusetts Institute of Technology
Stone R (1956) Quantity and Price Indexes in National Accounts. Paris
UNCSD (1996a) Indicators of Sustainable Development: Methodology Sheets. New York
UNCSD (1996b) Indicators of Sustainable Development: Framework and Methodologies. New York.

UNCSD (1996c) Sustainable Development Indicators: a pilot study following the methodology of the Commission of the United Nations on Sustainable Development. Luxembourg

UNEP (1994) An Overview of Environmental Indicators: State of the Art and Perspectives. Environmental Assessment Technical Report, New York

Uzawa H (1961) Natural Inventions and Stability of Growth Equilibrium. Royal Economic Society, London

Metrics for supply chain sustainability

Roland Clift

Centre for Environmental Strategy, University of Surrey, GUILDFORD, Surrey
GU2 7XH, Tel.: + 44 1483 689271, Fax: +44 1483 686671, E-mail address:
r.clift@surrey.ac.uk

Abstract

Most interpretations of sustainable development recognise that there are constraints on long-term human activities imposed by material and energy availability and by the capacity of the planet to accommodate wastes and emissions; inter- and intra-generational equity within these constraints is then an ethical principle underlying sustainability. This leads to identifying three dimensions of sustainable development: techno-economic, ecological and social. This paper reviews the development of indicators to reflect these three dimensions, applicable to industrial sectors, companies and broad groups of products or services. Indicators of environmental and economic performance are relatively well established. They can be combined to indicate the sustainability of products, services and supply chains. Indicators of social performance are more problematic, particularly indicators to describe the social value of products and services. Cases from the process, petroleum and petrochemicals, electronics and fast moving consumer goods sectors are reviewed, showing that social indicators must be developed through public participation.

1 Introduction: Sustainability, Equity and Constraints

"Right now, we've got freedom AND responsibility – it's a very groovy time" (Austin Powers)

"Sustainable Development" was placed centrally onto the international agenda by the Brundtland Commission on Environment and Development (WCED, 1987), which introduced the oft-quoted statement that sustainable development is "development that meets the needs of the present without compromising the ability of future generations to meet their needs". Most of the subsequent literature has developed the concept of equity, inherent in the Brundtland statement, as an ethical principle of equal access to opportunities and resources (see e.g. Reid 1995; Clayton and Radcliffe 1996; Clift 2000). Brundtland specifically highlighted inter-generational equity – the rights of future generations, sometimes expressed as the responsibility not to steal from our grandchildren. It is perhaps surprising that intra-generational equity – equal rights for all people inhabiting the planet at this time – was not part of the Brundtland definition; perhaps it was considered obvious that inter-generational equity is meaningless without intra-generational equity. Most subsequent attempts to make sustainable development more "operational" have recognised that it must embrace both dimensions of equity; for example, the UK Government states "At the heart of sustainable development is the simple idea of ensuring a better quality of life for everyone, now and for generations to come" (DETR 1999).

"Freedom" and "responsibility" come together when it is recognised that human activities are constrained by the capacity of the planet which is, in thermodynamic terms, a closed system : it exchanges energy but not matter with the rest of the universe (Jackson 1996). For example, it is ignorant, pointless or simply dishonest to state that all inhabitants of the planet should have the opportunity to achieve the level of consumption enjoyed at present in most of the industrialised countries, because to do so would require resources and carrying capacity many times larger than those with which the planet is equipped. Constraints arise from resources: material and energy availability; human and economic capital; or from the capacity of the biosphere to accommodate emissions and wastes from human activities, an obvious example being the effect of emissions on the global climate (which is emerging as the active constraint on the use of fossil fuels, rather than the long-term availability of those fuels (RCEP 2000)); and from the capacity of human beings and human social structures to respond to pressure and remain within the ecological constraints. Recognition of constraints thus lies behind the familiar three dimensions of sustainability: techno-economic, ecological and social (Clift 1998).

2 Sustainability Indicators

Indicators for sustainable development must therefore cover all three dimensions of sustainability, reflecting the significant constraints on human activity and conveying information about the level of inter- and intra-generational equity. Following the approach proposed by the Global Reporting Initiative (GRI), it is useful to proceed from broad *categories* through definite *aspects* to specific *indicators*, interpreted as:

Categories: broad areas or groupings of economic, environmental or social issues of concern to stakeholders;

Aspects: general types of information related to a specific category (e.g. greenhouse gas emissions, or donations to host communities);

Indicators: specific measurements of an individual aspect that can be used to track and demonstrate performance.

Although general indicator frameworks can be developed (e.g. Azapagic and Perdan 2000; IChemE 2002), it is commonly agreed that indicators need to be established on a sector-by-sector or even case-by-case basis. Sustainability indicators may be used at various levels: national economies, industrial sectors, companies, business areas or product groups, and specific products or services. The focus in this paper is on indicators which might be applied to sectors, companies and broad groups of products or services. Indicators are interpreted as parameters to show which aspects of performance must be improved and to indicate the direction of change, rather than measuring incremental improvements.

For the environmental and economic dimensions of sustainability, indicators are already available (Biswas et al. 1998; Lehni 1999). Environmental impacts can usefully be aggregated into the set of categories commonly used in Life Cycle Assessment (summarised in Table 1) leading naturally to quantitative indicators, such as atmospheric emissions weighted according to their Greenhouse Warming Potential to indicate the category of global climate change (Wright et al., 1997). A commonly used economic indicator for sectors, companies and business areas is Value Added (VA), defined as the value of sales less the cost of goods, raw materials (including energy) and services purchased[1]. Value Added also represents the contribution of an activity to Gross Domestic Product (GDP).

Table 1. Common Set of Environmental Impact Categories used in LCIA (Clift 2001)

Abiotic Depletion Potential	Extraction of non-renewable raw materials such as ores.
Energy Depletion Potential	Extraction of non-renewable energy carriers; can be included in Abiotic Depletion Potential.
Global Warming Potential	Contribution to atmospheric absorption of infra-red radiation leading to increase in global temperature.
Ozone Depletion Potential	Contribution to depletion of stratospheric ozone, leading to increase in ultraviolet radiation reaching earth's surface.
Human Toxicity	Contribution to human health problems through exposure to toxic substances via air, water or soil (especially through the food chain).
Aquatic/Terrestrial Ecotoxicity	Contribution to health problems in flora and fauna caused by exposure to toxic substances.
Acidification Potential	Contribution to acid deposition onto soil and into water.
Photochemical Oxidant Creation Potential	Contribution to formation of tropospheric ozone.
Nutrification Potential	Contribution to reduction of oxygen concentration in water (or soil) through providing nutrients which increase production of biomass.

---

[1] Variants on the simple Value Added have been proposed allowing, for example, for capital depreciation, but VA remains the most common metric for economic performance; see Azapagic and Perdan (2000).

In formulating and estimating the values of indicators, it is necessary to distinguish between application to "in-house" activities and to complete supply chains. The former is appropriate when describing the performance of a sector or a company (Behmanesh et al. 1993; Wright et al. 1997) while the latter (life cycle) approach is to be used when assessing products or services. Some authors, particularly in Life Cycle Assessment, advocate aggregating the environmental impacts in Table 1 into a single metric measured as "ecopoints" or damage costs. However, this level of aggregation loses information and transparency, and is therefore to be avoided for the kind of strategic assessment which is the topic of this paper.

It is also useful to use some form of normalisation to indicate the significance of indicators. For example, estimates of the environmental impact of an activity may be "normalised" by dividing by the total impact of human activities globally, or in the relevant region or country, or in the industrial sector to which the activity belongs. Normalisation using a combination of environmental and economic indicators can give valuable insights into the sustainability (or otherwise) of products and supply chains; see below.

Notoriously, it is the social dimension of sustainable development which presents the greatest difficulty. Some indicators give the impression of having been proposed because they are measurable (DETR 1999; see also Azapagic and Perdan 2000), without the discipline of examining what aspect of what category they represent. Possible approaches to developing social indicators will be explored later in this paper.

3 Economic and Environmental Indicators for Supply Chains

Overall Business Impact Assessment An indicator set which relates both environmental and economic performance has been developed by Unilever (Taylor and Postlethwaite 1996) in the approach known as Overall Business Impact Assessment (OBIA). OBIA was developed originally for businesses and groups of consumer products. The OBIA parameter measuring the performance of business or product group j in environmental impact category i is defined as:

$$\emptyset_{i,j} = \frac{[\text{Impact in category i / Value of business j}]}{[\text{Total anthropogenic contribution to impact category i / Total global economic activity}]} \quad (1)$$

where the environmental impact is evaluated over the whole life cycle and the "value" is taken as the total sales from the business.

Ecometrics like $\emptyset_{i,j}$ can be used to identify highly unsustainable activities, or to distinguish between discrete options or scenarios. The OBIA approach provides a means of screening product or business areas which should be targeted for environmental improvement or substitution. Figure 1 shows the screening process schematically. Business for which $\emptyset_{i,j}$ is around unity – such as 1 - 4, 6 and 8 in Figure 1 – show adequate environmental performance at the present levels of expectation. If the value of $\emptyset_{i,j}$ is much less than unity, then the performance in this environmental category is unusually good. However a product or business area for which $\emptyset_{i,j}$ is much larger than unity – such as 5 and 7 in Figure 1 – shows disproportionate environmental impacts in those categories for which $\emptyset_{i,j}$ is large.

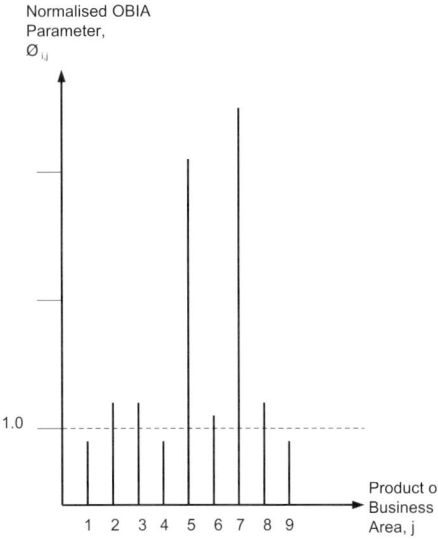

Fig. 1. Use of OBIA normalised metric to identify least sustainable products or business areas

Other variants on the OBIA approach have been reported. The World Business Council for Sustainable Development recommend an indicator which is essentially the reciprocal of $\emptyset_{i,j}$ (Lehni 1999). BASF are amongst the companies using this kind of approach to estimate eco-efficiency and hence select potential products and processes for commercial development (Saling 2002). Applied to complete industrial sectors, with $\emptyset_{i,j}$ evaluated for the sector rather than the complete supply chain, these indicators have been used to explore the implications of different economic development scenarios (e.g. Zakotnik and Radej 2002).

Application to Supply Chains OBIA has been extended to analyse the environmental and economic performance of supply chains (Jackson and Clift 1998; Clift and Wright 2000). The approach is shown schematically in Figure 2. The abscissa aggregates the Value Added along the supply chain; VA at any point is less than the sales price by the cost of energy and ancillary materials purchased, but this difference may be small by comparison with VA particularly for the later segments of the supply chain. The "environmental impact" ordinate refers to the quantified contribution to one of the impact categories in Table 1, or to some other category such as solid waste. Thus Figure 2 is a projection of a multi-dimensional surface in the different environmental impact categories, to avoid reducing the categories to a single metric. The origin represents primary resources, while point A represents the finished product. The gradient of the chord OA represents $Ø_{i,j}$ for the product. As indicated above, the overall eco-efficiency of the product is improved when this gradient is reduced.

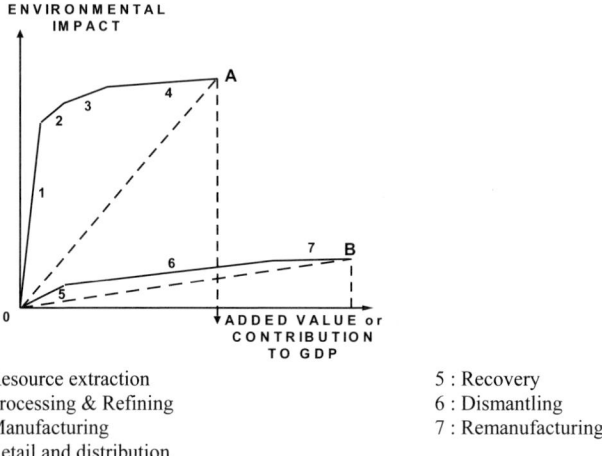

1 : Resource extraction
2 : Processing & Refining
3 : Manufacturing
4 : Retail and distribution

5 : Recovery
6 : Dismantling
7 : Remanufacturing

Fig. 2. Accumulation of economic value and environmental impact along the supply chain

The convex segmented curve shows the supply chain. Each segment represents one of the principal steps in production and distribution, with the gradient indicating $Ø_{i,j}$ for that specific step. The strong convexity is typical (Clift and Wright, 2000). It may be noted that Figure 2 actually understates the convexity. For example, for electronic equipment (including mobile telephones, the specific case examined by Clift and Wright (2000)) the quantity of solid waste produced per kg of final product is of order 200 kg in extraction, processing and refining and 20 kg in manufacturing (McLaren et al. 2000). The significance is that the primary resource industries incur disproportionately high environmental impact but receive disproportionately low economic benefit. Thus the convexity of the supply chain is an indicator of unsustainability: for equity along the supply chain, i.e. equitable distribution of impacts and benefits, the curve should be essentially straight. As well as reducing environmental impacts, the primary resource industries need to deliver much more economic and social benefit. The curve also demonstrates how companies and economies can "export unsustainability" by restructuring to concentrate on the later, high VA activities in the supply chain. This kind of analysis can help to illuminate the source of inequity, but leaves no indication that present economic structures can deliver equity. However, as a small example, the analysis illustrates why the "Fair Trade" movement, which aims to return more income to primary producers at the expense of profits in later parts of the supply chain (particularly retailing), is to be seen as a genuine attempt to develop more sustainably.

4 End-of-life Products

For many manufactured products, including vehicles and electronic or electrical equipment (see below), management at end-of-life is recognised as an environmental concern. In Europe, the issue has been framed as one of managing used products to ensure that toxic components do not escape into the environment. This has led to the banning of some components, but also to the introduction of "take-back" legislation which extends the responsibility of the manufacturer to liability for the product at the end of its service life. The real objective of the "take-back" approach is to promote re-use of components and recycling of materials, to reduce environmental impacts over the supply chain. Even so, legislated "take-back" has met much industry resistance, with claims that it will increase costs.

The extreme convexity of the supply chain curve also helps to explain why recovery and re-use or recycling of the product can appear to be uneconomic even if the environmental impacts are reduced. Figure 2 illustrates this case as curve OB, where the used product is treated as equivalent to virgin material; i.e. starting from O. Although Figure 2 is only schematic, again the form of OB is consistent with (proprietary) data, specifically for mobile telephones. Collection of the dispersed product (segment 5) and particularly dismantling so that components can be re-used or materials reprocessed (segment 6) are associated with high costs and VA. The re-manufactured product, at B, therefore turns out to be more expensive (and therefore economically uncompetitive) compared to the "new" product at A, even though it is associated with much reduced environmental impact. This simple conclusion helps to show why the European Union, for example, has found it necessary to introduce mandatory recovery targets for "end-of-life" manufactured goods rather than relying on economic incentives. In the longer term, "take-back" legislation is intended to drive manufacturers to design their products for ready dismantling, to reduce the economic cost of segment 6 even if this means increasing the economic cost of manufacturing (segment 4). However, this does nothing to improve equity along the supply chain, either of new or remanufactured products.

5 Categories, Aspects and Indicators for Social Benefits

5.1 Human Needs

Indicators for the social dimension of sustainability have, thus far, concentrated on a company's own activities (see Azapagic and Perdan 2000; IChemE 2002). Ethical indicators cover aspects such as stakeholder inclusion, preservation of cultural values and benefits to communities surrounding a company's operation; consistency of employment conditions and health and safety standards for all operations regardless of their location (intragenerational equity); and leaving the environment in a condition likely to be acceptable to future generations and not creating problems, such as toxic or radioactive wastes, for which solutions are not known (intergenerational equity). The "Sustainable Development Progress Metrics recommended for use in the Process Industries" developed by the Institution of Chemical Engineers (IChemE 2002) follows this approach, including social indicators which aim to reflect "the company's attitude to treatment of its employees, suppliers, contractors and customers and also its impacts on society at large". More contentiously, the IChemE indicators include the disparity of income and benefits between the company's direct employees.

However, none of these indicators addresses the social value of the products or services which a company provides. To take an obvious if extreme example, a company producing "weapons of mass destruction" might operate in a way which appears to be benign in terms of these social indicators, but that would not justify the company's activities. To take a less extreme example, if I were to buy a gun[2] and shoot a colleague, it would be no consolation to his family to know that its manufacturer operates to sound ethical standards; equally, it would be of little interest that I used "environmentally friendly" lead-free bullets with low values of $\varnothing_{i,j}$.

Therefore some attention must be paid to social indicators applicable to products and services. This is a relatively unexplored area. The following discussion is a very preliminary account of a possible approach. The discussion of social benefits will be strictly at an anthropocentric level[3], trying to relate social benefits of products and services to satisfying human needs. Starting at the *category* level, two attempts to categorise human needs are widely used. Perhaps the best known is that due to Maslow (1954), who articulated the hierarchy of human needs summarised in Figure 3. The common interpretation is that higher level needs remain latent until the lower level needs are satisfied (although there is some question as to whether Maslow himself intended this interpretation) (see Jackson 1996). An alternative categorisation, much discussed in the development literature, is due to Max-Neef and others (1991). Max-Neef identified a set – strictly not a hierarchy – of needs summarised in Table 2. Each of the nine needs is envisaged as applying in each of four "existential" categories – being, having, doing and interacting – leading to a matrix of 36 categories of fundamental needs. In addition, Max-Neef suggested a tenth need, transcendence, which might correspond to Maslow's "growth needs" (see Figure 1).

---

[2] It may be noted that the purchase, the funeral expenses and other resultant transactions would represent positive contributions to GDP – a caution against using conventional short-term economic indicators

[3] However, it may be noted that Banner (1999) has argued that the distinction between anthropocentric and ecocentric environmental concerns is false.

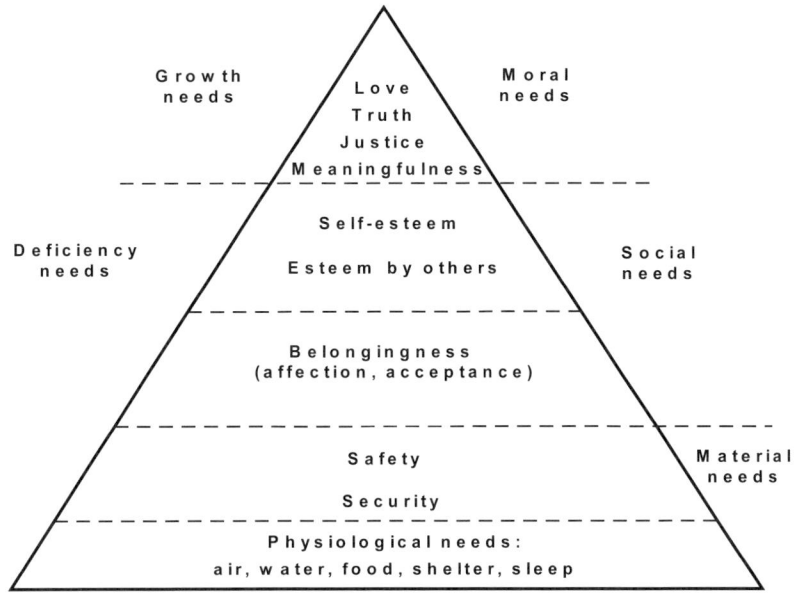

Fig. 3. Maslow's "Hierarchy of Human Needs"

Table 2. Max-Neef's "Fundamental Human Needs"

Material needs	Subsistence	Protection
Non-material needs	Affection	Understanding
	Participation	Identity
	Idleness	Creativity
	Freedom	(Transcendence)

A crucial distinction is also drawn between **satisfiers** which meet needs and **violators** which reduce or destroy the possibility of satisfying needs. "Weapons of mass destruction" are usually seen as violators. Inter- and intra-generational equity fit readily into the framework of human needs: in view of the growing evidence for the role of social inequality in reducing levels of public health (see e.g. Wilkinson, 2000), it might also be argued that social inequality can be interpreted as a violator. **Singular satisfiers** satisfy one need without inhibiting satisfaction of others; **synergic satisfiers** help to satisfy more than one need; **inhibiting satisfiers** satisfy one need at the expense of one or more of the other needs; **pseudo-satisfiers** bring a false and temporary sense of satisfaction. For examples, see Jackson (1996).

Investigation of products as satisfiers and violators provides a possible link from needs as categories, through aspects to indicators. Following the idea that indicators have to be developed from a common set of categories on a case-by-case basis, a very preliminary exploration is made here of possible ways to establish social indicators, using specific sectors as exemplars.

5.2 Process Sector

The Responsible Care initiative goes at least some way towards ensuring that products are used beneficially and responsibly. However, one of the characteristics of the process industries is that their products are generally used as inputs to final products (building components, manufactured goods etc.) rather than being the final products themselves. Given the close regulation of process plant throughout the industrialised world, it is increasingly recognised that the principal environmental impacts arise from chemical products rather than chemical production (RCEP 2003). Therefore the environmental impacts affect people who do not receive direct benefit from the industry or from use of products; the impacts are then solely violators.

Given the range of possible uses, many of which are unknown to chemical producers, it is difficult for companies in the process sector to answer the question "What human needs to your products satisfy?" Nevertheless, the sector is increasingly being required to an answer this question, in effect addressing the balance between social benefits and environmental impacts. The OECD (2000, 2002) is developing a "Framework for incorporating socio-economic analysis (SEA) in chemical risk management decision-making" as a structured way to estimate the balance between the benefits of a chemical product against the risks it poses to human health and the environment. The OECD argues that the indicators which emerge from the process should be expressed as monetary values, so that the process becomes a form of cost/benefit analysis. Other groups advocate a participatory approach to balancing risks and social benefits but with less emphasis than the OECD on monetary valuation and cost/benefit analysis. The Royal Commission on Environmental Pollution (1998, 2003) stresses that environmental issues and policy choices of this kind invariably raise question of value (as distinct from economic preference) which need to be integrated into each critical stage of decision-making, including the "framing" of the problem under consideration (which amounts to defining the categories and aspects of concern). All these approaches imply that formulation of social indicators needs to be an extensive and carefully structured process of public engagement, analogous to the kind of debate which has taken place in many European countries over the commercial use of Genetically Modified Organisms. One likely outcome is that many more chemical products than at present will be licensed for use only in applications with high recognised social benefit, satisfying genuine needs.

5.3 Petroleum and Petrochemicals

Additional considerations apply to the petroleum and petrochemicals sector: in common with some but not all of the process industries, it uses a non-renewable resource – fossil hydrocarbons. Long-term use of this resource is constrained, although by the capacity of the biosphere to accommodate the emissions rather than by the availability of the resource itself (RCEP 2000). The Shell Group, for example, has developed a Sustainable Development Management Framework which includes the aspects "engage and work with stakeholders" and "maximise benefits to the community" (see Cunningham et al. 2002), recognising the need for public engagement outlined above for the process sector. In terms of the categories discussed above, these aspects refer to both material and social or non-material needs because the intention is to support personal and community development rather than just providing employment as a route to satisfying material needs. As in the IChemE (2002) indicators, the Shell approach recognises the broader social responsibilities of the company, even though the environmental and economic indicators refer only to the company's own operations.

The general approach to defining indicators must also cover both products and projects. Cunningham et al. (2002) report two case studies to explore the development of indicators: for a specific product – biodiesel: rapeseed methyl ester (RME) - and for a specific project –development of a natural gas resource in Sakhalin (an area of Eastern Russia with rather special social and environmental characteristics). Biodiesel is of particular interest as an attempt to shift from non-renewable to renewable fuels, and contribute to the social goal of helping to stabilise rural populations. Sakhalin is a stark example of a case where a finite resource needs to be developed in a way which permanently enhances a community which is depressed both socially and economically. Table 3 summarises preliminary aspects and indicators of social benefit. The value of these indicators is currently being tested. They are attempts to quantify categories and aspects which are really qualitative, but the category - aspect - indicator framework helps to ensure that the indicators are significant, not merely reported because they can be measured.

Table 3. Examples of Indicators developed by Shell Group

Aspect	Indicators	
	Product-Biodiesel (RME)	Project - Nat.Gas Exploitation
Engage and work with stakeholders	Acceptability of fuel to consumer.	Plan for stakeholder dialogue in place.
	Effect of fuel on engine performance.	Frequency of meetings between company and local community.
Maximise benefits to the community	Level of employment generated.	Number of employees recruited within a 50 km radius.
	Amount of RME required not secured locally*.	Percentage of suppliers from the local area (radius 50 km).

* Note that this indicator is to be interpreted negatively or inversely.

5.4 Electronics

Like other manufactured items, electronic goods are final products. Unlike the process sector, it is therefore possible to examine the social impact of the product itself, by asking whether it satisfies any category of human need. The specific example of mobile telephones will be considered here. Mobile telephones can be regarded as synergic satisfiers, meeting needs which include safety, security and acceptance for Maslow; or protection, identity and participation for Max-Neef. (Their role as violators of some needs, for example by use on crowded vehicles, is recognised by some manufacturers who have promoted codes of practice or guides to courtesy in using mobile telephones). The fact that they require limited fixed infrastructure gives them a particular role as satisfiers in developing countries.

Perhaps because the social benefit of the product has not been seriously questioned, some companies in the mobile telecommunications sector have started address the question themselves. Social and environmental aspects of their business are being explored through systematic engagement with their own employees (e.g. Oxley Green et al., 2002) as a preliminary to engagement with a broader set of stakeholders including customers and suppliers. The process of identifying categories of social impact and proceeding through aspects to develop specific indicators then becomes, as for the other sectors discussed above, an open process to introduce corporate social responsibility into product development. This represents a notably innovative approach to the use of sustainability metrics.

5.5 Fast-Moving Consumer Goods

Two examples will be considered briefly of products in the Fast-Moving Consumer Goods (FMCG) sector. The first is domestic laundry detergent, usually considered necessary[4] to deliver the function "supply of clean clothes". This service is unusual in lying at the intersection of several supply chains (including detergent, machine, water, energy and fabric) with limited communication between the supply chains (Ransome and Clift 2002). However, it is universally regarded as a satisfier of needs: physiological and social for Maslow; material and non-material for Max-Neef. Companies in the FMCG sector are accustomed to stakeholder engagement in the form of consumer focus groups. Systematic work is now in progress to establish the environmental concerns of stakeholders, starting with company employees (Clarke et al. 2002). Much as for mobile telecommunications, the objective is to use structured social processes to elicit the categories and aspects of public concern, and hence to develop performance indicators which can be used to guide innovation and future product development. Whether this approach can also be used to inform communication between the intersecting supply chains remains to be investigated.

Other fast-moving consumer goods, such as cosmetics or hair styling aids, arguably meet none of the basic needs identified by Max-Neef or Maslow. At best, they might be regarded as pseudo-satisfiers for Maslow's "self-esteem" (which is consistent with the singularly repugnant advertising slogan "Because I'm worth it"). If purchase and use of such products has no identifiable social value, there might still be an argument that they are sustainable if purchase by the end user creates activity along a supply chain which helps to satisfy the needs of others,

---

[4] Washing clothes by hand may require less energy input, but avoiding this chore arguably satisfies needs such as Max-Neef's "idleness". Washing in unconfined conditions such as streams is arguably undesirable on environmental grounds.

most obviously by employment which is rewarding in both the economic and personal senses. This would require the curve describing the supply chain (cf. Figure 2) to be concave rather than convex, providing Value Added to the primary resource industries. We have yet to find a product for which this is the case...

6 Conclusions

Indicators for the direction of sustainable development need to represent all three dimensions: techno-economic, ecological and social. Aggregation across the dimensions, for example expressing ecological indicators in monetary form, is unnecessary and undesirable. However, normalised parameters which combine different dimensions can be informative. Indicators of economic and environmental performance are well established for products and services, and can be combined to identify activities which are significantly less sustainable than the average of human economic activities. They can also be used to reveal unsustainability in supply chains. Social indicators applicable to products and services are also needed, but are not generally available. To be valid, they need the kind of public acceptance which can only be achieved through well-structured participatory decision processes.

For indicators to be used effectively, it must be recognised that they will identify some economic activities which are so unsustainable that they must be discontinued. Industrial sectors which recognise the importance of equity but not the existence of constraints on human activities will be in the position of trying to ensure that everyone on the Titanic has access to a deck chair.

References

Azapagic A, Perdan S (2000) Indicators of sustainable development for industry: a general framework, *Trans. IChemE* **73B**: 243-261

Banner M (1999) Why and how (not) to value the environment. Chapter 5 in "Christian ethics and contemporary moral problems", Cambridge University Press, London

Behmanesh N, Roque JA, Allen D (1993) An analysis of normalized measures of pollution prevention, *Poll. Prev. Rev.* Spring 1993: 161-166.

Biswas G, Clift R, Davis G, Ehrenfeld J, Förster R, Jolliet O, Knoepfel I, Luterbacher U, Russell D, Hunkeler D (1998) Ecometrics: identification, categorisation, and life cycle validation. International Journal of LCA 3: 184-190.

Clarke L, Clift R, Wehrmeyer W, King H, McKeown P (2002) Addressing employees' concerns to facilitate environmentally conscious decision making innovation. 10th International Conference of the Greening of Industry Network, Göteborg, June

Clayton AHJ, Radcliffe NJ (1996) Sustainability – a systems approach. Earthscan, London

Clift R (1998) Engineering for the Environment: The New Model Engineer and Her Role. *Trans. IChemE* Vol. **76B**: pp. 151-160

Clift R (2000) Contribution to Forum on sustainability, Clean Products and Processes 2 : 67

Clift R (2001) Clean technology and industrial ecology. Chapter 16 (pp. 411-444) in R.M. Harrison, ed., "Pollution: causes, effects and control", Royal Society of Chemistry, London (4th ed.)

Clift R, Wright L (2000) Relationships between environmental impacts and added value along the supply chain. *Tech. Forecasting and Social Change*, **65**, 281-295

Cunningham B, Wehrmeyer W, Clift R, Brewer L (2002) Integrating social concerns into the decision-making process associated with the petroleum industry. 10th International Conference of the Greening of Industry Network, Göteborg, June

DETR (1999) A better quality of life. UK Depart. of Env., Transport and Rural Affairs, The Stationery Office, London

DETR (1999) Monitoring Progress: indicators for the strategy for sustainable development in the United Kingdom, UK Dept. of Env., Transport and Rural Affairs, The Stationery Office, London

ECTEL (1997) End-of-life management of cellular phones: an industry perspective and response, ECTEL Cellular Phones Takeback Working Group, London

IChem E (2002) The Sustainability Metrics – sustainable development progress metrics recommended for use in the Process Industries, Institution of Chemical Engineers, Rugby

Jackson T (1996) Material concerns – pollution, profit and quality of life, Routledge, London

Jackson T, Clift R (1998) Where's the profit in industrial ecology? *J.Ind.Ecol.* **2** : 3-5

Lehni M (1999) Measuring eco-efficiency with cross-comparable indicators. WBCSD, Geneva

Maslow A (1954) Motivation and personality. Harper and Row, New York

Max-Neef M, Elizade A, Hopenhayn M (1991) Human scale development – conception, application and further reflections. Apex Press, New York

McLaren J, Parkinson SD, Jackson T (2000) Modelling material cascades – frameworks for the environmental assessment of recycling systems. *Resources Conservation and Recycling* **31**: 83-104

OECD (2000) Framework for integrating socio-economic analysis in chemical risk management decision-making. Report ENV/JM/MONO (2000)5, Organisation for Economic and Cultural Development, Paris

OECD (2002) Technical guidance document on the use of socio-economic analysis in chemical risk management decision-making. Report ENV/JM/MONO (2002)10, Organisation for Economic and Cultural Development, Paris

Oxley Green AS, Wright L, Burningham K, Clift R (2002) Assessing the environmental views and concerns of Nokia employees as part of stakeholder participation. 10th International Conference of the Greening of Industry Network, Göteborg, June

Ransome T, Clift R (2002) The supply, use and waste management of domestic clothes washing. Appendix C (pp.205-230) in N. Wrisberg and H.A. Udo de Haes, eds. Analytical tools for environmental design and management in a systems perspective, Kluwer, Dordrecht

RCEP (1998) Setting Environmental Standards. 21st Report of the Royal Commission on Environmental Pollution, The Stationery Office, London

RCEP (2000) Energy: the changing climate. 22nd report of the Royal Commission on Environmental Pollution, The Stationery Office, London

RCEP (2003) Chemicals in Products. 24th Report of the Royal Commission on Environmental Pollution, The Stationery Office, London

Reid D (1995) Sustainable Development – an introductory guide. Earthscan, London

Saling P (2002) Realising more sustainable products and processes in different fields of business by application of the eco-efficiency analysis. NATO Advanced Research Workshop, Maribor

Taylor AP, Postlethwaite D (1996) Overall Business Impact Assessment (OB1A). *4th LCA Case Studies Symp.*, pp. 181-187, SETAC-Europe, Brussels

WCED (1987) World Commission on Environment and Development: Our Common Future, Report of the Brundtland Commission. Oxford University Press, London

Wilkinson R (2000) Mind the gap – hierarchies, health and human evolution. Weidenfeld & Nicolson, London

Wright M, Allen D, Clift R, Sas H (1997) Measuring corporate environmental performance: the ICI Environmental Burden system. Journal of Industrial ecology 1: 117-127

Zakotnik I, Radej B (2002) Environment as a factor of national competitiveness in manufacturing. NATO Advanced Research Workshop, Maribor

Quantifying technological aspects of process sustainability: a thermodynamic approach

S. Lems, H. J. van der Kooi, J. de Swaan Arons

Laboratory of Applied Thermodynamics and Phase Equilibria, Delft University of Technology, Julianalaan 136, 2628 BL Delft, The Netherlands

Abstract

Thermodynamic analysis has greatly helped to compare and to improve the energy efficiency of all kinds of technological processes, and recently we have also attempted to analyze some important biochemical processes under intracellular conditions. This work has pointed to some key strategies on sustainable process operation and to the exceptionally high thermodynamic efficiencies of chemical and solar energy conversion in living cells.

From this it was expected that the sustainability strategies of specific biochemical processes and those of the ecosphere as a whole could be of guidance to current technological processes, especially now that there is a growing demand from government and industry to effectively deal with sustainability aspects in process analysis. Our focus on this issue has led to methodologies to quantify technological aspects of sustainability by making use of thermodynamic principles.

Three indicators were constructed to express three technological aspects of process sustainability. First, an indicator for the sustainability of resource utilization considers the thermodynamic input and the availability the resources used in the process. Secondly, an efficiency indicator focuses on the conversion and loss of thermodynamic quantities in the process itself. Thirdly, an indicator for environmental compatibility takes into account the thermodynamic input required to prevent possible negative side effects of the process, such as global warming or acid rain. The three indicators are used to reflect on (un)sustainable characteristics of current technological processes compared to biochemical processes. Finally, we address the drawbacks of combining indicator values to express overall sustainability.

1 Introduction to thermodynamic analysis

The thermodynamic analysis of technological processes greatly helps to understand where and how inefficiencies in processes come about and how these can be diminished. The concepts of energy and energy efficiency prove to be inadequate to properly analyze processes, since they do not account for the inevitable loss of the quality of energy as dictated by the second law of thermodynamics. In this way, processes can appear quite efficient in terms of energy, while they are actually very wasteful from a thermodynamic point of view. Hence, energy efficiencies can be quite misleading, and a better way of determining and expressing the efficiency of processes is by including second law principles. Exergy analysis is the most comprehensive form of second law analysis (Rosen 1999) and central in this is the concept of exergy.

Exergy can be defined (see also Szargut et al. 1988) as the amount of work (e.g. mechanical shaft work) that can ideally be performed by a material or the amount of energy when it is brought from its actual conditions to the equilibrium conditions of the natural environment (see Fig. 1). The term ideally in this definition refers to the thermodynamic limit that can only be achieved theoretically in a completely reversible process, i.e. a process in which all losses due to irreversibility (i.e. entropy increase) are absent. Such a hypothetical process generates the maximum amount of work that can be obtained from a material flow or energy flow in a given reference environment. In exergy calculations, this environment is normally a reflection of the natural environment at the earth's surface, since this is where most technological processes occur.

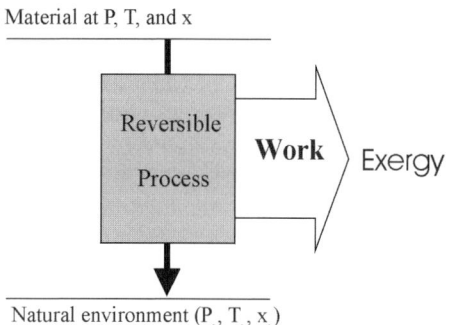

Fig. 1. Exergy (Szargut et al. 1988) is the maximum amount of work that can be obtained from an amount of energy or matter when it is brought from its actual conditions (P, T, x) to the conditions of the natural environment (P_0, T_0, x_0)

The exergy analysis of a technological process starts with calculating the exergy value of each material or energy flow to or from the process. The total exergy coming out of the process is always less the total exergy that went in, and the resulting loss of exergy is directly related to the total entropy increase S_{total} (of system and surroundings) and the temperature of the surroundings T_σ:

$$Ex_{loss} = T_\sigma \cdot \Delta S_{total} \qquad (1)$$

It should be stressed however that exergy is in no way equivalent or proportional to entropy. Although an increase of entropy also means a loss of exergy, entropy lacks the absolute reference of exergy to the conditions of the natural environment. In addition, some effective exergy losses do not show up as entropy increases in the process itself, which is important for practical process analysis.

Exergy analysis basically locates the thermodynamically inefficient process steps and by identifying the causes of the exergy losses there, valuable insights into the process are obtained and ways to significantly improve the thermodynamic efficiency may be found. Even those processes that have already been optimized with other tools, as for example pinch technology to design heat integration systems, can sometimes benefit from exergy analysis. Such optimization tools usually consider only one driving force, e.g. temperature difference, while exergy analysis accounts for all driving forces, whether it is a pressure difference, a mixing effect, or the driving force of a chemical reaction. Also, exergy analysis enables processes to be judged and compared to other processes in terms of their thermodynamic efficiency and, as we will see later, exergy values of process streams can also help to express other technological aspects relevant to process sustainability.

It should be noted at this point that sustainability in this paper refers strictly to the ability of a (technological) process to physically sustain its conversion of materials and energy. A complete description of sustainability would involve various other disciplines, such as economic and social sciences, but in this paper we focus on the technological aspects of sustainability only. At this point, we will first discuss some sustainable characteristics of biochemical processes, since these have been the basis for the quantification method presented in this paper.

2 Sustainable features of biochemical processes

Recently we have attempted to adapt the conventional method of exergy calculation and to analyze some important biochemical processes under intracellular conditions. This work has revealed the exceptionally high thermodynamic efficiencies of chemical and solar energy conversion in living cells, which is probably an important aspect of their sustainability. Nevertheless, high thermodynamic efficiency alone cannot explain the billions of years of existence of living cell processes; other major factors must also play a role and they seem to involve (1) the nature of the underlying driving force, i.e. the source of exergy, and (2) the closing of material cycles by continuous recycling of matter.

First, the ecosphere obtains nearly all of its exergy from solar energy, which seems an important sustainability strategy for several reasons. First of all, solar energy is thermodynamically quite potent and thereby capable of effectively driving various processes at the natural conditions on earth. Secondly, solar energy is available in large quantities and on a very long timescale, making it the most abundant and reliable source of exergy available. Finally, solar energy is an immaterial source of exergy, meaning that its use in processes does not require the net conversion of matter in the earth's system.

Secondly, the ecosphere operates within (nearly) closing material cycles, neither emitting nor extracting compounds from the natural environment on a net basis. This prevents basic-resource depletion and it also avoids man-made structural changes in the natural environment, for instance by accumulation of combustion products such as carbon dioxide in the atmosphere. Our current technological processes are largely based on the net conversion of material resources, creating exactly the problems that the ecosphere avoids. Ideally, future technological processes would operate within their own material cycle (i.e. separate from the material cycles of the ecosphere), preventing disturbances of natural material conversions on the planet

In view of the above, the use of abundant, immaterial sources of exergy, efficient conversion steps, and the closure of material cycles are important requirements for process sustainability. Even if these requirements may never be completely met in real processes, they point at 'ideal' process characteristics and are therefore meaningful starting points for the design of sustainability indicators. At this point, the key role of exergy in indicator design should be noted, as exergy provides valuable information on different technological aspects of sustainability.

3 Exergy and sustainability

There are strong links between exergy and sustainability, and the exergy values of material and energy flows are therefore particularly useful quantities to indicate technological aspects of process sustainability. All real processes must consume exergy to proceed, and this means that all our technological activities are ultimately limited by our ability to feed exergy to our processes. As discussed above, solar energy is likely to provide the most sustainable flow of exergy, because it is abundantly available on a very long time scale and because it is an "immaterial" source of exergy.

It should be realized however that our ability to practically obtain exergy from solar energy is limited. Although the total amount of exergy reaching the earth as solar radiation is enormous, this exergy is dispersed over a very large area and its effective harvesting is limited to relatively few sites, preferably near the point of its utilization. These sites are often also the areas needed for agriculture, living space, and industrial activity, and in addition the intensive harvesting of even solar exergy can in principle be disruptive to the natural environment (e.g. by changing the light reflection of the earth's surface or the natural heat and convective flows in the atmosphere). Obtaining a substantial sustainable flow of exergy is therefore far from straightforward, making exergy a basic scarce quantity in all (technological) processes.

Another link to sustainability is made by the fact that exergy is by definition relative to the earth's natural environment. For instance, the exergy value of hydrogen is about 236 kJ/mol (Szargut et al. 1988), which means that it takes at least 236 kJ/mol of work to produce 1 mol of hydrogen from components that are thermodynamically stable at the earth's surface, which in this case is water. It also means that at most 236 kJ/mol of work can be obtained when hydrogen is converted back to the stable form from which it originated. A direct connection is made in this way between the production and disposal of technologically relevant materials, and the natural environment on earth.

Given these unique features of exergy, the exergy of material and energy flows are valuable building blocks for the construction of indicators of process sustainability, which will be illustrated in the following sections.

4 Sustainability indicators

By making use of the exergy of process flows, three indicators of sustainability have been constructed. The three indicators aim to express the availability of resources utilized, the efficiency of the conversion of energy and materials, and the compatibility of the process with the natural environmental, respectively. The essence of each indicator's construction is discussed in three sections below.

4.1 Resource availability indicator

The first indicator focuses on the availability of the resources being used in the process, and its construction begins with defining a quantitative measure for resource depletion. Instead of classifying each resource as either renewable or non-renewable (DeWulf et al. 2000), we calculate the resource depletion time for each resource from its consumption rate $F_{m,consumption}$, its regeneration rate $F_{m,production}$, and the extent of its natural reserves $M_{reserves}$ as these are known at this point in time:

$$\tau = \frac{M_{reserves}}{F_{m,consumption} - F_{m,production}} \qquad (2)$$

Such a depletion time scale is better suited to express resource availability than the renewability concept, which distinguishes only so-called renewable and non-renewable resources; solar (and derived) energy is considered renewable because the dissipation of solar energy does not exceed its production, while resources such as fossil fuels (oil, gas, and coal), nuclear fuels, metal ores or other minerals are considered non-renewable because they are net consumed.

The first point is that the distinction between renewable and non-renewable resources is somewhat artificial. The sun is continuously depleting a finite amount of nuclear fuel and the material source of solar energy is therefore not renewable at all. Also, some 'non-renewable' resources, including even fossil fuels, cannot be considered completely non-renewable, because they are still being formed naturally to some extent. The depletion time τ conveniently takes into account such effects on resource availability.

The second point is that renewability is only one part of resource availability; the other part involves the natural reserves of resources. In fact, the high availability of solar energy is due to the sun's enormous reserves of nuclear fuel, and is not the result of any renewal of resources. Also, a temporary discrepancy between the consumption and the regeneration rate of a material resource within the earth's system does not necessarily threaten its availability when the resource is plentiful. The environmental effects of such net material emissions or extractions can be extensive, but these are beyond the scope of this indicator dealing with resource availability.

Depletion times more adequately express the availability of resources, and hence the depletion times of the resources used in a particular process can be used to construct a meaningful indicator of the sustainability of the process' resource utilization. By relating the depletion times to a common reference time scale, each ingoing flow of resources can be assigned a factor expressing availability on a scale of zero to one, and the exergy of the resource flows can be taken as weights, determining the contribution of each factor to the overall indicator value. These exergy values express the minimum amount of work required to (artificially) produce the flows of resources from compounds that are thermodynamically stable in the natural environment, and they are therefore particularly suited to express the relative importance of the different resource flows (much more so than the mass, volume, or energy content of the resource flows).

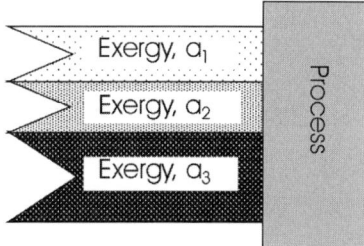

Fig. 2. The flows of resources to the process can be viewed as flows of exergy with different levels of availability in future processes

The indicator is taken as the product of the average availability factor (resource flows weighed by their exergy) and the minimum availability factor of the resources used in the process. The average factor takes into account the availability and exergy of all resource flows utilized, while the minimum factor only expresses the availability of the resource that is being depleted at the highest rate. This minimum availability represents the bottleneck in future resource supply, which explains its key position in the indicator construction, where it acts as a limit to the indicator value. For a more detailed description of this indicator's construction see Lems et al. 2002.

4.2 Thermodynamic efficiency indicator

The second indicator of process sustainability considered in this paper is the thermodynamic efficiency with which flows of materials and energy are converted in the process. As explained earlier, it is not straightforward to obtain a significant sustainable flow of exergy, and its efficient utilization in technological processes is therefore crucial to sustainability. The efficiency indicator is hence defined as the total exergy of all outgoing flows of the process divided by the total exergy of all ingoing flows (see figure 3).

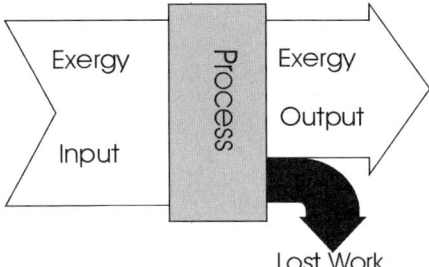

Fig. 3. The efficiency indicator expresses the total exergy output in the form of useful products as a fraction of the total exergy input

The thermodynamic or exergy efficiency of processes diminishes as a result of exergy losses, and two types of exergy losses can be distinguished. First, exergy is always lost as a result of irreversibility in the process itself, i.e. the internal exergy losses. In current technological processes, these losses can often be reduced by applying smaller driving forces, i.e. smaller gradients in temperature, pressure, concentration, or chemical composition. The more fundamental solution however is to improve the mechanism with which the exergy is transferred (Lems et al. 2003), because this allows higher thermodynamic efficiencies at all magnitudes of driving forces.

Secondly, exergy can also be lost when the exergy of outgoing process streams is allowed to decay rather than that it is utilized in other processes. Examples of these external exergy losses are the direct discharge of hot flue gasses or pressurized gas streams to the environment. This type of exergy losses can be quite significant and can be reduced by measures such as heat recovery, controlled expansion, and reuse of non-trivial chemical compounds.

The thermodynamic efficiencies of current technological processes are usually quite low. For example electricity generation in a conventional power plant achieves about 35% efficiency; the efficiency of distillation columns, although depending on the equipment and operation, is usually below 10%, while the efficiency of cryogenic distillation is even considerably lower than that. Simple chemical conversions such as the partial oxidation of natural gas can sometimes have relatively high thermodynamic efficiencies (up to 50–60%), but as soon as more process steps are involved and more waste streams are produced, as is notoriously the case in for example the production of fine chemicals, thermodynamic efficiencies drop drastically.

Biochemical processes score substantially better on thermodynamic efficiency, and some of our other work on biochemical energy conversion (Lems et al. 2003) has indicated that the reasons for this lie in the basic kinetic features of the conversion and leak mechanisms. This suggests that fundamentally different mechanisms may be required to achieve significantly higher thermodynamic efficiencies in technological processes.

4.3 Environmental compatibility indicator

The third indicator of sustainability designed by DeWulf et al. 2000, relates the exergy input required to run the process in the proposed or existing way to the exergy input required to run the process in an environmentally benign way. The latter includes the extra exergy required for abating potentially harmful effects on the natural environment (see Figure 4), which is taken as a measure for the lack of the process' compatibility with that environment. Hence, a large exergy requirement for abatement relative to the exergy requirement of the process itself leads to a low indicator value.

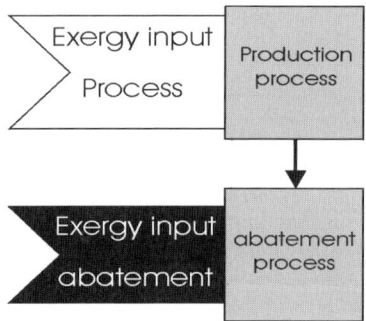

Fig. 4. The indicator views the exergy input to the process in relation to the exergy input required for abating of the process' harmful effects.

It must be noted that abating negative effects can be a very broad concept. In the first place it is considered to apply to all measures required to close material cycles. In the case of carbon dioxide (CO_2) emissions, strictly this would mean reconverting the CO_2 to the original fuel, which immediately shows that burning fossil fuels can never really be a sustainable source of exergy; the second best alternative would be to at least close the carbon cycle and sequestrate the formed CO2, e.g. in empty oil or natural gas wells. The consequence of using fossil fuels is then the extra exergy required for the CO2 sequestration, which diminishes the indicator value.

Apart from closing material cycles, abatement can also apply to various other effects such as thermal pollution, erosion of soil after deforestation, or subsidence of ground after extraction of natural gas or oil from underground deposits. The measures taken against these effects again involve the input of exergy, which like before decreases the value of the indicator on environmental compatibility.

5 Improving the indicator values

The indicators on resource availability and environmental compatibility usually turn out to be quite low for current technological processes, mainly because these processes rely heavily on the net consumption of relatively scarce materials such as oil as main source of exergy and also because these processes do not recycle the remaining material resources that are not necessarily exergy providers. As a result, on the one hand materials are being extracted from the environment and consumed as resources, and on the other hand materials are being emitted to the environment in the form of used products and emissions. The ecosphere in contrast does not convert material on a net basis, but instead operates within closed material cycles.

The first and main step to improve the values of these two indicators is hence to find a highly available source of exergy. This would preferably be an immaterial source like solar energy (as used by the ecosphere) because that would in principle allow the closure of all material cycles, improving both indicators at the same time; strictly in terms of improving the resource availability indicator however the exergy source could in principle also be material in nature if the resource is very abundant. An extreme example of the latter could be the use of hydrogen atoms as fuel for the process of nuclear fusion, although up to now this has not been a technically feasible option.

The second step to improve the two indicators is to recycle those materials that are not in the first place exergy providers in the process. Simple inorganic materials like metal and glass, but also more complex organic molecules can in principle be endlessly regenerated and reused, given that the exergy is available to run the regeneration processes. It should also be realized that certain elements could be so scarce that there is simply not enough to support a significant material cycle, as may be the case in the production of certain types of solar cells. It is best to avoid rare elements and to make use of the more abundant elements.

Living cell processes show that a very large diversity of materials can be obtained mainly from very common elements like carbon, hydrogen, nitrogen, and oxygen. Not only are the building blocks of such organic molecules widely available (mainly water, carbon dioxide, nitrogen, and oxygen), the interconnecting chemical bonds are only moderately energetic, allowing easy degradation and conversion into other products while still being quite stable at the natural conditions on earth. This is in contrast to the chemical bonds in many man-made inorganic materials, which usually require very high temperatures to be broken.

The efficiency indicator of current technological processes is also mostly characterized by low values. The first reason for this is that the exergy of outgoing (waste) process streams is often not utilized, and a first step in improving the indicator value is therefore to prevent these exergy losses. Preferably the material or energy flow is reused in the process (or in another) either directly or with a minimum of degradation; otherwise at least as much of the exergy should be recovered from these flows before they are discharged into the environment. Living cells indeed utilize every part of the chemical exergy present in the substrates they use, and even if conditions do not allow a substrate to be processed any further (e.g. glucose fermentation to alcohol), it can usually be utilized by another organism operating under other conditions.

The second major reason for the high exergy losses lies in the operation of the process itself. Although the losses are often attributed to the application of too large driving forces, e.g. the very spontaneous chemical reactions occurring during combustion, the real underlying cause is rather the ineffective coupling mechanisms with which energetically favorable and energetically unfavorable processes are coupled. Currently, most energy coupling systems in technological processes use raw heat as an intermediate carrier of exergy and much exergy is already lost during the generation of such a disordered form of energy. In biochemical processes this coupling is achieved in a much more subtle way by directly coupling chemical reactions in enzyme complexes. The enormous catalytic power of these enzymes allows for exceptionally high thermodynamic efficiencies in processes operation at very mild process conditions (Lems et al. 2003).

6 Evaluating multiple aspects of sustainability

The indicators discussed are deliberately designed to express different aspects of process sustainability, and they should also be interpreted separately. It may seem appealing to numerically combine the three indicator values to a single sustainability parameter, but this has major drawbacks.

First of all, one sustainability parameter falsely suggests that sustainability is one-dimensional and thereby undermines the idea that there are different aspects to a complex concept as sustainability. Secondly, valuable information is lost when merging the individual indicator value into one overall value; the combination basically means that concrete and meaningful expressions become part of an overall expression without a tangible meaning. Finally, the numerical method of combining the indicators value is highly subjective. Many methods can be used, but there is no objective reason to choose one method over the other; each method unavoidably involves a value judgment on the relative importance of one indicator over the other.

Hence, the sustainability of a process is best evaluated by considering the set of three independent indicator values, which express three fundamentally different aspects of process sustainability.

7 Conclusions

Thermodynamic principles are essential elements of a quantitative description of the sustainability of technological processes. This is because the proceedings of all processes and the availability of all resources ultimately come down to the availability of exergy (not energy), which by definition views all energy and matter relative to the natural environment.

Living cell processes contain essential strategies on sustainability, and these can be the starting point for the construction of meaningful sustainability indicators. By making use of exergy flows to and from processes, three indicators can be defined, which quantitatively express resource availability, thermodynamic efficiency, and environmental compatibility, respectively.

Valuable insights on process sustainability can be obtained from this set of indicators. First, a major step toward sustainable process operation is clearly to efficiently obtain exergy from a highly available, preferably immaterial source; solar or solar-derived energy (e.g. wind energy) is most likely to best fulfill these requirements. Also, the indicators point at the importance of improving energy coupling mechanisms and that of closing material cycles by continuous recycling of materials; also scarce elements should be avoided and the use of organic molecules instead of inorganic ones could prove to be more sustainable in the end.

Finally, individual sustainability parameters should not be merged into one overall sustainability coefficient, because this destroys the meaningfulness and the objectiveness of the quantification method. The parameters address fundamentally different aspects of process sustainability, and their interpretation should therefore be independent from each other.

References

DeWulf J, van Langenhove H, Mulder J, van den Berg MMD, Van der Kooi HJ, de Swaan Arons J (2000) Illustrations towards quantifying the sustainability of technology. Green Chem 2:108–114

Lems S, Van der Kooi HJ, de Swaan Arons J (2002), The sustainability of resource utilization. Green Chem 4:308–313

Lems S, Van der Kooi HJ, de Swaan Arons J (2003) The optimisation of energy transfer in chemical reaction systems. Chem Eng Sci 58:2001–2009

Rosen MA (1999) Second law analysis: approaches and implications. Int. J. Energy Res. 23:415–429

Szargut J, Morris DR, Steward FR (1988) Exergy analysis of thermal, chemical and metallurgical processes. Hemisphere, New York

Defining and Measuring Macroeconomic Sustainability – The Sustainable Economy Indices

Jochen Gassner

Graz University of Technology, Institute for Chemical Engineering Fundamentals and Process Engineering, Resource Efficient and Sustainable Systems Working Group, Inffeldgasse 25, 8010 Graz, Austria, tel.: ++43 316 8737977, fax: ++43 316 873 7963, e-mail: gassner@rns.tugraz.at, web: http://rns.tugraz.at/new_web/

Abstract

Throughout this paper, a set of indices for the measurement of the sustainability of national and regional economic systems is developed. The basic concept for the development of the indices sees economic sustainability co-determined by the natural and social environment of economic activity as well as by the operational principles of economic systems which are found in actions of payment.

Accordingly, indices of economic sustainability for the interfaces between economic systems and their social (Consumption Surplus) and natural environment (Ecological Sustainability) as well as for the exchange of money between economic subsystems (Solvency) are developed. Together with auxiliary indices, a set of indices for the measurement of the sustainability of national and regional economic systems is developed.

The calculation of the indices is based on an extended version of Supply and Use Tables Including Environmental Accounts. The extended SUTEA contains accounts for the ecological valuation of flows of resources and residuals as well as a novel classification of economic activity (Survivability, Production and Consumption Surplus).

1 Introduction

During the last decades, sustainability and sustainable development have spread geographically and at the same time found their way into a number of disciplines. Sustainability has become part of political talk, scientific research and even leisure time attitudes. Natural, technical as well as social sciences have sustainability on their research agendas. From this diversity of technical and cultural backgrounds accrues ineluctably a multitude of viewpoints on sustainability. This multitude of viewpoints – in the form of concepts – provides a variety of backgrounds for sustainability evaluation and measurement. Different conceptual starting points result in different indicator systems which in turn necessarily point to different "sustainabilities" and foster different forms of sustainable development.

In the following, the principal concepts of economic sustainability are scrutinised, conceptual extensions are proposed and subsequently a system of indices for the measurement of economic sustainability at the national and regional level (The Sustainable Economy Indices) is developed. Importance is attached to the blueprint of the indices discussed – the determinants and the reference systems for sustainability measurement.

2 Determinants of Economic Sustainability

Knowledge about and speculations on scarcity represent a major impetus to theoretical sustainability considerations and more practical efforts. The central scarcity that sustainability researchers and practitioners live on is the apparently limited availability of natural resources. This primordial scarcity and the ensuing proto-fear that an overuse of natural sources and sinks might endanger the survival of nothing less than mankind has a fundamental influence on all activity in the field of sustainability. In consequence, sustainability efforts are based on guidelines that refer to the use of natural resources (efficiency, sufficiency), on concepts that differ according to the importance attributed to natural resources (strong sustainability, weak sustainability) and on considerations about the intertemporal and social distribution of natural resources. (Ott 2001) So, it does not give cause for surprise that the main concepts of economic sustainability accord pivotal positions to the availability, the management or the exploitation of natural resources as well. (Solow 1986; Pearce et al. 1989)

But limited supply of natural resources is not the only scarcity economic systems are confronted with and therefore is not the sole determinant of economic sustainability. A second scarcity appears at the interface between economic systems and their social environment.

The final aim of every economic activity is the provision of goods and services for final consumption by households. This amount of goods and services is not unlimited. Although throughout phases of economic growth the value of commodities for final consumption is usually growing, it can be decreasing in times of economic crises. The measuring rod for the assessment of the scarcity of goods and services for final use (the need for commodities) lies in the expectations and preferences of the households. Unlike for the exchange between economic systems and their natural environment, here scarcity is not a fact determined by the (social) environment alone, but by a combination of external (environmental) and internal (economic) factors. The expectations of the social environment of the amount of value provided by the economic system for final use by the households is on the one hand co-determined by the social environment to economic activity (consumer expectations) and on the other hand influenced by the economic system (that creates consumer expectations and needs) itself. This second scarcity is reflected by concepts that see sustainability determined by non-declining utility or consumption. (Solow 1974a, 1974b; Hartwick 1977, 1978)

Most existing concepts of sustainable economic development are built on the common ground of ethical concern about intergenerational equity which is inherent to nearly all sustainability efforts in other disciplines as well. All concepts formulate constraints for economic activity to be called sustainable. Without going into detail, mainly two different types of reference systems that serve as backbone for sustainability metrics can be distinguished.

The first is applied to cultural determinants such as utility and consumption and is widely used in economic analyses. It is based upon "the actual state" as a reference for sustainability measurement. The requirement of non-declining utility or consumption claims first and above all that some future amount of utility or consumption must not fall behind the actual (present) amount. The ongoing discussions about prospects of limited growth of the global economy rely on the same reference. (In the field of sustainability) the background of the reference of the actual state is mainly ethical. The imperative of intergenerational equity requires that future generations must be at least as well off as today's.

From the concepts determining sustainability by means of natural elements (natural resource services, the natural capital stock, ecosystem resilience and stability) reference systems of non-ethical nature are usually derived. A non-declining stock of natural capital or a sustainable yield of natural resource services implies respect for reproductive and absorptive capacities of natural systems. The respective sustainability references, though still based on ethical ground, are determined by naturally given factors (e.g. flows of natural resources) or, more precisely, our scientific knowledge about these factors. Sustainability is assessed by referring anthropogenic elements to their natural counterparts.

A third determinant can be detected in existing concepts of economic sustainability - the scarcity of internal economic elements (production opportunities). The requirement of the maintenance of production opportunities implies the management of resources such as physical (man-made) capital, intellectual capital and human capital. (Perman et al. 1999) The reference system for the measurement of the sustainability of the management of internal elements is usually the actual state. Non-declining amounts of the different forms of capital (or of overall capital when the possibility of aggregation is implied) are required.

It has been shown that these internal elements (physical, human and intellectual capital) are not of genuinely and exclusively economic nature and that their use as determinants for economic sustainability and guidelines for the measurement of the same is critical in some respects. (Gassner 2002) In contrast, we propose a different (third) determinant that arises from the internal operations of economic systems.

Social systems theory sees social systems constituted by communication. Communication (the media and the codes) is what distinguishes one differentiated social system (e.g. the political system) from another (e.g. the economic system). Communication in economic systems is carried out through actions of payment. Money is exchanged between subsystems of the global economic systems. Actions of payment constitute and perpetuate economic systems. The need for and the scarcity of other forms of capital (physical, human and intellectual) is only derived from the primordial scarcity of economic activity – the scarcity of money.

Ein Verständnis von Wirtschaft, das bei Zahlungen als den Grundoperationen des Systems ansetzt, kann alles, was sonst als Grundbegriff der Wirtschaftstheorie fungiert, - also etwa Produktion, Tausch, Verteilung, Kapital, Arbeit – als derivativen Sachverhalt behandeln.[1] (Luhmann 1988)

Participation in economic activity requires solvency (money, and not necessarily produced or other capital) from the participating economic subsystems. This holds true for economic actors at every level (households, firms, nations). Solvency perpetuates economic activity, or in other words, it assures economic sustainability. Insolvency results in exclusion from economic activity, or in other words, the unsustainability of the economic actor.

In diesem System ermöglichen Zahlungen Zahlungen. Dadurch ist eine im Prinzip unbegrenzte Zukunft eingebaut. Alle Dispositionen im System sichern zugleich die Zukunft des Systems. Jenseits aller Ziele, aller Gewinne, aller Befriedigung geht es immer weiter. Das System kann sich nicht beenden, da der Sinn des Geldes im Ausgeben des Geldes liegt. Die Gewährleistung des Ausgebenkönnens (zu Bedingungen, die die Annahme von Geld als lohnend erscheinen lassen) gibt eine abstrakte, in "Warenform" allein gar nicht mögliche Zukunftssicherheit, und die Schätzung von Eigentum, Kapital, Arbeitsplätzen und Versorgungsberechtigungen wird diesem Ziel untergeordnet.[2] (Luhmann 1988)

We can conclude that three essential scarcities determine the sustainability of economic systems. The externally given limit of the availability of natural sources and sinks, the limited amount of value of goods and services for consumption by the households, that is co-determined by the social environment and the economic system itself and the genuinely internal scarcity of money that results from the basic systemic operations of economic systems (actions of payment).

---

[1] An understanding of economic activity built on payments as basic systemic operations can treat everything that usually serves as basic concept of economics (e.g. production, exchange, distribution, capital, labour) as derived elements. (translation by the author)

[2] Within this system [the economic system, the author] payments enable payments. Thereby, an, in principle unlimited, future is inherent to the system. All systemic dispositions assure at the same time the future of the system. Beyond all aims, beyond all profits and all satisfactions, the system goes on. The system cannot stop its own operations because the meaning of money lies in spending money. The guarantee of the possibility to spend money (under conditions that make appear the acceptance of money worthwhile) gives an abstract security for the future that cannot be given in the form of goods. The appreciation of property, capital, jobs and the right on the supply [of goods and services, the author] are subordinated to this aim. (translation by the author)

3 The Sustainable Economy Indices

The Sustainable Economy Indices (Gassner 2002) comprise two sets of indices. The first set of indices is based on our reflections on scarcities. It comprises distinct measures for the exchange between economic systems and their natural, social and economic environment. It is assumed that each scarcity is a determining factor of economic sustainability in its own right. Trade-offs between the different determinants (e.g. substitution of man-made for natural capital) are methodologically excluded as no aggregate measures of (e.g. stocks of natural plus man-made capital) are formed. This underlines the conviction that for main natural resources (e.g. climate, assimilation capacities) no adequate substitutes exist. The exchange between economic systems and their natural environment is measured in physical terms, while the communicative exchanges between economic systems and their social environment (the households) and between economic systems and their economic environment are expressed in monetary terms (Gassner 2002).

3.1 Measuring Scarcity 1: Natural Sources and Sinks

Within the Sustainable Economy Indices, the Sustainable Process Index (Krotscheck 1995) is used to measure sustainability of the interactions between economic systems and their natural environment. For an explanation of our choice we refer the reader to (Gassner 2002).

The SPI is a measure of the sustainability of the exchange between anthropogenic systems and their natural environment. It measures the pressure exerted on natural systems by human activity. The main pressure inducing elements of the interaction between social systems and their natural environment are flows - flows of resources from natural systems to anthropogenic systems and flows of residuals from anthropogenic systems to natural systems. The basic idea of the SPI is to calculate the area that would be needed to embed these anthropogenic flows in a sustainable manner and to compare it to the area available. A process[3] is rated sustainable when the area needed to supply its flows from natural sources and to dissipate the flows to natural sinks does not exceed the area available. The area available is usually a concrete geographical (land and sea) area. The area needed is calculated by conversion of flows to area.

For the SPI the determinants of ecological sustainability are natural flows (local assimilation capacities, geogenic flows, quality and quantity of global material cycles and local fertility). From these determinants is directly derived the sustainability reference system applied within the SPI:

- For renewable resources the reference flow is the flow of resources that can be supplied per area given the local fertility. The conversion of the

---

[3] The SPI, though originally developed for the evaluation of industrial processes, is applied to a wide variety of human activities, such as single productive, consumptive processes, value chains in LCAs or regional and national economic systems.

anthropogenic mass flow of a renewable resource to area is straight forward.

$$A_{RR} = F_R/y_R \qquad (1)$$

In this expression, A_{RR} is the area needed to supply the flow of a renewable resource [m^2], F_R is the anthropogenic flow of the resource [kg/a], y_R is the yield of this resource given the natural fertility [kg/m^{2}a] (the mass of the resource that can be harvested in one year).

- For fossil resources the reasoning is similar. Seen over geological periods fossil resources are renewables, the main natural sink within the global carbon cycle being sedimentation in the oceans. It follows that the reference flow for fossil resources is derived from the rate of renewal of fossils per area. The conversion of the process related mass flow of fossil resources is calculated as

$$A_{FR} = F_F/y_F \qquad (2)$$

where AFR is the area needed to supply the flow of a fossil resource [m^2], FF is the anthropogenic flow of the resource [kg/a] and y_F is the yield of fossils given the sediment setting per m2 of ocean floor [kg/m^{2}a]. The value for y_F is approximately 0,002 kg/m^{2}a. (Krotscheck 1995)

For both fossil and renewable resources the sustainability determinant of quality and quantity of global material cycles applies.

- This is different for mineral resources. Most mineral materials do not form global cycles. It follows that for mineral resources no natural reference flows exist. It follows further that the use of mineral resources is inherently dissipative. Within the SPI it is assumed that the environmental pressure from dissipation of flows from mineral resources (emissions) is far more significant than that of the extraction of these materials. Therefore, the area needed to supply mineral resources is omitted except direct area use. Instead, the sustainability determinant of local assimilation capacities has to be applied. Direct area consumption and area used to absorb dissipative flows is explained below.

From an interpretative point of view it should be noted that the omission of area calculations for mineral resources allows for depletion of these resources and substitution by other natural or man-made resources.

- Calculations of the area needed to assimilate dissipative flows from processes are based on the assumption that every flow emitted is dissipated in the environment and that every flow has a corresponding area to assimilate this flow in a sustainable manner. To calculate this corresponding area, it is assumed that, analogous to the regeneration rates for renewable resources, an environmental compartment has a rate of renewal. For water this rate is determined by the seeping to the ground water body, for soil by the process of composting. Composting biomass can be used to replenish soil. The mass of composting from 1 m2 of fresh biomass per year is the calculation basis in this case. Together with natural concentrations of substances in the environmental compartments, these renewal rates yield natural reference flows for anthropogenic emissions.

$$A_{ci} = F_i/(R_c * c_{ci}) \qquad (3)$$

In this equation A_{ci} denotes the area needed to absorb substance i in compartment c, F_i denotes the emission flow of substance i from a process [kg/a], R_c the rate of renewal of compartment c [kg/m²a] and c_{ci} the concentration of substance i in compartment c. It is assumed that the same area of an environmental compartment can absorb emissions of different substances, so that the overall area to absorb emissions for a compartment calculates as the maximum of the single substance areas.

- The calculation of direct area consumption (sealed surface due to installations and other constructions) is straight forward. It is simply the area used by the respective infrastructure. Usually direct area consumption is insignificantly small in relation to the area from conversion of mass flows.

Within the system of the Economic Sustainability Indices, the SPI will be called Ecological Sustainability Index (ESI). For a regional/national economic system it calculates as the total domestic area consumed by domestic productive and consumptive activities (the economic system and the households) plus foreign productive and consumptive activities (through transboundary flows of residuals) divided by the geographical surface of the domestic economic system (land area plus sea area).[4] (Krotscheck 1995; Gassner 2002)

$$ESI = (A_{Decon} + A_{Fecon} + A_{Dhh} + A_{Fhh}) / S_D \quad [m^2/m^2] \qquad (4)$$

where ESI denotes the Ecological Sustainability Index, A_{Decon} the domestic area consumed by domestic production, A_{Fecon} the domestic area consumed by foreign production, A_{Dhh} the domestic area consumed by domestic households, A_{Fhh} the domestic area consumed by foreign households, and S_D the geographical surface of the domestic economic system.

An ESI between 0 and 1 stands for ecological sustainability of an economic system (and the domestic households). An ESI value bigger than 1 represents unsustainability, the area consumed exceeds the area available.

3.2 Measuring Scarcity 2: Consumption Surplus

The second index within the Sustainable Economy Indices, the Consumption Surplus Index (CSI), measures the sustainability of the exchange between economic systems and their social environment (the households). The final aim of every economic activity, is the provision of goods and services for final use by consumers. The existing sustainability concepts based on the requirement of non-

---

[4] For background information on SPI calculations see (Krotscheck 1995). For a more detailed description of SPI calculations within the framework of the Economic Sustainability Indices see (Gassner 2002).

declining consumption or utility focus on this form of exchange. Unlike suggested by these previous concepts, the CSI does not measure overall consumption (the value of all goods and services provided by the economic system).

Generally speaking, the CSI follows the Index of Sustainable Economic Welfare (ISEW) (Cobb and Cobb 1994) by excluding certain (mainly defensive) goods and services from the calculations of the consumption surplus. Thereby, it is assumed that the amount of value of these commodities should not necessarily be non-declining over time.

Within the concept of the Sustainable Economy Indices (Gassner 2002) the goods and services excluded from the calculations of sustainability of the economy-society interface are derived from the underlying notion of a "survivability level of consumption". This originally biological approach (the survivability level of consumption comprises all goods and services needed for biological survival and the reproducibility of a population) is given a new (cultural, social) interpretation. Social survival means the survival of functions, structures and processes of anthropogenic systems. The survivability level of consumption comprises all goods and services necessary not only for biological, but social survival. The commodities needed for social survival comprise organisational (e.g. administration), material (e.g. food, energy) and social (human and intellectual) resources.

There is no single absolute level of social survivability valid for every anthropogenic system. The survival of industrialised societies with complex administrative structures, a large number of energy consuming artefacts and a higher minimum level of education requires more material, organisational and human resources than the survival of agricultural societies. We suggest that the survivability level be determined at the level of the single product and service. Then, all "survivability products and services" can be excluded from the calculation of the CSI. The remaining goods and services (more accurate: their monetary value) for final use by the households yield a surplus, they make the "consumption surplus".

It has been shown that food, energy, government expenditure on public administration, health and social work services, education services and sewage and refuse disposal services are products for social survival. (Gassner 2002)

Much like for the social survivability level, there is no absolute level of sustainability for the CSI. We have already mentioned that the sustainability reference for the economy-society exchanges is not given by "objective" factors, such as carrying capacities for the nature-economy interface, but is determined by preferences and wants of the consumers (the households) and ethical arguments of intergenerational equity. Preferences and wants are different for different societies and change with time when societies evolve. If we assume that today's wants are formed by yesterday's satisfaction of wants, we can use the amount of value provided by the economic system at a point in time as a reference state for sustainability. In other words, we assume that generations do not want to fall behind anterior generations, which means that they demand at least as much value as provided to the anterior generations. For the CSI this means that the value of goods and services for final use by the households (except goods and services for survivability) must be non-declining.

Commodities for final consumption are provided by means of production or trade. Domestically produced as well as imported goods conduce to the fulfilment of (domestic) wants. Therefore, the CSI calculates as the value of produced minus exported plus imported goods.

$$CSI = C_{tot} - C_{surv} \qquad [\$, €] \qquad (5)$$

where CSI denotes the Consumption Surplus Index, C_{tot} total consumption (domestically produced plus imported minus exported) and C_{surv} the survivability level of consumption (domestically produced plus imported minus exported). The sustainability criterion is:

$$dCSI/dt \geq 0 \qquad (6)$$

3.3 Measuring Scarcity 3: Solvency

Sustainability of the third system-environment interrelation is measured by the Economic Exchange Index (EExI). Economic systems exchange goods, services, unrequited transfers and (financial) capital. The exchange of goods and services is accompanied by a reversal exchange of money.

Domestically produced goods and services can be (domestically) consumed or exported. In the same way, goods and services produced by the economic environment (the rest of the world economies) can be imported. Flows of imported commodities cause an outflow of money to the economic environment. Respectively, exports cause inflows of money. When monetary outflows exceed inflows (when the value of imports is bigger than the value of exports) an economic system is incurring debts. Such outflows of money can be counteracted by investment of foreign capital. Thus, a deficit in the exchange of goods and services does not necessarily lead to a lack of solvency. (Remember that solvency is our third determinant of economic sustainability.)

However, constant dissavings represent a threat to sustainability as exchange deficits might not be financed by creditors and investors forever. There is no agreement among economists on a precise definition of an "unsustainable" current account deficit. Generally, it can be said that sustainable current accounts do not "trigger a sharp hike in domestic interest rates, a rapid depreciation of the domestic currency, or some other abrupt domestic or global disruption". (Holman 2001) Crucial in this respect is of course the current account deficit - GDP ratio. Some argue that a current account deficit exceeding 4,2 % of GDP is to be judged unsustainable. (Mann 1999) Regardless of where the unsustainability level for current account deficits has to be set, a precautionary approach to measuring the scarcity of the medium of internal economic operations (the scarcity of money) must be based on the principle that an economic system must not consume and invest more than it produces. It follows that a possible threat to the solvency of an economic systems arises from a negative balance of trade and services.

$$\text{EExI} = V_{exp} - V_{imp} \qquad [\$, €] \qquad (7)$$

where EExI denotes the Economic Exchange Index, V_{exp} the value of goods and services exported and V_{imp} the value of goods and services imported. The economic exchange criterion is:

$$\text{EExI} \geq 0 \qquad (8)$$

For interpretations it is of course useful to look at the EExI – GDP ratio rather than the absolute value of EExI.

4 Interpretations

Prima facie, our determinants of economic sustainability appear overly strict and inflexible. First, one might argue – and with good reasons – that most economic systems are very far from complying with at least one of the sustainability requirements today and will be so in the future. Case studies show that virtually all developed economies have Ecological Sustainability Indices of more than 25, which means that they consume more than 25 times the area available. An SPI between 0 and 1 seems unachievable in most cases. Even profound changes in technology and lifestyles will most probably not assure ecological sustainability of economic systems (according to our criteria). The other criteria are violated at least from time to time. Even developed economies show current account deficits – let alone developing countries that usually cannot induce economic growth without significant amounts of foreign investment. Decreasing amounts of consumption surplus are possible in times of economic crisis or less dramatic phases of economic downturn or if a shift from products for final consumption to survivability products is made necessary by e.g. war or natural disasters.

Strictly following our determinants, all such economic states would have to be rated unsustainable. But sustainability cannot be detected by looking at single points in time and isolated points in space (single economies). Sustainable development is a process and has to be evaluated over time. Sustainable development involves the interactions of different (economic) actors. Sustainability is relative with time and space.

Concerning the apparent ecological unsustainability of virtually all developed economic systems it can be argued that, first, the criteria underlying the SPI reflect precautionary considerations and that, second, the imperative must be to get somewhat nearer to a state of sustainability and that having the right goal and the right reference system is more important than whether the sustainability requirements are attainable in the near future or not.

From a long-term perspective, phases of non-compliance with sustainability criteria can be unavoidable or even desirable. Present exchange deficits can be used to "finance" future sustainable development. Periods of investment in foreign technology can be necessary to allow for future economic growth. Periods of technological change may imply increased consumption of natural resources.

When future technologies are more environmentally sound such increased present consumption can be desirable (in the absence of irreversible effects).

From a geographically wider perspective, the unsustainability of single economic systems may reveal beneficial to the sustainability of a greater whole. The hinterland of an agglomeration is usually not sustainable in terms of the Economic Exchange Index but is relatively more sustainable in terms of the Ecological Sustainability Index. Its supply of natural sources and sinks as well as of labour may guarantee a relatively more sustainable city-hinterland system than evenly distributed production and consumption.

These arguments do not principally discredit the sustainability determinants and criteria. They simply claim that these be applied as guiding principles in the evaluation of a development situated in its temporal and spatial context rather than irrefutable measuring rods for the sustainability of single economic systems at single points in time.

5 Efficiency Indices

Making visible the functioning of economic systems in the light of the three scarcities is one essential aspect of the measurement of economic sustainability. States and development paths of economic systems are assessed against the backdrop of the possibly conflicting constraints of the limited availability of natural resources, the increasing demand for value of consumer goods and the constant necessity of solvency. The ESI, the CSI and the EExI intend to inform about which dimensions of an economic system are in a state of sustainability or not and whether the system is approaching sustainability or not.

Linking the different determinants of economic sustainability to form indices of efficiency yields additional information for the assessment of economic systems on the way to sustainability. While our indices of effectiveness determine the position of an economic system in relation to sustainability reference states, indices of efficiency refer one factor of sustainability to another. Thereby, they indicate the relative strengths and weaknesses in economic functioning and may point to possibilities for correction and improvement.

Both the achievement of the final aim (the provision of consumption surplus) and the perpetuation of the operational principle of every economic activity rely on the use of primary (natural) economic resources. The linkage of the use of natural resources (the input to economic activity) with the two other determinants of sustainability (solvency and consumption surplus) is shown by our efficiency indices.

The first efficiency index, the Economic Efficiency Index (EEI), is calculated as the ratio of consumption surplus produced per SPI area used. While the CSI comprises domestically produced consumption surplus and the balance of imported and exported consumption surplus, the EEI – as it intends to measure the functioning of domestic production – considers domestically produced consumption surplus only. To take into account the total SPI area used for the production of this consumption surplus, SPI area used in foreign economies to produce goods and services that are used for economic purposes in the domestic economy have to be added to domestic use of SPI area (by the economic system and not the households). Analogously, SPI area used for the domestic production of goods and services that are used in rest of the world economies (but not by rest of the world households) are deducted.

$$\text{EEI} = \text{CS}_{dom} / (A_{Decon} + A_{Pimpecon} - A_{Pexpecon}) \qquad [\$/m^2], [€/m^2] \qquad (9)$$

where EEI denotes the Economic Efficiency Index, CS_{Sdom} the domestically produced consumption surplus, $A_{Pimpecon}$ the area incorporated in products for economic use imported and $A_{Pexpecon}$ the area incorporated in products for economic use exported.

A pair of efficiency indices unifies two crucial question of economic sustainability. How does trade affect the solvency of economic systems? And how does trade affect the ecological sustainability of economic systems? We have drawn the attention to the importance of economic exchange (trade) by introducing the basic systemic operations (actions of payment) as one determinant of economic sustainability. We are now reinforcing the importance of trade for economic sustainability by claiming that not only balanced monetary exchange but also balanced exchange in natural resources (in terms of SPI area) can be needed to assure economic sustainability. Or - to stress the nature of efficiency ratios – that an exchange deficit in one magnitude (monetary, natural) must be offset by a surplus in the exchange in the other magnitude. The Import Efficiency Index (IEfI) calculates as value added imported per SPI area imported, the Export Efficiency Index (EEfI) as value added exported per SPI area exported.

$$\text{IEfI} = VA_{imp} / A_{imp} \qquad [\$/m^2], [€/m^2] \qquad (10)$$

where IEfI denotes the Import Efficiency Index, A_{imp} the area incorporated in goods and services imported and VA_{imp} the value added incorporated in goods and services imported.

$$\text{EEfI} = VA_{exp} / A_{exp} \qquad [\$/m^2], [€/m^2] \qquad (11)$$

where EEfI denotes the Export Efficiency Index, Aexp the area incorporated in goods and services exported and VAexp the value added incorporated in goods and services exported.

6 The System of Accounts Underlying the Sustainable Economy Indices

A comprehensive overview of the state of the art in environmental-economic accounting is given in the System of Environmental Economic Accounts (SEEA) handbook. (The London Group on Environmental Accounting 2001)

Among all the accounting systems discussed within the framework of the SEEA the supply and use tables including environmental accounts (SUTEA) serve our purpose best. A full SUTEA contains all necessary data on monetary and physical flows needed to calculate our indices of economic sustainability. A SUTEA is based on conventional monetary supply and use tables. Monetary tables are brought together with physical supply and use tables to yield an integrated system of physical and monetary accounts. For the calculation of our indices, the standard SUTEA has to be slightly rearranged and extended.

6.1 Economic Functions – Survivability, Production and Consumption Surplus

At the disaggregate level a SUTEA uses conventional classifications of industries (General industrial classification of economic activities within the European Communities – NACE) and products (Classification of products by activity - CPA). These remain unchanged, because monetary as well as physical data are collected according to these classification systems by national statistical offices.

For structural analysis of economic systems at a highly aggregated level it is useful to form economic sectors (e.g. in the System of National Accounts: non-financial corporations, financial corporations, general government, households, non-profit institutions serving households). Within the framework of the Sustainable Economy Indices it is useful to form aggregates of value added and SPI area consumption per sector.

From our specific point of view, it is interesting to see which share of overall value added and SPI area is directly used for the provision of the Consumption Surplus and which share is spent for survival. A functional classification that is derived directly from the notion of the consumption surplus can give information about structural strengths and weaknesses and about efficiency in the fulfilment of the final economic aim.

The functional classification does not allocate industries to sectors but starts by classifying final uses (products and services). On basis of this classification of final uses into economic functions, intermediate flows (in use tables) can be ascribed to economic functions as well. Thereby, we can ascribe values that are recorded and calculated for activities (and not products) such as value added or generation of residuals and use of natural resources (and the derived consumption of SPI area) to the economic functions.

The first economic function is the assurance of survival (Survivability). It comprises the final uses of all survivability products (see p.7) regardless of the final uses sections in the SUTEA (and all the intermediate flows necessary for the provision of these products).

The second function is the provision of the Consumption Surplus. It comprises all products in the final uses sections FCE, Dwellings and Valuables except survivability products. Parts of investment (gross fixed capital formation – GFCF) consist of goods and services for households as well. These goods and services are to be included in CS.[5]

The third economic function comprises all products for intra-economic use. Products for intra-economic use are products in the final uses section GFCF (except dwellings). Again, survivability products are excluded. This third function shall be called Production.

Table 1. Allocation of final uses to economic functions

Survivability	Production	Consumption surplus
Food, energy and products of defensive nature	GFCF/Other buildings and structures (except Survivability)	FCE by households (except Survivability)
	GFCF/Machinery (except Survivability)	FCE by government (except Survivability)
	GFCF/Transport equipment (except Survivability)	FCE by NPISH (except Survivability)
	GFCF/Other GFCF (except Survivability)	GFCF/Dwellings (except Survivability)
	Changes in inventories (except Survivability)	Valuables

6.2 SPI Valuation Accounts

The second extension that we are going to discuss in this manuscript is the inclusion of SPI valuation accounts.[6] The first step of the calculation of the SPI yields weighting factors for natural resources, ecosystem inputs and residuals to air water and soil. On the input side, these factors represent the inverse of the mass of a resource per area and year. Accordingly, their dimension is m^2a/kg. On the output side, the factors are calculated as the inverse of the mass of a residual that can be absorbed per area and year. They are of the same dimension as the input factors. In a second step, the SPI factors are multiplied by flows of inputs from the environment (resources) and outputs to the environment (residuals) [kg/a] recorded in physical units in the SUTEA.

---

[5] Consumption and investment can represent goods and services available to society once survivability requirements are met. Both (parts of) consumption and (parts of) investment are results of economic activity made available to society. It follows that the consumption surplus must comprise consumption goods as well as investment goods that are made available to the social environment of economic systems and that are not for survivability purposes.

[6] For additional extensions and calculation instructions we refer the reader to (Gassner 2002).

The product of weighting factor and flow gives the area needed to contain a resource or residual flow. In order to integrate SPI calculations with physical-monetary supply and use tables, the physical flows of resources, ecosystem inputs and residuals are multiplied by the respective SPI factors. The overall SPI area of an industry is calculated as the sum of the single areas needed to supply flows of resources and dissipate flows of residuals. Aggregation of flows of resources is straight forward. For flows of residuals, it is assumed that the same area of an environmental compartment (air, water and soil) can absorb flows of different substances at the same time. Thus, to calculate the overall SPI area for the three environmental media single substance areas are not added. Overall SPI area for a medium calculates as the biggest single substance area for that medium. The (biggest) single substance area is thought to contain all other single substance flows as well. The sum of SPI areas per industry yields the total SPI area of an economic system. The SPI area of an economic system plus the SPI area for flows of resources and residuals due to consumptive activities (the SPI area of the Households) gives the overall SPI area of a given nation or region.

7 Résumé

The Sustainable Economy Indices surpass in mainly four aspects the current state of the discussion of economic sustainability measurement:

1. The issue of money flows (solvency) is introduced to the field of sustainability accounting and indicators. An element that has always been of utmost importance in conventional economics is thereby accorded its legitimate place in economic sustainability. Together with the money flows related to trade, the flows of incorporated environmental pressure in trade flows are for the first time calculated within a standardised accounting framework. This "environmental-economic balance of payments" shifts the focus of sustainability analysis from single economic systems to the economic and environmental interrelations between economies. This allows for an extended analysis of co-operations and symbioses of economic systems against the backdrop of economic sustainability.
2. A novel classification of economic activities is proposed. The formation of three economic functions (Survivability, Consumption Surplus and Production) allows for the analysis of structural particularities of economic systems against the backdrop of sustainability. Better than standard classifications, it allows analysing the share of economic activity (and the related consumption of natural resources and value added) that contributes directly to meeting the demands of the social environment. The formation of the functions Survivability and Production show how much value added and how much natural resources are consumed for "auxiliary" economic activity.
3. Standard accounting instruments and methods are brought together with aspects of alternative measures of sustainability (such as the ISEW). Thereby, an alternative view on sustainability is used and at the same time the usual difficulties

in calculation of alternative measures are avoided. Standard conceptual views are widened while standard methodology is assured.
4. A measure of the ecological footprint family (the Sustainable Process Index) is integrated into a standardised system of environmental economic accounts. Thereby, a widely acknowledged and widely used method of ecological sustainability measurement is made part of a concept of economic sustainability and environmental-economic accounting. At the same time, evaluations of ecological sustainability for regional and national economic systems are facilitated and standardised.

References

Cobb CW, Cobb JB (1994) The Green National Product - A Proposed Index of Sustainable Economic Welfare. University Press of America, Maryland

Gassner J (2002) The Sustainable Economy Indices – Measuring the Sustainability of National and Subnational Economic Systems. Technische Universität, Graz

Hartwick JM (1977) Intergenerational Equity and the Investing of Rents From Exhaustible Resources. American Economic Review 67: 972-974

Hartwick JM (1978) Substitution Among Exhaustible Resources and Intergenerational Equity. Review of Economic Studie 45: 347-354

Holman JA (2001) Is the Large U.S. Current Account Deficit Sustainable? Economic Review of the Federal Reserve Bank of Kansas City, First Quarter: 5-23

Krotscheck C (1995) Prozessbewertung in der Nachhaltigen Wirtschaft. Technische Universität, Graz

London Group on Environmental Accounting (2001) System of Environmental Economic Accounting 2000, Vorburg Draft
www4.statcan.ca/citygrp/London/publicrev/pubrev.htm

Luhmann N (1988) Die Wirtschaft der Gesellschaft. Suhrkamp, Frankfurt

Mann CL (1999) Is the U.S. Trade Deficit Sustainable? Institute for International Economics, Washington

Ott K (2001) Eine Theorie "Starker" Nachhaltigkeit. Natur und Kultur 2/1: 55-75

Pearce DW, Maradya A, Barbier EB (1989) Blueprint for a Green Economy. Earthscan, London

Perman RYM, McGilvray J, Common M (1999) Natural Resource and Environmental Economics. 2nd Edition, Longman, Essex

Solow RM (1974a) The Economics of Resources or the Resources of Economics. American Economic Review 64: 1-14

Solow RM (1974b) Intergenerational Equity and Exhaustible Resources. Review of Economic Studies Symposium: 29-45

Solow RM (1986) On the Intergenerational Allocation of Natural Resources. Scandinavian Journal of Economics 88: 141-149

Environment as a factor of national competitiveness in manufacturing

Bojan Radej[a,b], Ivanka Zakotnik[a]

[a] Institute of Macroeconomic Analysis and Development of the Republic of Slovenia – IMAD; [b] University of Ljubljana, E-mail addresses: bojan.radej@siol.net (B. Radej), ivanka.zakotnik@gov.si (I. Zakotnik)

Abstract

This paper examines the changes in environmental quality that have occurred in Slovenia's economic transition using estimates and projections of pollution levels and natural resources content of manufacturing goods exports for the period 1992–2006. The empirical questions addressed are two. The first is whether the transition from a centrally planned to a market-driven economy exacerbates environmental degradation, or instead encourages environmental improvement. The second question under study is whether the export growth projected in the Strategy of Economic Development for Slovenia is compatible with environmental sustainability. The study used export structure data and calculated relative changes in environmental pressures using international databases that link industrial structure with pollution emissions and use of natural resources. We found that in the early phase of transition (1992 to 1998), there was a clear decoupling between export patterns and environmental pressure. After 1998, this trend is no longer clear. Moreover, analysis of export projections show that pollution emissions linked to exports would increase by 22% between 2000 and 2006 while the natural resource intensity of goods exports would fall by 0.7 structural points. Based on these results, in particular rising pollution emissions, we conclude that there is a conflict between the Slovene Strategy's stated goal of sustainability on the one hand and the increasing environmental burden of exports on the other.

Keywords: competitiveness, environment, market transition, development patterns

1 Introduction: background and thesis

Accelerated globalisation and a greater contribution to the international division of labour by ten EU accession central and east European transition countries (CEECs) are the key features characterising the world economy of the 1990s (Aiginger et al. 1998). From this statement, two research issues arise concerning the environmental sustainability of economic development in CEECs: (i) the link between economic transition and environment; and (ii) the link between globalisation and environment.

The first empirical question addressed is whether the transition from a centrally planned to a market-driven economy exacerbates environmental degradation, or instead environmental improvements complement economic transition. The literature on explicit linkages between the environment and the socio-economic transition from central planning to market economies tends to be based on scarce and incomplete evidence (Pearce and Warford 1993; OECD 1995; Zylicz 1997; World Bank 1998). In theory at least, the transition from more to less market failure is consistent with an overall improvement in environmental quality. The conclusion would most likely hold in the very long run. However, in the short and medium run, the consequences of transition are far from obvious. This is especially true since even in the most developed and mature market systems, environmental media are the last areas where property rights are established and market failures alleviated (Vukina et al. 1999). Zamparutti and Gillespie (2000) distinguished two groups of transition countries regarding their success to decouple economic growth from additional environmental pressures in the 1990s. In the first group there was a synergy among economic reforms, democratic development, and environmental improvement. These countries, now in the second phase of transition, face a variety of challenges to deepen integration between the environment and sectoral policies. In the second group of countries (particularly countries of the former Soviet Union[1]), pollution reductions initially resulted from declines in economic production, but many of these countries now face renewed environmental problems linked to renewed economic growth.

The second issue under study here is the environmental effect of globalisation and internationalisation. These have diverse environmental consequences which are creating new problems and challenges for national and international economic policy (Zamparutti and Gillespie 2000). In theory, every country should tend to optimise its growth pattern relative to its availability of resources (environmental, social, and economic).

Globalisation raises many issues, including the appropriate roles of national economic and environmental policies. At the least, all economic inputs—and environmental ones in particular—should be closely monitored as a precondition for policies to improve the environmental performance of national competitiveness.

After joining the EU, the CEECs will no longer be national economies in which the government controls all economic policies, but will become regional econo-

---

[1] Not including the three Baltic States: Estonia, Latvia, and Lithuania.

mies in which at least some policy responsibilities are transferred upward to EU administration. Domestic policies will no longer be able to protect national welfare with measures that erode competition in general and price competition in particular (IMAD 2002). There are other problems increasingly aired in connection with the globalisation issue, which, incidentally, also heightens pressures to harmonise national economic policies across industrialised countries (Aiginger et al. 1998; Zamparutti and Gillespie 2000) in areas such as humanitarian issues, development co-ordination, social and labour standards, ecological issues, and unemployment.

Most CEECs are small open economies—their exports average about 45% of GDP (Aiginger et al. 1998) and reach almost 60% for Slovenia (IMAD 2002). In the pretransition (pre-1989) era, industry was the highly favoured sector, its share in GDP was much higher than in comparable Western economies and its structure was different as well. The following main features characterised pretransition industries: a general overhang of heavy industry, a relative surplus in labour-intensive industries and a pronounced structural deficit in sophisticated technologies. This pattern was accentuated through the central role played by natural resources. The major reason for this structural outcome was that production patterns were decided by planning authorities instead of the market. This led to a widening gap with the technological progress and associated structural changes in the West, as well as systematic inattention to environmental side effects. Improvement in the development pattern in the first period of transition (early 1990s) was mostly a by-product of defensive economic restructuring (Simoneti et al. 2000). In the second half of the nineties, dynamic economic growth continued and in the more advanced group of transition countries the social component of welfare improved (Seljak 2001), particularly in terms of reduced unemployment and relative poverty. With regard to the environmental component, however, the situation deteriorated as a result of restored economic growth without sufficient parallel environmental improvement (IMAD 2002). Accordingly, the transition to market-driven economies generated a pressing need for environmental adjustments in industry as one of the priority elements of environmental integration (Zamparutti and Gillespie 2000). In industry, output has declined more than national GDP and a fundamental restructuring has been foreseen with important consequences for the further development (Aiginger et al. 1998).

Slovenia is one of the smallest of small CEECs. It is poor in competitive stocks of mineral resources and fossil fuels, but rich in environmental services[2], a consequence of its geographical diversity. Nonetheless, Slovenia's ecological footprint[3] is twice as big as its territory. The corresponding figure for all countries in the world is 1.2 (indicating that humanity exceeds the Earth's ecosystem carrying capacity by 20%), for the USA 5.6, for CEECs between 1.1 (Bosnia) and 2.7 (Estonia; Wackernagel et al. 2002). The ecological footprint estimates suggest that in the

---

[2] The capacity to absorb, disperse or otherwise neutralise anthropic emissions to the environment.
[3] Over the long-term, an ecological footprint larger than 1 for the whole planet is not achievable without deteriorating conditions for economic development in the future.

long run the global pattern of economic growth needs to change in order to preserve the level of sustainability and economic activity already achieved. At the national level, however, economic development is not necessarily unsustainable if one country relies on the surpluses of economic resources of another—or if it succeeds in covering welfare debts to present and future generations at home and abroad (Tietenberg 1996). Nonetheless, Slovenia's current footprint indicates that economic trends need to be closely monitored for their environmental impacts and long-term sustainability.

For a small economy, an open domestic market together with a strong export sector can substantially compensate the lack of competition at home and increase its attractiveness as a location for investment, with growing significance for economic performance and for the country's global (national, complex)[4] competitiveness (IMAD 2002). In addition, environmental factors can be an important element in global competitiveness (Kovačič 2001). For Slovenia, however, several studies have identified Slovenian exports of goods as relatively high in terms of their content of natural resources and environmental emissions (UNECE 1997; Vukina et al. 1999). The energy intensity of Slovenia's GDP is almost twice that of the EU average and investment has further increased energy-intensive economic activity (IMAD 2002). These and other studies indicate that the share of exports based on natural resources and environmental pollution is too high (World Bank 1998; Aiginger 1998).

Following Dunning (1992), with higher economic development competitiveness is increasingly determined by created factors, such as technology- or human resource-intensive exports, while previously decisive non-created factors of competitiveness, such as an availability of commercially attractive supplies of natural resources or cheap labour are diminishing in their importance for national competitiveness. The growing role of created assets in competitiveness also means that social and the environmental capital have ceased to represent restrictions, but have themselves become important created assets for integral economic development.

In addition, it would be reasonable to assume that countries with strong environmental regulatory regimes (government provisions) generally show strong competitiveness and GDP performance, as is confirmed by Kovačič (2001). Competition alone does not always lead to the optimum results and thus government should provide for the creation of socially desirable "efficient competition" (IMAD 2002). Governments in general share an essential responsibility for designing the profile of national competitiveness.

The Strategy for the Economic Development of Slovenia 2001–2006 (SEDS; IMAD 2001a) established a new development paradigm based on the principles of

---

[4] A measurable capacity of manufacturing internationally competitive products and services that ensures sustainable development and sufficiently high living standards (the Institute for Management and Development; IMAD 2000). Another and theoretically equivalent definition sees a nation's competitiveness as the degree to which the country can, under free and fair market conditions, produce goods and services that meet the test of international markets and that simultaneously maintain and expand the real incomes of its people over the long term (WEF 2002).

sustainability. The Strategy also identifies export growth as one of the main components for Slovenia's economic growth. Our research investigated the compatibility of these two goals, through an analysis of current and projected manufacturing export patterns.

Natural resources and environmental services contribute to economic development and welfare in two ways: first by use of natural resources in increased economic activity and second by absorption of additional waste or pollution. The identification of the relationship between trade and environment has to take into account both impacts: i.e., the natural resource factor content of exports and the pollution load linked to exports.

In the section Working methodology and data we elaborate our working methodology. Then we try to identify the nature of the relationship between environment and economic growth for Slovenia's exports of manufacturing goods. We investigated (i) the content of natural resources in exports and (ii) the composition of goods exports in terms of total pollution load. We conclude by identifying the problematic sectors of Slovene goods exporters in terms of sustainability. General conclusions and strategic policy reforms, in the light of the SEDS's new development paradigm of sustainability (IMAD 2001a), are proposed at the end.

1.1 Theory

One of the central questions in economic trade theory is that of specialisation and thus, explaining the international trade that is actually observed. The Law of comparative advantage states that countries engaged in trade specialise in the production of those goods in which they have a relative, not necessarily an absolute, cost advantage over their trading partners. In the simplest setting of the Ricardian model, there are two countries, two goods and only one factor of production (labour; Wolfmayr-Schnitzer 1998). In such a setting, the countries' factor endowments are bound to be identical and cross-country differences in production technologies are the main source of comparative advantage. Differences in technology are thus at the centre of explanations of specific patterns in international trade. The Heckscher–Ohlin theory of trade, or factor-proportions theory, assumes that production technologies are the same across countries, thereby ruling out productivity differences which were of central importance in the original Ricardian model. In the two-factor economy of the Heckscher–Ohlin theory, comparative advantages are then entirely due to differences in factor endowments or the relative abundance of factors of production. Thus, under the assumption of given and basically identical technologies across countries, the Heckscher–Ohlin model predicts that countries will tend to specialise in the export of goods whose production is intensive in factors with which they are abundantly endowed. In the early 1980s a new generation on trade models, drawing heavily on the insights of the new endogenous growth theory by Romer (1986) and his followers, began to emerge. In these models, technological change driven by investments in R&D had a major role to play. In a certain way, trade theory returned to the basic Ricardian idea that the trade pattern is largely driven by international differences in technology rather than factor endowments (Wolfmayr-Schnitzer 1998).

With technology playing a decisive role in the formation of international trade patterns, it should be possible to draw conclusions on the technological competitiveness and to that end the qualitative competitiveness of a country from an analysis of observed trade specialisation (Wolfmayr-Schnitzer 1998).

According to the new trade theories, which explain trade in terms of technology, technology diffusion/adjustment lags and continuous innovation processes, CEECs would specialise in the export of old, mature goods. The production processes for these goods become routine and less skilled labour has to play a greater role, either because technology for the production of the new products is not available (Posner 1961) or because, while technology is available, the production process is intensive in the use of skilled labour, which is relatively rare in CEECs (Hirsch 1967) compared to the OECD, or because CEEC markets do not represent the (high-income) markets where new products are first demanded (Vernon 1966).

Based on the existing literature on growth, trade, and environment, several operational hypotheses pertaining to the transition and post-transition periods can be developed. First, one can argue that the environment in the economies in transition would be cleaned up quickly because of rising energy prices, penalising energy-intensive activities (Hughes 1991). In addition, foreign direct investment should introduce less polluting technologies not domestically developed, and Klavens and Zamparutti (1995) found, through a survey of multinational corporations, that most foreign companies looked thoroughly at environmental issues when it came to making decisions about investing in CEECs.

Second, one can argue that the abundance of skilled labour in CEECs may result in a new allocation of resources towards cleaner industries induced by international market discipline, as shown to be the case for Mexico (Beghin et al. 1995; Grossman and Krueger 1993). These optimistic views may be invalidated by the fact that many transition economies have lax environmental regulations and are "abundant" in environmental endowment (Rauscher 1995). If this is true, then one could potentially observe worsening pollution due to an increased specialisation in pollution-intensive industrial activities.

Some of these questions can be addressed by looking into the relationship between the pollution intensity of an economy and its openness. The literature on this question presents conflicting evidence. Lucas et al. (1992) found that pollution intensities for outward-oriented and fast-growing economies were decreasing rapidly in the 1980s, and further, that a negative relationship could be established between an index of openness and pollution intensity of value added. This finding suggests that an outward orientation is environmentally less damaging than an inward orientation. However, Rock (1996) found little econometric evidence to support the relationship between outward orientation and environmental improvement.

Analysis of foreign trade data is one of main techniques for the assessment of competitiveness[5]. It is based on the assumption that the share of intra-industry foreign trade and the terms of trade are indicators of value-added growth from foreign trade, and value-added growth reflects the changing competitive position of the country. If the share of natural resource-intensive and highly polluting industries in value added and in goods exports drops, this indicates improved environmental performance of goods exports and this is interpreted as an improvement of the competitive position of a particular country (industries, niches). Following this line of thought, our study assesses environment as the qualitative (non-price) aspect of Slovenian competitiveness in goods exports. Slovenia's transformation curve and production function are considered, in a range of studies, as sustainable, reflecting a relatively clean environment (WEF 2001), high biodiversity, and relatively good human resources (Seljak 2001; IMAD 2001a, 2002, 2003; Hanžek 1998; World Bank 1998; Radej 2001). Relative prices of energy are above average for CEECs and environmental legislation has steadily improved over the last decade (MOPE 2002; IMAD 2001b, 2002). These are clear preconditions for an ambitious sustainable development policy in Slovenia and specifically for successful implementation of SEDS (IMAD 2001a). However, linking environment to foreign trade issues is successful only when economic policy recognises and supports preconditions for the sustainability of economic growth. This is not necessarily taking place in transition countries because of persisting structural problems, such as the exemption of large energy consumers from eco taxes and a large gap in implementation of environmental policy (Radej 2001). Therefore, from both an economic and an environmental perspective, it is reasonable to study environmental issues in Slovenia via trends in goods exports over the 1990s and the first half of this decade, to determine whether or not Slovenia specialises in production that is intensive in relatively abundant economic resources.

2 Working methodology and data

2.1 Estimating integrated emissions linked to exports

For pollution intensity estimates we used a database called The Industrial Pollution Projection System (IPPS) developed by the Economics of Industrial Pollution Control Research Project at the World Bank (IPPS 1994). Emissions intensities in the IPPS database have been calculated from the U.S. Manufacturing Census, using detailed information on more than 100 different effluent types collected from approximately 200,000 plants. This analytical approach rests upon the assumption

---

[5] In addition to benchmarking, surveying stakeholders on factors of competitiveness (as undertaken by as IMD and WEF) and consultations among businesses and government are undertaken (Kovačič 2001).

that industrial emissions are heavily affected by the scale of industrial activity, its sectoral composition, and the technologies employed in production.

At the broadest level of pollutant aggregation, IPPS intensity estimates are available for the sum of all toxic pollutants released to all media (air, water, land), an indicator called "integrated pollution". At the narrowest level, separate intensities have been estimated for air, water, and land release of over 100 toxic pollutants. Access to these emissions and economic data presents a unique opportunity to develop a comprehensive picture of the environmental risks associated with industrial development, enabling a reasonable estimate to be made of the emissions associated with any given level of activity, in any specified industrial sector. Conceptually, such estimates can be presented as an index of "emissions (or pollution) intensity", expressed as a ratio of emissions per unit of manufacturing production or value added.

Cross-country variations in regulatory, economic, and technological conditions clearly impose limitations on the applicability of the pollutant intensity indices derived in this study for other countries (IPPS 1994). Since research work that could enable adjustments of IPPS parameters for national differences in regulatory regimes, factor prices, and availability of technology in Slovenia is largely limited by the availability of the national data, we rely on the original IPPS database as an approximation. Results based on IPPS data have already been made for various countries[6] where insufficient data on industrial emissions has been an impediment to establishing pollution control strategies and prioritising environmental measures. Zylicz (1997) as well as Vukina et al. (1999) also reported severe data and statistical limitations when researching inter-temporal and inter-country trends in pollution emissions in former CEECs.

For Slovenia, previous research has shown that the IPPS database provides a ranking of industrial sectors by integrated pollution that, at least for the highest level of aggregation, is closely correlated to the ranking based on their actual emissions (Mani et al. 1997). While the IPPS database provides a measure of integral pollution, it also indicates that pollution-intensive sectors ("dirty industries") differ markedly across pollutants. The implication for applied work is clear. As pollution intensities are very sensitive to differences in the composition of production within a particular sector, pollution projections should always be done with the most disaggregated data available. The resulting gains in accuracy are often quite striking.

We used industrial activity data at the three-digit International Standard Industrial Classification (ISIC) disaggregation level[7] and the applicable IPPS intensity estimates for 14 effluent types, grouped into four subsets which contain a set of

---

[6] IPPS has been used to produce the first comprehensive cross-country estimates of toxic pollution in *World Resources 1994–95* (Table 12.4), published by the World Resources Institute. Work on trade and the environment by the OECD has also been based on IPPS, most notably the paper by David Roland-Holst and Hiro Lee: "International Trade and the Transfer of Environmental Costs and Benefits" (OECD, December 1993).

[7] The three-digit level was the most detailed aggregation available for data on economic activity in Slovenia.

conventional pollution parameters including (i) toxic chemicals released in air, water, and soil; (ii) bio-accumulative metals in air, water, and soil; (iii) air pollutants (nitrogen oxides, sulfur oxides, volatile organic compounds, suspended particulates, fine-particulates, carbon monoxide); and (iv) water pollution (biological oxygen demand and total suspended solids).

Intensities are expressed in pounds of pollutant per million U.S. dollars of value added at 1987 prices.

We have estimated the baseline integrated pollution levels of manufacturing production in Slovenia simply by multiplying the pollution intensity coefficient for each pollutant (from IPPS estimates for individual production sector) by Slovenian data on value added of production for each industry. Baseline estimation of integrated pollution levels linked to exports for each manufacturing sector was derived from the estimated sector pollution load using the share of exports in production of the sector.

2.2 Factor content of exports

According to distinct relative factor intensities, we cluster commodity groups of exports into four groups. This breakdown of exports is derived from the International Trade in Commodity Statistics (ITCS) database using a common conversion key (based on UN classification work) which maps product groups according to Standard International Trade Classification SITC Rev.3 (three-digit breakdowns).[8] The four groups are: (i) natural resource-intensive products, (ii) unskilled labour-intensive products, (iii) technology-intensive products, (iv) human capital-intensive products.

The first two groups represent lines of production characterised by low value added, high natural resource intensities, and simple technologies. The first group consists of food, beverages, crude materials, mineral fuels, animal and vegetable oils, leather, plywood, mineral manufactures, diamonds, and non-ferrous metals. The second group includes commodities with the lowest value added per worker (textiles, garments, furniture, glass, etc.)

The line dividing technology- and human capital-intensive groups is fuzzy: both contain products requiring more sophisticated inputs than found in the first two groups. In the third group, technology-intensive products, are found goods with the highest ratios of R&D (research and development) expenditures to value added, whereas the human capital-intensive group contains goods with the lowest ratios of R&D expenditures to value added. The third group includes chemicals (plastics, fertilisers, etc.) some capital equipment, telecommunications equipment,

---

[8] The classification was developed by the United Nations Conference on Trade and Development (UNCTAD) in SITC Rev.2 and updated in SITC Rev.3. The product composition of exports is based on trade data reported to the database COMTRADE of the United Nations Statistical Division (UNSD; http://www.Intracen.org/menus/countries.htm). The COMTRADE database covers more than 90% of world trade.

medical, scientific, and measuring equipment, and photographic supplies. The fourth group includes such goods as paints, rubber, paper, TV and radio sets, etc.

Finally we decompose estimated changes in the factor content of exports. However, this is not possible in the same way as in the case of decomposition of emissions from dirty industries (see the section Medium-term projections of integrated pollution levels linked to exports). The reason is methodological–the definitions of factor groups of products which we adopted mean that products are not allowed to move between intensity groups, setting their per unit intensities as constant (1). For example, installing technologically sophisticated environmental protection devices in a previously natural resource-intensive business might call for it to be reclassified into the technology-intensive group. Therefore, the aggregate of changes in the factor content of exports of goods can be decomposed only into changes of exports (in real terms) (i) by scale effect (subgroups) and (ii) by composition effect (within a subgroup). Scale increases in the export of goods by 2006 are of course the same for pollution intensity and for a factor contents study and are thus estimated from UNCTAD grouping and SEDS projections of export growth by manufacturing subindustries (IMAD 2001a). From the estimation of scale increases by subgroups and the structure of exports in the previous period, the change in structure of exports by prevailing factor was calculated. In the final step we assume that changes in the structure of exports by prevailing factor reflect (are equal to) the changes in factor composition of exports within the subgroup.

2.3 Medium-term projections of integrated pollution levels linked to exports

We analysed the dynamics and structure of manufacturing exports, emissions, production, and value-added growth for the period between 1992 and 2006 using data sets from statistical sources and the latest IMAD macroeconomic forecast.

The projected improvements in technical/technological processes over time which enable decreases in environmental degradation per unit of production were also estimated. The expected improvements in technological process of production in the next medium-term period were derived from a separate study of development in manufacturing for the period between 1995 and 2006 (Gmeiner 2000). Drawing on the findings presented in this study, we have corrected the U.S.-based estimates of pollutant intensities (originally applied in the baseline estimation of integrated pollution for the beginning year) by the projected index of quality changes in manufacturing's production function. Gmeiner (2000) assumes that technological improvements as a whole will be reflected in environmental improvements. He assumes that new technologies (environmental investment) in manufacturing will accumulate depending on the ratio between index of value-added growth and index of physical production growth in manufacturing. Pollution intensity for each year is obtained simply by multiplying this calculated ratio for the year by the pollution intensity in Slovenia in the base year.

Finally, an aggregate pollution projection was calculated as:

$$P_2/P_1 = \sum_i (p_i/Pr_{si}r_{ui})E_2/E_1 \qquad (1)$$

where P_2/P_1 denotes the change in total pollution (P) linked to exports in period 2 to 1; p_i/P denotes the share of the ith sector pollution (p) in total pollution (P) linked to exports; $r_{si}=s_{2i}/s_{1i}$ denotes the change (r) in share (s) of sector i in total exports in period 2 to 1; $s_{1i}=e_{1i}/E_1$ denotes the share of exports of sector i (e_i) in total exports (E) in year 1 ($E_1=\sum e_{1i}$); $s_{2i}=e_{2i}/E_2$ denotes the share of exports of sector i (e_i) in total exports (E) in year 2 ($E_2=\sum e_{2i}$); $r_{ui}=u_{2i}/u_{1i}$ denotes the change in the pollution intensity factor (u) of sector i; $u_{1i}=p_{1i}/e_{1i}$ denotes the integrated pollution level of sector i (p_i) per unit of exports of sector i (e_i) in year 1; $u_{2i}=p_{2i}/e_{2i}$ denotes the integrated pollution level of sector i (p_i) per unit of exports of sector i (e_i) in year 2; E_2/E_1 denotes the change in total exports (E) in real terms. Change in pollution linked to exports (P_2/P_1) can be finally decomposed to composition effects, $\sum_i (p_i/Pr_{si})$, change in pollution intensity (r_{ui}) and scale effects (E_2/E_1).

3 The results

3.1 The structure of Slovenia's exports of goods by factor content in the 1990s

In the period of market transition, the structure of goods exports by prevailing factor content (inputs in production) changed significantly. The changes in the relative share in goods exports that occurred to each group of products in the period 1992–2001 are presented in Table 1

Technology-intensive and human capital-intensive products recorded the largest increases (created or derived factors of competitiveness). The same pattern is seen in the Czech Republic, Hungary, and Slovakia (see Appendix). The total share of these products in Slovenia's exports increased from 54.7% in 1992 to 64.7% in 2000 (higher shares were recorded by Germany, Austria, Finland, Ireland, the U.K., France, and Sweden, as well as the Czech Republic and Hungary); the share of technology-intensive products rose from 21.3% in 1992 to 27.6% in 2000, while the share of human capital-intensive products increased from 33.4% to 37.1%.

The export of goods that intensively use non-created factors of competitiveness (natural resources and unskilled labour) accounted for 35% of Slovenia's exports of goods in 2000, drawing close to the shares of Belgium, Italy, the Netherlands, and Spain. The lowest shares in the EU were recorded by Austria, Finland, Ireland, Sweden, and the U.K. Unskilled labour-intensive products represented 19.8% of Slovenia's exports of goods in 2000. Their share has fallen steadily since 1994 (they accounted for 24.8% of exports in 1994), but is still substantially higher than in most EU member states and was higher than in some transition countries in 2000 (Czech Republic, Hungary, and Slovakia). EU members that recorded larger shares than Slovenia in 2000 were Italy, Greece, and Portugal. The main groups of labour-intensive manufactures exported from Slovenia were furniture, men's coats, women's coats, footwear, and other clothing. These five groups accounted for up to 60% of Slovenia's unskilled labour-intensive exports.

Table 1. The structure of Slovenia's exports of goods by factor content, %

Primary factor (intensity) groups	1992	1993	1994	1995	1996	1997	1998	1999	2000	2001
Natural-resources	20,6	20,9	17,7	16,7	16,3	16,6	15,6	15,3	15,5	15,4
Unskilled-labour	24,7	26,4	24,8	23,3	22,7	21,6	21,3	21,4	19,8	20,1
Technology	21,3	21,9	24,4	24,9	25,7	26,4	25,9	26,6	27,6	28,4
Human-capital	33,4	30,8	33,1	35,1	35,4	35,5	37,2	36,7	37,1	36,1

Source: ITC COMTRADE database of the United Nations Statistics Division, SORS, calculations by Zakotnik, 2000a.

Natural resource-intensive products represented 15% of Slovenia's exports of goods in 2000, and here Slovenia was closest to Austria, Finland, Hungary, and the Czech Republic. The main groups of natural resource-intensive products in Slovenia's exports of goods were food, beverages, crude materials, mineral fuels, animal and vegetable oils and fats, leather, veneers and other wood in the rough or roughly squared, and ferrous and non-ferrous metals.

In 1995–1999, the share of exports produced with a high content of created factors of production rose by about 0.8 of a percentage point per year. Share of exports produced by natural resource-intensive industries fell proportionately. In 2000, structural shifts towards more intensive use of created factors increased, with the share of exports involving created factors of production rising by 1.4 percentage points; the share of unskilled labour-intensive exports dropped by 1.6 percentage points, while the share of resource-intensive exports increased by 0.2 percentage points.

These results echo previous studies. Vukina et al. (1999) reviewed CEEC manufacturing trends in terms of their pollution emissions, and estimated that in Slovenia the composition effect of manufacturing output resulted in consistent environmental improvements in the early and mid-1990s for most pollution emission types, except for volatile organic compounds and bio-oxygen demand.

Thus, the favourable trends of decreasing shares of unskilled labour-intensive and natural resource-intensive exports and the rising shares of human capital-intensive exports stagnated or worsened after 1998, while the trend of rising shares of technology-intensive exports continued. This suggests a divergence from SEDS goals, according to which Slovenia should become more competitive in both technological and human created factors of production (IMAD 2001a).

Overall, in the past decade significant favourable changes in the composition of Slovenia's exports by factor intensity occurred. The shares of natural resource-intensive and labour-intensive products have gradually declined. In 2000, technology-intensive products accounted for 27.6% of Slovenia's exports, compared with

21.3% in 1992. Human capital-intensive products have the highest share (37.1%) in Slovenia's exports (2000); nine years ago their share in exports was 33.4%. The share of natural resource-intensive exports has decreased (from 20.6% of total exports of goods in 1992 to 15.5% in 2000).

3.2 Export projections through 2006

In the medium-term projections for 2001–2006 (Table 2), we expect further improvements in line with the orientation towards sustainability. The share of created factors of export competitiveness (technology, human capital) against natural ones (unskilled labour force, natural resources) is expected to increase by 3.3 structural points. The greatest increase is expected in the share of human capital-intensive exports (by 2.5 structural points). It is expected that major investments in the improvement of product quality together with a significant modernisation of the organisation of business activities will result in a large increase of production of transport equipment, most of which will be exported (through co-operation with strategic partners abroad). The share of technology-intensive products in exports is expected to increase by 0.8 structural points. This takes into account benefits realised through higher R&D expenditures and other advantages of the specialised research (Gmeiner 2000). In the medium term the greatest technological improvements are expected in the production of machinery and equipment, in particular electrical and optical equipment, transport equipment, and rubber and plastic products. The share of natural resource-intensive exports is expected to decrease by 0.7 structural points. The expected decrease is mainly derived from projected lower shares of exports of pulp and paper products, basic metals, and fabricated metal products.

The high and increased shares of food, beverages, and wooden products exports-natural resource-intensive sectors-by 2006 is not necessarily problematic in terms of sustainability. These exports are based on domestic renewable or reproducible natural resources: wood, water (for beer, beverages), and land (agricultural production). If these natural resource-intensive goods exporters produce with low emissions, they can be treated as sustainably favourable, green exports. The same conclusion applies to predicted increases in the export of wood and wood products, which are also renewable and abundant natural resources in Slovenia. The share of basic metals and products in exports is projected to fall, which would be environmentally favourable.

Following SEDS projections (IMAD 2001a), the natural resources content in goods exports would increase by 37.7% by 2006 (Table 3), compared to year 2000. Thus, the medium-term increase is projected to be substantially higher than that of the preceding period (24.9% increase). This would result from the combined effect of growth in natural resource-intensive exports (a predicted increase in volume of 44.4%) together with a decreased content of natural resources, 4.6%, resulting from environmentally favourable restructuring, with the outcome of the latter being much lower than in the previous period. Overall, the increase in natural resources content in goods exports is far too high - in terms of Slovenia's relatively poor endowments (World Bank 1998). Indeed, the increase surpasses labour-intensive export of goods' projected growth by 2006. Moreover, the projected increase is a concern in terms of sustainability, considering that Slovenia's ecological footprint already exceeds its national carrying capacity (Wackernagel et al. 2002).

Table 2. Structure of exports of goods by product origin and factor content, shares in %

Export product groups	Exports of goods		Exports of goods by structure of factor content							
			Natural resources		Unskilled Labour		Technology		Human capital	
	2000	2006	2000	2006	2000	2006	2000	2006	2000	2006
Exports of goods (1+2)	100.0	100.0	15.5	14.8	19.8	17.2	27.6	28.4	37.1	39.6
1. Exports of goods from manufacturing, in which:	98.9	98.9	14.4	13.7	19.8	19.5	27.6	28.4	37.1	37.3
Food, beverages, tobacco	3.5	4.1	3.5	4.1	0.0	0.0	0.0	0.0	0.0	0.0
Textiles and textile products	7.4	5.4	0.5	0.4	6.9	5.0	0.0	0.0	0.0	0.0
Leather and leather products	1.6	1.6	0.1	0.1	1.5	1.5	0.0	0.0	0.0	0.0
Wood and wood products	3.5	3.8	3.4	3.6	0.1	0.2	0.0	0.0	0.0	0.0
Pulp, paper and paper products; media; printing	5.4	4.6	0.3	0.1	0.0	0.0	0.0	0.0	5.1	4.5
Chemicals, products, fibres	11.0	10.8	0.0	0.0	0.4	0.2	7.5	7.5	3.1	3.1
Rubber and plastic products	4.5	4.9	0.0	0.0	1.4	1.4	0.6	0.9	2.5	2.7
Other non metallic mineral products	2.6	2.8	1.4	1.4	1.2	1.4	0.0	0.0	0.0	0.0
Basic metals and products	12.1	11.2	4.7	3.9	0.0	0.0	0.0	0.0	7.3	7.3
Machinery, equipment n.e.c.	13.9	13.9	0.0	0.0	0.6	0.4	7.6	7.8	5.7	5.7
Electrical, optical equipment	12.2	12.2	0.0	0.0	0.6	0.2	11.0	11.1	0.6	0.9
Transport equipment	13.2	15.7	0.0	0.0	0.1	0.1	0.5	0.6	12.6	15.0
Other goods, n.e.c.	7.9	8.0	0.3	0.3	7.0	6.8	0.4	0.5	0.2	0.4
2. Exports of goods from non-manufacturing activities	1.1	1.1	1.1	1.1	0.0	0.0	0.0	0.0	0.0	0.0

Source: Zakotnik, 2000a.

Table 3. Changes in factors content of exports

	Period	Natural-resource	Unskilled-labour	Techno-logy	Human capital
Unit effect	2000/1995	1.000	1.000	1.000	1.000
	2006/2000	1.000	1.000	1.000	1.000
Composition effect	2000/1995	0.926	0.849	1.109	1.059
	2006/2000	0.954	0.872	1.029	1.066
Scale effect	2000/1995	1.349	1.235	1.611	1.540
	2006/2000	1.444	1.313	1.556	1.610
Changes in factors content of exports of goods	2000/1995	1.249	1.048	1.786	1.631
	2006/2000	1.377	1.145	1.601	1.716

3.3 Dirty industries and pollution-intensive exports

As can be seen in Table 4, the contribution of exports to the integrated pollution level of the manufacturing sector in the observed year (1995) was greater (65.0% of total estimated pollution of manufacturing sector) than their economic welfare, measured by output value (the share of exports in total production in 1995 accounted to 50.7%). More structured insight into aggregate figures shows distinctive differences between contributions of domestic and foreign demand to the national aggregates of production and pollution. This indicates that the export of goods from dirty industries is a relevant problem for economic sustainability in Slovenia.

Results presented in Table 5 show that exports from dirty industries are very concentrated. The greatest contributions to integrated pollution levels of the manufacturing sector are observed in water and air pollution. Export product groups with the greatest contributions to water pollution are as follows: steel and iron, paper and paper products, industrial chemicals, and non-ferrous metals. Exports of these four groups of products account for over 95% of the total water pollution linked to exports. The greatest part of contributions of exports to air pollution can be linked with exports of steel and iron and of industrial chemicals; exports of these two groups are also mainly responsible for toxic metal pollution.

Table 4. Integrated emissions from manufacturing, 1995

	Cumulative Production	Integrated emissions, structure in%				
		Total	Toxic	Toxic metal	Water[1]	Air[2]
Manufacturing	100.0	100.0	11.9	1.4	38.7	47.9
Production linked to foreign demand	50.7	65.0	9.0	1.0	26.1	28.9
Production linked to domestic demand	49.3	35.0	2.9	0.4	12.6	19.0

Source of data: Zakotnik, 2000a.
Notes: [1] BOD and total suspended particles; [2] NO_2, PM10, SO_2, CO, particles, VOC.

"Dirty industry" exports represent the greatest part in the export-related burden on Slovenia's environment. Dirty industries in Slovenia are (Fig. 1): steel and iron, non-ferrous metals, industrial chemicals, non-metal mineral products, paper and paper products, and other chemicals.

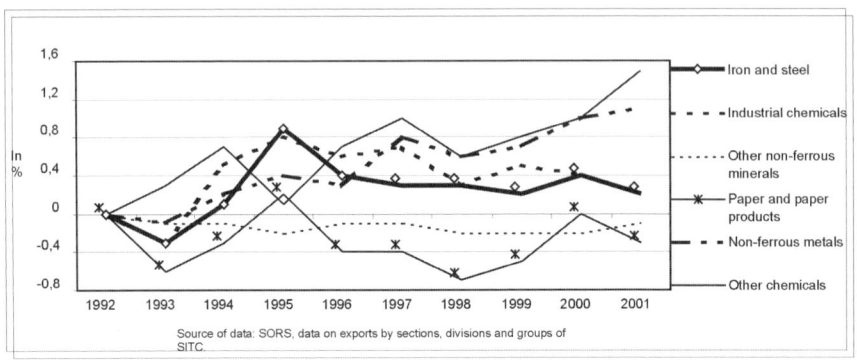

Fig. 1. Changes in the shape of pollution-intensive industries in merchandise exports

Dirty industry exports contribute about one-fifth of total goods exports (21.7%); the contribution of these exports to the integrated pollution generated by Slovenian manufacturing is over 56%. Production volumes of dirty industries increased more (3.1%) than manufacturing's total production volumes (2.5%) in 1995–2000 (also in 2001, with increases of 5.4% and 2.8%, respectively). Dirty industries are concentrated in approximately 900 companies with more than 40,000 employees, generating around a fifth of value added in manufacturing.

Figure 1 shows that dirty industries have increased their shares in Slovenia's exports of goods since 1998. Total exports of goods rose by 12.8% in 2000 (6.6% in 2001) and the proportion of dirty industries in total exports of goods remained stable at about 22% in the last 2 years.

Our aggregate projection of integrated pollution levels linked to goods exports take into account three different types of changes: changes in the composition of exports, changes in their scale (volume), and, last but not least, changes in pollution intensity per unit (of production, value added or per employee).

Implementation of SEDS projections for goods exports by 2006 (IMAD 2001a) would result in relatively high increase (about 22.5%) in emissions (Table 6). Looking at the decomposition of such a high growth in emissions, it is noticed that growth of exports of goods is projected to be more than twice as high (by 51.2% between years 2006 and 2000). On the other hand, composition effects (4.5% drop) and decrease of pollution intensities per unit of production (14.5% drop) are environmentally favourable. If all three effects are observed it becomes obvious that structural and per unit of product improvement would not be sufficient to neutralise environmental pressures from scale changes in export of goods. As a result, the environmental component of national competitiveness in goods exports would be deteriorated by 2006. It is obvious from the aggregate results that two SEDS goals are in conflict: the SEDS projected growth path for the export of goods from manufacturing (IMAD 2001a) is not in line with the SEDS sustainability ambitions.

Table 5. Structure of integrated emissions from exports of goods from manufacturing, 1995[1] in structural points

	Exports of goods	Integrated emissions				
		Cumulative Pollution	Toxic pollution	Toxic metal pollution	Water pollution	Air pollution
Exports from manufacturing = A+ B	100,0	65,0	9,0	1,0	26,1	28,9
A. Exports from dirty industries	21,7	56,6	7,2	0,8	25,4	23,2
Iron and steel	3,6	27,8	0,7	0,4	20,0	6,7
Industrial chemicals	3,6	15,9	5,6	0,2	1,8	8,3
Other non-metal mineral	1,3	1,7	0,0	0,0	0,0	1,7
Paper and paper products	4,7	3,7	0,2	0,0	1,5	2,0
Non-ferrous metals	3,2	4,4	0,4	0,2	1,6	2,2
Other chemicals	5,3	3,1	0,3	0,0	0,5	2,3
B. Exports from non-dirty industries	76,0	8,4	1,8	0,2	0,7	5,7

Source: BS, SORS; recalculations by Zakotnik, 2000b.
Note: [1] Integrated pollution from manufacturing (linked to domestic and foreign demand) = 100

Table 6. Change in integrated pollution linked to exports of goods

		2000/1995	2006/2000
Change in pollution intensity per unit	r_{ui}	0,869	0,858
Composition effects	$\sum_i (p_i/P * r_{si})$	1,031	0,945
Scale effects	E_2 / E_1	1,464	1,512
Change in integrated emissions linked to exports	P_2 / P_1	1,312	1,225

4 Conclusions

Our research aimed to identify structural changes in Slovenia's goods exports, focusing on the influence of export competitiveness and specialisation on the environment. We attempted to test the consistency between observed and projected export patterns (structure and dynamics) and the SEDS's stated orientation towards the sustainability of economic development (IMAD 2001a). Another SEDS priority, reducing the gap in GDP with EU countries, is obviously taking place. The problem is that GDP growth is achieved by increasing value added *and* environmental degradation (IMAD 2003). Export expansion would further widen the existing unsustainability of economic development if additional environmental measures are not introduced to compensate increases in production volumes with lower pollution intensities or environmentally favourable structural changes. In their study Vukina et al. (1999) found that the scale effect virtually dominates composition effects in all CEECs for all pollutants. GDP growth rates in transition years have been impressive in many CEECs, not only in Slovenia. The estimated industrial labour productivity improved strongly, and there have been marked micro efficiency gains (Aiginger 1998, p 25): industrial production is expanding with fewer or a constant number of workers employed, the estimated level of CEECs' industrial labour productivity is already much higher than in the pre-reform period (Aiginger 1998). Vukina also found that various composition patterns, observed across pollutants and CEECs, suggest diverse patterns of specialisation induced by transition. Thus, the results of empirical investigations of international trade specialisation in CEECs during transition are essentially in line with the predictions of trade theory. Specialisation patterns of most CEECs are still strongly biased towards energy- and labour-intensive manufacturing (Wolfmayr-Schnitzer 1998).

Although the general pattern is similar in different CEECs, there are significant differences between them in the intensity of the changes in environmental factors of competitiveness (Zamparutti and Gillespie 2000). Slovenia performs relatively well with respect to specialisation patterns in that its trade is relatively more specialised in the sophisticated product markets of human capital-intensive industries than that of the other CEECs (Wolfmayr-Schnitzer 1998). Results obtained from IPPS (1994) also indicate that dirty industries differ markedly across pollutants, as noted by Vukina et al. (1999).

A comparison of Slovenia's structure of exports and imports to and from OECD countries reveals a comparative specialisation in labour- and resource-intensive industries, in particular the former. Thus, Slovenia has a gap in exports of human capital- and technology-intensive exports, in particular for high-tech exports (Wolfmayr-Schnitzer 1998).

Analysis of growth factors for Slovenia (Bovha and Mayer 2002) reveals that the main structural weaknesses of the Slovene economy are to be found in poor technological development and innovation. This conclusion can be supported by a finding that Slovenia lags behind other transition countries in introducing production programmes that require new investments, technological solutions and a generally active approach to corporate restructuring and development (Gmeiner 1999). This is the result of the thus far slow restructuring process and failures in establishing of an efficient ownership structure in enterprises as well as a slow inflow of foreign direct investment (IMAD 2002). Following competitiveness studies, it has been noticed that Slovene enterprises mostly neglect non-price factors of competitive advantages (IMAD 2002). Exports are concentrated on non-differentiated products and services with less value-added, higher environmental impacts, indicating lower non-price competitiveness. In terms of national competitiveness, Slovenia lags behind EU member states and certain other candidate countries in the efficiency of the Government. Governmental policy is particularly weak in terms of environmental regulation and management (WEF 2001).

Our results appear generally consistent with those of Lucas et al. (1992), World Bank (1998,), Aiginger et al. (1998), Wolfmayr-Schnitzer (1998), Zamparutti and Gillespie (2000), and Vukina et al. (1999). Lack of sufficient improvement in the created factors of competitiveness and excessive reliance on labour-, natural resource-, and energy-intensive exports and industries have been reported as a central issue in all studies available to us.

With regard to SEDS and the sustainability of economic development, we conclude that there are problems related, first, to the relatively high exports of natural resource-intensive goods, and second, more significantly, to environmental degradation from exports of dirty industries.

Acknowledgements

The authors are grateful for the support provided by IMAD, the Institute for Economic Analysis, (financial) and in particular to Dr. Gmeiner (Institute of Macroeconomic Analysis and Development of the Republic of Slovenia - IMAD) and Anthony Zamparutti for their comments on an earlier version of the text and to Meta Žigman (IMAD), for technical assistance.

Appendix: Structure of goods exports by factor intensities

Country	Factor intensities (1999)					Average annual change (from 1995 to 1999), in structural points				
	Natural resources Primary	Others	Labour	Technology	Human Resources	Exports of goods	Natural resources Primary	Others	Labour	Technology	Human Resources
Austria	10	5	13	38	33	2	4	-7	-1	2	4
Belgium	13	11	12	33	29	1	2	0	3	6	1
Czeck republic	10	5	18	32	35	7	-6	-1	5	11	12
Denmark	28	3	16	35	13	0	-1	-3	3	4	1
Finland	9	6	7	43	32	1	-5	0	-4	7	-2
France	15	3	10	43	28	1	-3	-5	0	4	2
Greece	35	10	22	14	10	-2	-5	-3	-4	6	-1
Croatia	17	6	40	22	9	-2	-3	0	2	-5	1
Ireland	11	1	3	69	12	14	-6	5	0	22	9
Italy	8	6	23	37	24	0	0	-2	-1	1	-1
Hungary	12	3	14	46	23	22	-4	6	11	38	32
Germany	6	3	8	45	30	1	-4	-1	-1	2	2
Netherlands	24	3	8	44	17	-1	-5	-4	1	3	0
Norway	57	8	7	13	6	0	-1	-2	3	5	-1
Poland	16	9	27	19	26	5	-3	0	5	7	11
Portugal	13	7	33	20	27	1	-5	3	-2	2	9
Romania	15	6	44	15	19	7	0	10	12	1	7
Slovakia	12	6	15	23	40	8	-2	-2	5	5	13
Slovenia	6	9	21	25	37	1	0	-1	-1	3	3
Spain	17	5	12	24	38	5	2	2	7	6	4
Sweden	8	3	6	45	30	2	-3	-1	1	6	-1
Switzerland	4	6	8	58	24	-1	-4	1	-2	1	-3
Great Britain	11	5	8	51	22	3	-3	-1	0	5	2

Source: ITC COMTRADE database of United Nations Statistics Division

References

Aiginger K, Havlik P, Wolfmayr-Schnitzer Y (1998) The world economy, economic growth and restructuring in transition countries. In: The competitiveness of transition economies, OECD Proceedings, Austrian Institute of Economic Research (Wifo), Vienna Institute for Comparative Economic Studies (Wiw), Organisation for Economic Co-operation and Development (OECD), pp 15–41

Beghin JC, Roland-Holst D, van der Mensbrugghe D (1995) Trade liberalization and the environment in the Pacific Basin: coordinated approaches to Mexican trade and environment policy, Am J Agric Econ 77:778–785

Bovha PS, Mayer HP (2002) Sources of GDP growth, potential output and the output gap in Slovenia: mid-term projection. Manuscript, IMAD, Ljubljana

Dunning JH (1992) The global economy, domestic governance, strategies and transnational corporations: Interactions and policy implications. Transnatl Corp 1(3):7–45

Gmeiner P (1999) Slovenija in mednarodne primerjave konkurenčne sposobnosti nacionalnih gospodarstev s posebnim ozirom na tehnološki razvoj. Urad za makroekonomske analize in razvoj, Ljubljana

Gmeiner P (2000) Projekcije razvoja po področjih dejavnosti, njihov okoljski pomen ter posledice. Raziskava 'Strategija gospodarskega razvoja Slovenije—Okolje kot razvojni dejavnik v pogojih notranjega trga'. Urad RS za makroekonomske analize in razvoj, Ljubljana, 11 pp

Grossman GM, Krueger AB (1993) Environmental impact of a NAFTA. In: Garber P (ed) The US–Mexico Free Trade Agreement. MIT Press, Cambridge, Mass.

Hanžek M (ed) (1999) Human development report—Slovenia 1999. United Nations Development Programme, Institute for Macroeconomic Analysis and Development, Ljubljana

Hirsch S (1967) Location of industry and international competitiveness. Oxford University Press, Oxford

Hughes (1991) Hughes G. (1991) "Are the Costs of Cleaning up Eastern Europe Exaggerated?" Oxford Review of Economic Policy 7(4) (1991): 106-135

IMAD (2001a) Slovenia in a new decade: sustainability, competitiveness, EU membership: the strategy of economic development of Slovenia 2001–2006. Institute for Macroeconomic Analysis and Development, Ljubljana, 136 pp

IMAD (2001b) Autumn report 2001. Institute for Macroeconomic Analysis and Development, Ljubljana, 127 pp

IMAD (2002) Development report. Institute for Macroeconomic Analysis and Development, Ljubljana

IMAD (2003) Development report. Institute for Macroeconomic Analysis and Development, Ljubljana

IPPS (1994) The industrial pollution project system, H Hettige, P Martin, M Singh, D Wheeler, The World Bank Group

Klavens J, Zamparutti A (1995) Foreign direct investment and environment in central and eastern Europe: a survey. 70 pp

Kovačič A (2001) Merjenje globalne konkurenčnosti držav ter pomen varstva okolja. IB revija za strokovna in metodološka vprašanja gospodarskega, prostorskega in socialnega razvoja Slovenije, 30(2001)4, str. 53–65

Lucas REB, Wheeler D, Hettige H (1992) Economic development. Environment regulation and the international migration of toxic industrial pollution: 1960–1988. In Low P (ed) International trade and the environment. World Bank Discussion Papers 159, Washington, D.C., pp 67–86

Mani, Muthukumara, Wheeler (1997) In search of pollution havens? Dirty industry in the world economy, 1960–1995. World Bank, Poverty, Environment and Growth

MOPE (2002) Povzetek Poročila o stanju okolja 2001/2002, Ministrstvo za okolje in prostor, februar 2003, 27 str. dopis št. 354-01-58/2002 z dne 28. 1. 2003

OECD—Centre for Economies in Transition (1995) Environmental funds in economies in transition. OECD, Paris

Pearce WD, Warford JJ (1993) World without end. Economics, environment, and sustainable development. Oxford University Press, New York

Posner M (1961) International trade and technological change. Oxford Economic Papers, (13), pp 232–341

Radej B (2001) Inštrumentarij okoljske integracije energetskega razvoja. Ministrstvo za okolje in prostor. Interno gradivo, prispevek k pripravi Nacionalnega energetskega programa, 33 pp

Rauscher M (1995) Trade law and environmental issues in central and east european countries. In: Winters LA (ed) Foundations of an open economy. Trade laws and institutions for eastern Europe. CEPR, London

Rock MT (1996) Pollution intensity of GDP and trade policy: can the world bank be wrong? World Dev 24(3): 471–79

Romer PM (1986) Increasing returns and long-run growth. J Polit Econ, (94), pp 1002–1037

Seljak J (2001) Kazalec uravnoteženega razvoja. Ljubljana: Urad za makroekonomske analize in razvoj. Zbirka Analize, raziskave in razvoj, p. 195

Simoneti M, Rojec M, Rems M (2000) Poslovanje slovenskega podjetniškega sektorja v razdobju 1994–1998 z vidika lastniške strukture. IB revija, 34(2):5–20

Tietenberg T (1996) Environmental and natural resource economics, 4th edn. Scott Foresman, Glenview, 614 pp

UNECE (1997) Environmental performance reviews: Slovenia. environmental performance reviews, Series No. 2. United Nations' Economic Commission for Europe, Geneva, 177 pp

Vernon R (1966) International investment and international trade in the product cycle. Q J Econ, (80), pp 190–207

Vukina T, Beghin JC, Solakoglu EG (1999) Transition to markets and the environment: effects of the change in the composition of manufacturing output. Environ Dev Econ 4:582–598

Wackernagel M, Monfreda C, Deumling D (2002) Ecological footprint of nations– November 2002 update: how much nature do they use? How much nature do they have? Sustainability issue brief, November 2002, 14 pp, www.redefiningprogress.org

WEF (2002) Environmental performance measurement, the global report 2001–2002. World Economic Form, Davos

Wolfmayr-Schnitzer Y (1998) Trade performance of CEECs according to technology classes. In: The competitiveness of transition economies, OECD Proceedings, Austrian Institute of Economic Research (Wifo), Vienna Institute for Comparative Economic Studies (Wiw), Organisation for Economic Co-operation and Development (OECD), pp. 41–71

World Bank (1998) Slovenia: trade sector issues. WB: European and Central Asia Regional Office–Poverty Reduction and Economic Management Unit

Zakotnik I (2000a) Umazane industrije v predelovalnih dejavnostih. Raziskava "Strategija gospodarskega razvoja Slovenije do leta 2006". Urad RS za makroekonomske analize in razvoj, Ljubljana, 23 pp

Zakotnik I (2000b) Intenzivnost blagovnega izvoza glede na naravne vire. Raziskava "Strategija gospodarskega razvoja Slovenije–Okolje kot razvojni dejavnik v pogojih notranjega trga". Urad RS za makroekonomske analize in razvoj, Ljubljana, 25 pp

Zamparutti A, Gillespie B (2000) Environment in the transition towards market economies: an overview of trends in central and eastern Europe and the new independent states of the former Soviet Union. Environ Plann B: Plann Des 27(3):331–47

Zylicz T (1997) Environmental policy in economies in transition. Warsaw Ecological Economics Center, Warsaw University

Indicators for sustainable energy development from a negentropic perspective

Jordan Pop-Jordanov

Research Center for Energy, Informatics and Materials (ICEIM), Macedonian Academy of Sciences and Arts (MANU), P.O. Box 428, Skopje, Macedonia, Phone +389 2 3114 200, Fax: +389 2 3114 685, E-mail: jpj@manu.edu.mk

Abstract

In this paper, the indicators for sustainable energy development derived recently by several international organizations and national teams are reviewed, in the light of a negentropic perspective related to the non-material human needs.

Specifically, indicators for efficient use of mental resources and capacities (in addition to energy and material ones) are derived, an approach which is lacking in the present lists of sustainability indicators. As a result, some additional indicators concerning organizational attention deficit, occupational stress, corruption pressure and brain-drain are introduced and discussed.

1 Introduction

Already the Agenda 21 (UNCED 1992) has emphasized the necessity for identifying indicators of sustainable development that can measure and monitor relevant important changes. Simultaneously, the energy sector is recognized worldwide as significant for sustainable development, and it has even been stated that "sustainability is dependent on the evolution of energy technologies" (CAETS 1996). However, as inferred recently by another competent institution, "despite some progress, no comprehensive set of indicators for sustainable energy exists" (IAEA 2002).

Additional motivation for the present study is the fact that, although scientific and technological knowledge is largely proclaimed as an important factor for sustainability, it is still not recognized in practice as a decisive factor. A fresh example is the just published "Key Outcomes of the Summit" in Johannesburg (UN/DESA 2002), where science and technology are mentioned only at the end of the seventh outcome (of the total eight). This means that we are still challenged by the warnings of the National Academies (USA) expressed few years ago: "Discussions of the role of science and technology have not been central to the last decade's debates on sustainable development... Even Rio's 'Agenda 21'... devotes only 3 out of 40 chapters to science and technology... In consequence, societies approach the 21st century with little in the way of a useful strategic appraisal of how to identify and create the knowledge and know-how most crucial to achievements in sustainable development" (NAS 1999, p. 29).

In this study an attempt is made to review the sets of indicators for sustainable energy development (ISED), determined recently by several international organizations and national teams (IAEA-IEA 2001; UN ECE 2001; IAEA 2002). The main purpose is to propose eventual improvements in the existing indicators, as well as to introduce some additional ones, attempting to comply with the above-mentioned warnings related to the role of science and technology (S&T).

2 Negentropic Perspective on Sustainable Development

The original definition of sustainable development[1] inevitably concerns the depleting of natural resources and the environmental degradation. Looking at the energy and material flows within the societal metabolism (Fig. 1, full lines) it can be noted that the overall direction of world economic growth, expressed in terms of resource depletion and waste emissions, is toward increasing entropy, as a measure of disorder and degradation (Georgescu-Roegen 1971; Noorman 1995). From this it can be inferred that sustainable development is compromised.

However, in contrast to the finiteness of physical resources and capacities, knowledge and human capacities (leading to GDP "dematerialization") can be and are actually being accumulated, rather than depleted. Consequently, a noetic[2] negentropic extension to the standard societal metabolic scheme could be introduced, adding a supplementary negentropy resource related to science and technology (Pop-Jordanov 1998). In this way, the creation of new forms of order provides new stocks of negentropy able to compensate for the overall entropy increase in time, making sustainable growth achievable in principle (Fig. 1, dotted lines).

Actually, there is a wide spectrum of perspectives on sustainable development, related to various world-views, from anthropocentric to ecocentric and from possibilistic to conservative (O'Riordan and Turner 1983; De Vries 1989; Schwartz and Thomson 1990). On the other hand, opinions about sustainability specific to S&T depend on the background of related disciplines (Renn 1994; Renn and Kastenholz 1996; Valenduc and Vendramin 1997):

- For an economist, sustainability is at first related to new economic models of growth and regulation, taking into account not only the traditional quantifiable components of welfare, but also a lot of environmental "externalities" and qualitative assets.

---

[1] "Development that meets the needs of the present without compromising the ability of future generations to meet their own needs." (WCED 1987)
[2] Noetic *a.* relating to intellect (from: nous *n.* mind)

- For an ecologist, sustainability means the use of natural resources to the extent that the carrying and regenerative capacities of the ecosystems are not jeopardized.
- For a physicist, sustainability means the ability of biological systems to fight against degradation of energy and resources (entropy) by creating new forms of order (negentropy) using the various inputs of solar energy.
- For a chemist or an engineer, the challenge of sustainability is to complete material and energy life cycles created by human activities, through new techniques for material design, re-use, recycling and waste management.
- For a social scientist, sustainability implies the social and cultural compatibility of human intervention in the environment with its images constructed by different groups within society.

The common denominator of all above opinions is that, although they treat S&T as one important factor supporting sustainable development, they still do not view it as a decisive and noetic negentropic factor. In fact, when (and if) dealing with entropy and negentropy in the context of sustainable development, the standard considerations are mainly concerned with abiotic and biotic spheres respectively, underestimating the non-material components.

In the next sections we will reconsider the ISED in the light of the described noetic perspective, attempting to introduce some improvements and extensions.

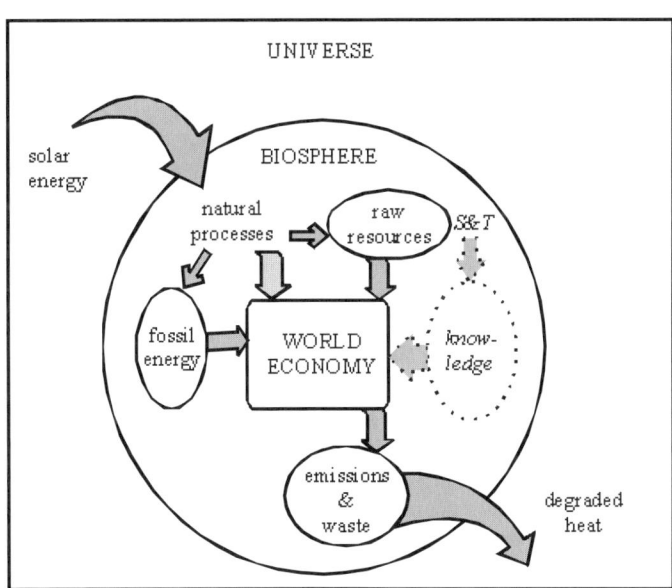

Fig. 1. Societal metabolic pathways
Full lines: standard scheme (Norman 1995).
Dotted lines: noetic negentropic extension (Pop-Jordanov 1998; 2000)

3   Consideration of Existing ISED

Indicators for Sustainable Development (ISD) could be defined as quantifiable parameters to measure and monitor changes (progress or degradation) in respect to sustainable development, thus signaling challenges or alarms. These parameters are usually considered in pressure/driving force–state–response (PSR/DSR) framework, i.e. cause–effect indicators and corrective response policy actions for improvement (NAS 1999).

Recently, a collaborative project on indicators for sustainable energy development was realized, involving participants from seven international organizations and 15 states. As a result, a full set of 41 ISED and a core set of 23 ISED were distilled (IAEA-IEA 2001; UNECE 2001; IAEA 2002). Reviewing the mentioned sets of ISED we can conclude that they are mostly concerned with the traditional natural source and sink limitations, underestimating the non-material indicators.

However, when measuring and monitoring the changes with respect to sustainable development, it should be taken into account that such a development in essence means increasing quality of life by satisfaction of both material an non-material human needs. Consequently, economic growth (both material and non-material) is only one of the means of development. Material economic growth is confronting the natural source and sink limitations, while non-material economic growth, i.e. the recently increasing dematerialization of the economy, includes expanding the share of services sector and "the higher energy - and resource - efficiency of the new and emerging knowledge-intensive technologies" (Gallopin 2001, p. 10).

When developing national energy efficiency strategies, as a rule, only efficient energy use is considered, while efficient energy supply is either underestimated (restricting it to fossil fuels only, as in the mentioned IAEA-IEA indicator sets) or completely neglected. However, our analysis of the core ISED shows that almost all of them can be positively affected by efficient energy supply (Fig. 2).

On a global scale, within the open system of the biosphere, evolution as a process of increasing negentropy importation in complex systems acts against the second law of thermodynamics (Schrödinger 1944; Prigogine and Stengers 1984; Laszlo 1996). This holds for renewable energy as well, whose ultimate source is outside our biosphere. (In this context, the mentioned noetic extension views the mind as a negentropic source in a way comparable to the sun).

Again, in the quoted IAEA-IEA full set comprising 41 ISED, the renewables are represented by two indicators only (related to hydropotential and fuel wood). The new renewables (solar, wind, etc) are not explicitly mentioned (they could be understood as incorporated in bulk indicators "energy mix" and "indigenous energy production").

The importance of energy supply efficiency for sustainability can be also illustrated by some technological innovations concerning renewable energy, elaborated recently at ICEIM. These improvements affect positively most of the indicators from Figure 2, and are based on (a) enhancing hydropower efficiency by reversible pump storage (Todorovski et al. 2002), and (b) increasing efficiency of solar cells using advanced quantum resonance modeling (Markovska et al. 2002).

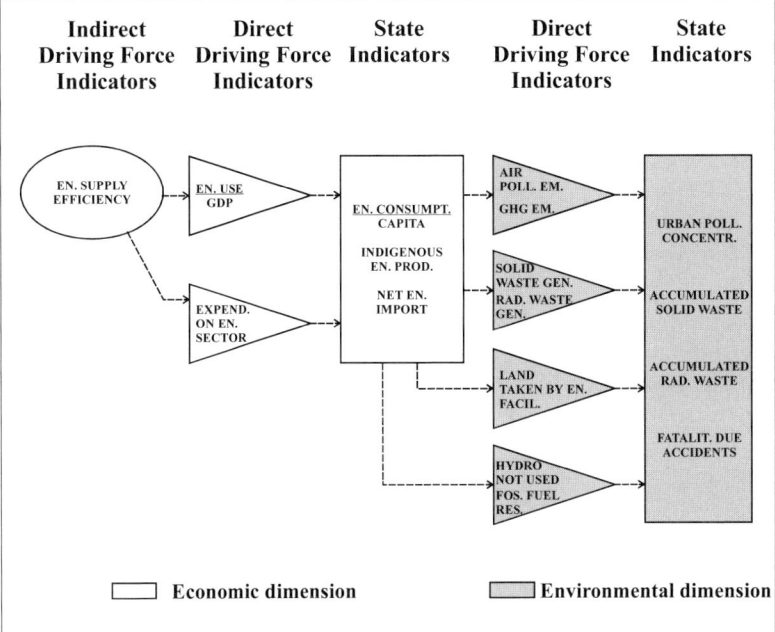

Fig. 2. Energy supply efficiency with positively affected core ISED

4 Introduction of Mental Indicators

4.1 Mental Resources and Capacities

Beside the mentioned non-material economic growth expressed, among others, in the higher energy efficiency of the new knowledge-intensive technologies, the non-material social growth should also be taken into account when considering the sustainability indicators. Its neglect in the present lists of indicators of sustainable development, both the general sets of UNCSD and the energy related sets of IAEA, IEA and UNECE, could also be attributed to the underestimation of S&T, correlated with the noetic perspective on sustainability.

As is known, modern society is characterized by a shift from material and energy resources to mental ones, i.e. from physical labor to mental labor (or more generally, from matter to mind[3]). Following the noetic negentropic ("dematerialized") extension of the societal metabolic pathways, indicators related to efficient use of mental (mind, intellect) resources and capacities (in addition to energy and material ones) are derived here, an approach that is missing in the present lists of sustainability indicators.

Consequently, some additional response actions with corresponding positively affected ISED are proposed (Fig. 3). As can be seen, the term "mental" is used here in a broader sense, to include cognitive, emotional and even moral components.

4.2 Information Overflow and Organizational Attention Deficit

The modern information age, characterized by an overflow of daily messages and data, induces the so-called organizational attention deficit disorder, with the following symptoms (Davenport and Beck 2001):
- Increased likelihood of missing key information when making decisions
- Diminished time for reflection on anything but simple information transactions such as e-mail and voice mail
- Difficulty in holding others' attention
- Decreased ability to focus when necessary

Let us recall that even the Chernobyl accident was largely caused by occupational attention deficit.

An additional contribution to organizational attention deficit is made by computer-related disorders, as a new phenomenon within the information society (Pepper and Gibney 2000).

The EEG biofeedback technique for adjustment of brain wave frequency is becoming a response measure competitive with traditional pharmaceutical therapy for attention deficit disorder. Some recent results of our team, showing the improvement of attention by brain wave operant conditioning, are displayed in Fig. 4.

4.3 Occupational Psychosomatic Disorders

The existing lists of ISED include only the accidental health risks, while the health effects from normal plant operation are unjustifiably disregarded. Moreover, stress related psychosomatic diseases, induced by nuclear and coal power plants, have been shown to be non-negligible (Pop-Jordanova and Pop-Jordanov 1996). Some our results for the Slovenian nuclear power plant at Krsko and the Macedonian coal power plant at Bitola are displayed in Fig. 5. Here again, the

---

[3] This could be considered as another manifestation of overcoming the traditional mind/matter separation, expressed by an old play on words: "What is mind?" / "No matter" / "What is matter?" / "Never mind" (Jibu and Yashue 1995).

non-pharmacological biofeedback techniques are shown to be an effective response for treating stress related disorders (Pop-Jordanova and Pop-Jordanov 2002).

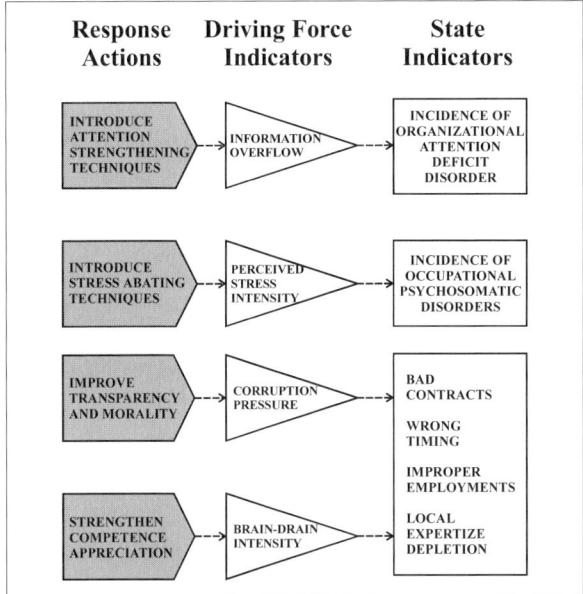

Fig. 3. Proposed response actions and corresponding positively affected ISED for efficient use of mental resources and capacities

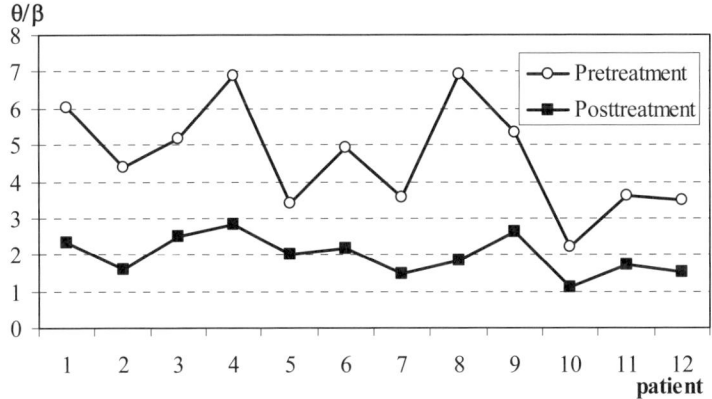

Fig. 4. Mean theta/beta ratio as attention deficit indicator

(EEG biofeedback treatment, 12 patients, 40 sessions)

4.4 Transparency and Morality

Although it enables sustainability, knowledge alone is not enough. In both research and decision making, the genuine concept of sustainability involves (or at least should involve) moral behavior, as motivation from within, complementary to legislative and economic motivations. Moreover, a recent World Energy Council study states that "ethics may become the new decisive factor in orienting and managing development [energy] strategies that will be truly sustainable" (WEC 2001).

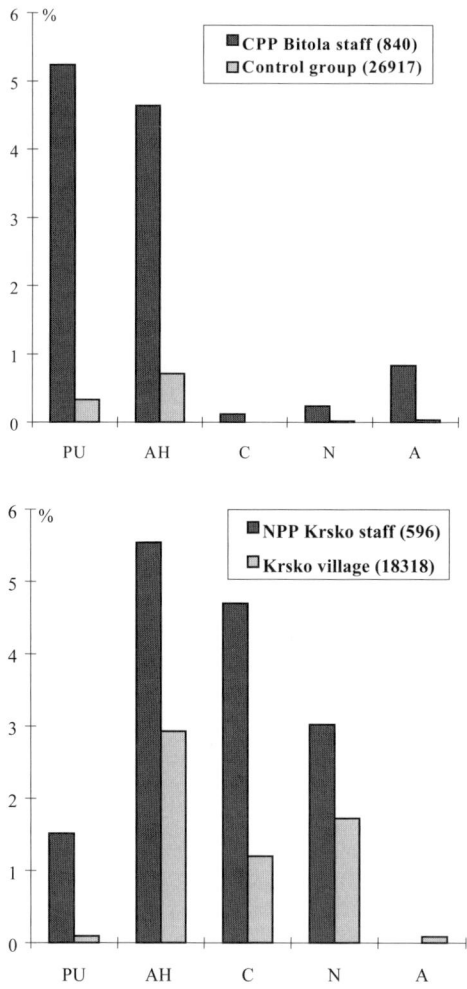

PU - Peptic ulcer   AH - Arterial hypertension   C – Colitis   N – Neurodermatitis   A - Asthma

Fig. 5. Incidence of primary psychosomatic diseases (% of the total number of patients in the group)

Ethical norms and transparency are especially important for efficient and timely restructuring and privatization of the energy sector. This holds particularly for countries where considerable corruption pressure (as an indicator of what could be called "moral entropy") takes place.

The correlation between S&T, ethics and sustainability could be considered in a broader sense to include the basic socioeconomic relations. In a recent pertinent publication of the World Innovation Foundation, established by a group of nobelists, it is stated (WIF 2001):

"Currently the creation of wealth and the management of that wealth are overridingly treated as separate entities... It was not Bill Gates who invented the software to run micro-computers, which marked the foundations of Microsoft itself, but those scientific enthusiasts who primarily conceived the technology that produced the initial software operating system ... and where today they are not even worth a mere 1,000th of Bill Gates's fortune... This and other matters is where things in the future world have to change, for if those who had initially invented the technology, that has spawned all the wealth, had been conveyed the wealth, many more technologies would have ensued... Therefore our politicians have to fully comprehend over the next few years that only scientists and engineers have the ultimate means to determine our existence or not. Consequently our presence is not as what is presently perceived in the hands of our political masters but in the hands of the world's scientists and engineers".

On the other hand, the different forms of modern terrorism, as an extreme manifestation of inhumanity and amorality, may jeopardize the transition to energy sustainability as well.

4.5 Brain-drain

Depleting local expertise, the brain-drain phenomenon could endanger sustainable development, especially in developing and transition economies.

The main reasons for the brain-drain, according to opinions of respondents - students of energetics and informatics at Skopje University - are ranked in Table 1 (Pop-Jordanov and Markovska 1994).

Table 1. Main reasons for brain drain

	Total	Energetics	Informatics
	Rank	Rank	Rank
Lack of professional progression	I	III	I
Egalitarianism	II	I	II
Boring work	III	II	V
Lack of organization	IV	IV	IV
Poor income	V	V	III

Consequently, an important response action should be strengthening of competence appreciation versus mediocrity, partisanship and corruption, i.e. noetic negentropy versus moral entropy.

5 Conclusions

As warned by competent scientific institutions, the role of S&T for sustainable development was underestimated in defining and following up the UN Agenda 21. As inferred here, this state of mind is also reflected in developing the concept of indicators of sustainable development and in the subsequent identification of such indicators (ISD by UNCSD and ISED by IAEA, IEA and UNECE). This situation, which conceptually could be attributed to the neglect of a noetic negentropic perspective on sustainable development, contributed among others to the underestimation of energy supply efficiency (in particular of renewables) within the economic dimension, and to the neglect of non-material needs (in particular the mental ones) within the social dimension.

Specifically, one can conclude that due consideration should be given to indicators and response actions for efficient use of mental (cognitive, emotional or moral) resources and capacities (in addition to energy and material ones), an approach which is neglected in the present PSR/DSR lists. In this respect, the mental indicators and response actions concerned with the organizational attention deficit, occupational psychosomatic stress, corruption pressure and brain-drain are particularly important for energy sustainability, as well as for sustainable technological development in general.

References

CAETS (1996) A declaration of the Council of Academies of Engineering and Technological Sciences (CAETS). In: Creating wealth in harmony with the environment. Royal Swedish Academy of Engineering Sciences (IVA), Stockholm, p 110

Davenport HT, Beck CJ (2001) The attention economy. Harvard Business School Press, Boston, Massachusetts

De Vries HJM (1989) Sustainable resource use. An enquiry into modeling and planning, Ph.D. Dissertation, Univ. Groningen

Gallopin G (2001) Science and technology, sustainability and sustainable development. LC/R. 2081, ECLAS, Economic Commission for Latin America and the Caribbean, pp 1-30

Georgescu-Roegen N (1971) The entropy law and the economic process. Harvard University Press Cambridge, Massachusetts

IAEA (2002) Indicators for sustainable energy development. FS Series 2/02/E, IAEA, Vienna

IAEA-IEA (2001) Indicators for sustainable energy development. Joint IAEA and IEA Contribution to CSD-9, New York

Jibu M, Yasue K (1995) Quantum brain dynamics and consciousness. John Benjamins Publ Co, Amsterdam/ Philadelphia, p XIII
Laszlo (1996) Evolution: The general theory. Hampton Press, New Jersey
Markovska N, Pop-Jordanov J, Solov'ev E (2002) Resonant effects in heterojunctions and thin films with possible impacts on solar cell efficiency. In: Proceedings of the World Renewable Energy Congress, Elsevier Publ, Cologne, Germany
NAS (1999) Our Common Journey. National Academy of Sciences, National Academy Press, Washington, D.C.
Noorman KJ (1995) Exploring futures from an energy perspective. D.Sc. Dissertation, Univ. Croningen, Netherland
O'Riordan T, Turner RK (1983) An annotated reader in environmental planning and management. Pergamon press, Oxford
Pepper E, Gibney KH (2000) Healthy computing with muscle biofeedback. BFE, Woerden, Netherlands
Pop-Jordanov J (1998) Science, technology and sustainable development. In: Science and culture for the joint future of South Eastern Europe, Macedonian Academy of Sciences and Arts (MANU), Skopje, pp 27-34
Pop-Jordanov J (2000) R&D networks as negentropic tools toward sustainability. In: Strategies of the international scientific cooperation in South-East Europe, IOS Press, Amsterdam-Berlin-Oxford-Tokyo-Washington, pp 148-153
Pop-Jordanov J, Markovska N (1994) Brain drain in energy and informatics. Encyclopaedia Moderna, 14, 4, pp 308-313; Also in: (1995) The energy economy in CEE in transitions, WEC, London, pp 139-145
Pop-Jordanova N, Pop-Jordanov J (1996) Psychosomatic and substitutional effect: comparative health risks from electricity generation. In: Electricity, health and environment, IAEA, Vienna, pp 177-187
Pop-Jordanova N, Pop-Jordanov J (2002) Psychophysiological comorbidity and computerized biofeedback. The International Journal of Artificial Organs, Vol. 25, No. 5, pp 429-433
Prigogine I, Stangers I (1984) Order out of chaos. Bantam, New York
Renn O (1994) A regional concept of qualitative growth and sustainability. Bericht Nr 2, Akademie für Technikfolgenabschätzung in Baden-Württemberg, Stutgart
Renn O, Kastenholz (1996) Ein regionales Konzept für nachhaltiger Entwicklung. In: Gaia ecological perspectives in science, humanities and economics. Spektrum Akademischer Verlag, Basel, pp 88-89
Schrödinger E (1944) What is life. Cambridge University Press, Cambridge
Schwartz M, Thomson M (1990) Divided we stand, redefining politics, technology and societal choice. Harvester Wheatsheaf, New York
Todorovski M, Bosevski T, Pop-Jordanov J, Causevski A (2002) Enhancement of efficiency in using national hydro potential and the consequent environmental benefits. In: Proceedings of the World Renewable Energy Congress, Elsevier Publ, Cologne, Germany
UNCED (1992) Report of the United Nations conference on environment and development Rio de Janeiro: UN. (Annex I: Rio Declaration; Annex II: Agenda 21)
UNECE (2001) Indicators for Sustainable Energy Development, A Collaborative Project, ENERGY/2001/8
UN/DESA (2002) Key Outcomes of the Summit, Johannesburg Summit 2002, United Nations Department of Economic and Social Affairs

Valenduc G, Vendramin P (1997) Science, technological innovation and sustainable development. International Conference "Science for a Sustainable Society", Roskilde

WCED (1987) Our common future. World Commission on Environment and Development, Oxford University Press, Oxford

WEC (2001) Values added: ethical experiences in the energy sector. World Energy Council, London

WIF (2001) Global inequality. Scientific Discovery, The Newsletter of the World Innovation Foundation, Vol. 5, No. 3

Sustainability indicators for anticipating the fickleness of human-environmental interaction

Janne Hukkinen

Helsinki University of Technology, Laboratory of Environmental Protection, PO Box 2300 (Otakaari 8), 02015 HUT (Espoo), Finland, Tel.: +358 9 451 3975, E-mail address: janne.hukkinen@hut.fi

Abstract

The development of environmental indicators is dominated by the so-called pressure-state-response (PSR) model. The PSR contains a set of indicators measuring anthropogenic pressure (P) on the environment, the state (S) of the environment resulting from such pressure, and the societal response (R) to ease the pressure. The strength of the PSR is its acknowledgement of the causal relationship between the state of the environment and human activity. Its major weakness, however, is the lack of sophistication of the mathematical and cognitive models representing the causal relationship. As a result, current indicator systems based on the PSR fail to take into account contingencies in human-environmental interaction that make the future state of the system difficult to ascertain. Recognizing the fickleness of human beings and nature will result in very different indicators from those traditionally developed. In particular, the article identifies the following important areas of indicator development: (1) Indicators of ecosystem impacts of production, which measure changes in production outputs and environmentally significant inputs; (2) indicators of bounded carrying capacity, which utilize alternative scenarios of human-environmental interaction to specify the ecosystem-specific limits that societies might pose on industrial production; (3) indicators of congruence between ecosystems, institutions and production, which measure the agreement between the functions of an ecosystem and the institutional rules governing its management; and (4) indicators of technological and institutional path dependence, which observe and potentially strengthen lock-ins in human-environmental interaction. These development challenges imply that sustainability indicators should be considered more as vehicles for improving communication between different communities of experts on the sustainability of a particular system of human-environmental interaction, and less as universal measures of sustainability.

Keywords: sustainable development indicator, human-environmental interaction, Baltic Sea, emergency response, participation, nonlinear dynamics, catastrophe, scenario, carrying capacity, path dependence

1 Introduction

Imagine that you survived the Black Year of the Baltic Sea 2012. It began with some of the worst winter storms in memory. In March, an oil tanker from the rapidly expanded Russian oil harbors on the Gulf of Finland collides in bad weather with a Helsinki-Tallinn passenger ferry, resulting in 125 drowned and hundreds of kilometers of black, lifeless shoreline. Following a hot early summer, August brings with it exceptionally intense blue green algae blooms. A new algae species is discovered, much more poisonous than the earlier ones, resulting in the death of 13 children from acute intestinal bleeding plus dramatic fish and bird kills along the shores of Finland, Estonia and Russia. After the holiday season, the fall begins with yet another disaster, when the safety system of the brand new mega-harbor of Helsinki, which handles virtually all of Finland's container traffic, fails to detect a container with poisoned breakfast cereal. Within two days the country's efficient logistics system has delivered the deadly breakfast to 21 individuals before the cereal is withdrawn from shop shelves.

Then come the tough questions from angry citizens. What happened certainly did not meet the goals of improving people's living and working conditions, managing natural resources in a sustainable way, or protecting the environment, as articulated in the Baltic 21 sustainable development program (Baltic 21 Secretariat 2000). Why did not the indicator system of pressure, state and response (PSR), such as the one found in Finland's sustainable development indicators (Rosenström and Palosaari 2000) and also applied in the Baltic Sea, do its job by giving an early warning of the pressures that eventually lead to the accidents, of the state of the environment that was so vulnerable to the accidents, or of the response mechanisms that were so inadequate to react to the accidents?

In retrospect, the existing indicator systems had measured the wrong pressures, states and responses. According to the critics, since several chains of events had taken place well before 2012 that should have been observed as early warnings, the sustainability indicators should have measured the pressure, state and response levels related to these chains of events. The early warnings included the potential for increasing storm frequencies, as discussed by climate change modelers (Broecker and Hemming 2001; Grassl 2000); the potential for oil disasters as a result of the oil harbor developments in the Russian coast of the Baltic that began already in the 1990s (Perttu 2002); the potential consequences of the dramatic increase in passenger traffic between Helsinki and Tallinn that began in the early 1990s; the potential for major marine accidents resulting from the combined effects of all of the above (Taivalkoski and Kaurala 2002); the potential for acts of terrorism in a region that was rapidly developing its container traffic (Pyykkönen 2002; The Economist 2002); or the potential for adverse consequences of the introduction of new species, which was evident already in the late-1990s (Finnish Environment Institute 2002). Instead, the indicators measured things like life expectancies and mortalities in Baltic countries, GDP growth, unemployment, income distribution, level of public participation, greenhouse gas emissions, acidifying air emissions, nutrient emissions to the sea, consumption of ozone depleting substances, energy intensities of nations, state of the fisheries, forest management, agriculture, environmental management systems in industries, and tourism. The final irony after the major marine transport accidents of 2012, and one which the critics did not hesitate to dwell on, was that the transport sector indicators for the Baltic region dealt with road traffic alone (Baltic 21 Secretariat 2000).

Officials responsible for the sustainable development indicators felt the critics were being unfair. After all, the indicators they were referring to were long term sustainable development indicators, when at issue was the adequacy of emergency response. The critics remained unconvinced and pointed out that the impacts of the catastrophes would persist for decades to come, and should therefore be considered as part of the sustainability agenda.

This article asks the question what went wrong with the indicators for sustainable development in the Baltic region in this hypothetical scenario. There is nothing wrong with the PSR framework as such, since it effectively focuses policy and management measures by laying out the causal chain of events from human action to environmental impact (OECD 1993). However, there is much to improve in the sophistication of both the mathematical and cognitive models that the causal chain is based on in current PSR indicator systems. Current PSR indicator systems are based on linear mathematical models that portray systems of human-environmental interaction responding to gradual change in a smooth way. In reality, studies of many systems of human-environmental interaction have shown that smooth change can be interrupted by sudden drastic switches to a contrasting state, better described by nonlinear models (Scheffer et al. 2001). Policy and management intervention is further complicated by the fact that stakeholders have different cognitive models about environmental issues and therefore react strategically to whatever policies are implemented. Such strategic reactions make all policy interventions based on formal mathematical models potentially unstable, despite attempts to include strategic reactions in the formal models (Hukkinen 1993; 1999; Flyvbjerg 2001).

The article also asks how might sustainability indicators be improved to prevent the failures envisioned in the Baltic Sea scenario. I will argue that recognizing the fickleness of human-environmental interaction will result in very different indicators from those traditionally developed within the PSR framework. I recommend that sustainability indicators be considered more as vehicles for improving communication between different communities of experts on the sustainability of a particular system of human-environmental interaction, and less as universal measures of sustainability.

2 Critique of current applications of the PSR

To understand the poor performance of PSR-type sustainability indicators in the hypothetical Baltic Sea case, I will compare the PSR framework with the challenges of sustainability. The list of sustainability challenges is based on my interpretation of what have been identified as key challenges in the academic and policy literature on the subject. In articulating the challenges, I have accepted the often-repeated statement originating in the Brundtland Commission's report of 'meeting the needs of the present without compromising the ability of future generations to meet their own needs' (World Commission on Environment and Development 1990). Although many academics have struggled with the poor translatability of that statement into anything meaningful in their own disciplines, there is a body of research that can offer interesting insights to the policy objectives of sustainability. Much of this research falls under the fields of environmental governance, environmental history and systems ecology. These fields of inquiry have dwelled on the conditions under which human communities have historically been able to manage a relatively long-lasting relationship with the ecosystem surrounding them (many inquiries have also considered the conditions under which communities have not been able to manage such a relationship). Despite their relevance to key issues of sustainability, these lines of inquiry have had a weak impact on sustainability indicators. Consider, for example, one of the most widespread indicator frameworks currently in use, namely, the PSR model.

The PSR model contains a set of indicators that assess the extent of anthropogenic pressure (P) on the environment, the state (S) of the system of human-environmental interaction resulting from such pressure, and the societal response (R) to ease the pressure (OECD 1993). As such, the PSR is a laudable effort to help decision makers to better understand the causes and effects of environmental stresses and the effectiveness of human efforts to respond to them. The problem is the simplistic application of the PSR in real life situations. A recent report on the Baltic Sea sustainability indicators illustrates this point (for a broader critical overview of the practical applications of the PSR, see Hukkinen 2003).

In the Baltic Sea case, all indicators have been selected on the basis of simplicity and availability of statistical data, without regard to the complex functional linkages between different indicators of pressure, state and response (Baltic 21 Secretariat 2000, p. 9). The Baltic indicator report also implies that measurements of the state of the environment should be made with respect to a fixed carrying capacity of the Baltic Sea ecosystem, by noting that past economic downturns have reduced pollution whereas recent economic activity poses a substantial 'risk of going back to unsustainable production and consumption'. Furthermore, state and response indicators in the report are based on the assumption that there is a single, agreed-upon trajectory of sustainability, as illustrated by the conclusion that 'the Baltic Sea Region has entered the road towards but is still far from sustainable development'. Although the report does recognize that there are 'different development paths' for different countries in the region, it makes no effort to develop response indicators that would measure the degree of path dependence in the development. Finally, the Baltic indicator report makes the universalistic assumption that despite regional differences in development, the measurements can be 'summarized for the region' to conclude that 'several important positive trends are visible' but that 'a number of fundamental economic, social and environmental criteria for a sustainable society are not met' (Baltic 21 Secretariat 2000, p. 43). Each one of these assumptions is problematic when viewed in light of literature on environmental governance, environmental history and systems ecology.

First, despite their broad coverage of ecological, economic and socio-cultural dimensions of sustainability (see, e.g., Rosenström and Palosaari 2000), current PSR indicator systems suffer from a lack of <u>functional linkages between indicators</u>. By functional linkages I mean not only linkages among the various indicators of, say, the ecological state of the environment, but also between the ecological, economic and socio-cultural indicators. Indicators should, for example, tell us how the total volume of production and its environmental inputs has changed. They should also tell us the implications of the changes in production inputs and total production for the dependent ecosystems. And finally, indicators should tell us to what extent the change in production input configuration conforms with or resists existing or emerging institutions (Hukkinen 1999). In the Baltic case, existing PSR indicators do tell us about changes in the volume of agricultural and industrial production and their implications for the Baltic nutrient balance, since agriculture and industry have traditionally contributed the main pollution loads. However, the indicators tell nothing about the potential for combinations of completely novel kinds of production in unexpected contexts to influence the state of the Baltic Sea. An example would be the combined effects of intensified marine transport and oil production under extreme storm conditions.

My second criticism of current PSR applications has to do with their simplified notion of <u>carrying capacity</u>. Many environmental indicators are based on the assumption that the more widespread and materially intensive our actions, the greater the risk of causing irreversible ecological disturbance and exceeding the carrying capacity of the Earth (Hinterberger et al. 1997; Georgescu-Roegen 1971). Yet due to contingencies in human innovation and biological evolution, there is no fundamental ecological reason why the Earth could not remain viable under a technological regime which utilizes radically more of the Sun's energy to circulate material flows much more intensively (Arrow et al. 1995; Ayres 1997). The bell-shaped population curve of ecology may hold for the special case of a particular ecosystem (such as the famous example of the Labrador-Newfoundland cod population reported by Longhurst 1998 and Kurlansky 1998). But it holds only under the specific boundary conditions of that particular ecosystem. There is nothing wrong with the PSR approach as such, but there is clearly a need to develop ecosystem-specific indicators within the PSR framework.

Third, current applications of the PSR approach assume there is a <u>single trajectory of development</u> in human-environmental interaction. The indicators of pressure, state and response are interpreted with respect to this trajectory. Yet the evolutionary path taken by a particular system of human-environmental interaction at any particular point in time is contingent upon chance events (Holling and Sanderson 1996; Gould 1990; Ehrlich 2000). Contingency refers to the problem of not knowing what developmental trajectories in the present will turn out to determine future events. While unpredictable, contingent developments are not random ones. A historical explanation of the unfolding of events in biological evolution, for example, does not emerge from deductions from laws of nature, but from sequences of preceding states, where major and often randomly triggered changes at any stage of the sequence can alter the final outcome (Gould 1990). Despite the random influences, outcomes can in retrospect be explained in causal terms because they are dependent on what happened before. Similarly, future evolution is highly unpredictable and final outcomes must be considered to be contingent (Bruun et al. 2002). The single trajectory assumed by the PSR is therefore too narrow an approach for understanding the richness of possible future trajectories. To return to the Baltic case, when a Russian supertanker collides with an Estonian passenger ferry in rough seas somewhere between Helsinki and Tallinn, causing trauma and pollution for decades, what good is it to know how well the regulatory systems of the Baltic states are geared to dealing with an increasing nutrient level caused by intensified agricultural activity? Clearly, our knowledge base is not the limitation to imagining and preparing ourselves for many other possible futures in the Baltic Sea than just eutrophication.

Fourth, current PSR applications fail to take into account various ecological, technological, institutional and cognitive path dependences that have a key influence in the outcomes of human-environmental interaction. Path dependence here refers to the tendency of past decisions or developments to constrain our choices in the future. Path dependence has been identified within many fields of inquiry relevant to sustainability issues (it has also been criticized; see Liebowitz and Margolis 1990 for a critical view of the market lock). In technology studies, commitment to large scale technological systems has been found to significantly delimit future choices (Hughes 1987). In institutional studies, earlier commitment to economic institutions has been found to develop into persistent lock-ins from which it is difficult for a society to divorce itself (North 1992). Decision theorists and cognitive scientists have recorded instances where individuals find it impossible to alter their mental models, even if they found such models unsatisfactory, because of the institutional support the mental models get from peers and society at large (Hukkinen 1999). And ecologists describe biological evolution as a path determined by its earlier stages and random events (Gould 1990; Ehrlich 2000). Sustainability indicators should be able to measure the degree of such path dependence. In the Baltic Sea, for example, the inability to incorporate in the indicator system measurements of major changes in marine transport technology, Baltic marine ecosystems, and Baltic Sea environmental regulations and dominant theories underlying such regulations, will only increase the tendency to treat marine accidents, algae blooms, and the reluctance of Baltic Sea states to participate in environmental cooperation as 'surprises' beyond human foresight.

The unique character of surprising events brings me to my final point of criticism against current applications of the PSR approach, namely, the assumption of universality in the face of case specificity of human-environmental interaction. Historical studies of systems of human-environmental interaction highlight the uniqueness of the solutions that human beings in different parts of the world throughout history have devised for coping with environmental challenges (Redman 1999; Ostrom 1994; McNeill 2000). The system of environmental governance, for example, which works in one place at a particular point in time does not necessarily work elsewhere (Ostrom et al. 1999). Although much research on sustainability indicators does acknowledge context as important (see, e.g., Holmberg 1995), some of the most influential PSR-type environmental indicators today make claims for universal applicability and comparability. Take eco-efficiency, for example, which has inspired environmental policy makers in government and industry worldwide as a concept that concisely articulates their concerns in environmental management. Proponents of eco-efficiency have argued, if not explicitly, then implicitly, that eco-efficiency can be applied around the globe, regardless of local ecological and social contexts (OECD Policy Brief 1998; World Business Council for Sustainable Development 2000; Commission of the European Communities 2001). I will argue in the next section that instead of universalism, the developers of sustainability indicators should strive for generalism, in which observations become meaningful only within the particular context of observation (Geertz 1993; Hukkinen 2001). The Baltic disaster case exemplifies such generalism: the early warning indicators of a mid-sea collision would most likely not be applicable anywhere else except the Baltic marine transport corridors between Finland, Estonia and Russia.

In sum, the PSR is an appropriate framework for getting a handle on the state of human-environmental interaction, and on the responses to reduce the pressures that may lead to failures in that interaction. However, several factors must be considered if we are to better anticipate the failures. The most influential PSR indicators today do not take these factors into account (Hukkinen 2001, 2003). In the next section, I will outline a set of proposals for developing sustainability indicators that face the challenges of functional linkages between indicators, carrying capacity, multiple trajectories of development, path dependence and ecosystem specificity.

3 The challenge of sustainability for indicator design

The PSR indicator framework can be improved at each stage of the causal chain of pressure, state and response. At the pressure stage, improved indicators of the impacts of production on ecosystems are needed. State indicators should be improved with indicators that contextualize carrying capacity and indicators that describe the congruence between ecosystems, institutions and production. Finally, the response stage would gain an intellectually broader base with indicators of technological, institutional and ecological path dependence. I will deal with each of these improvements in turn.

To address the lack of functional linkages between different pressure indicators, improved indicators of ecosystem impacts of production are needed. At least three broad categories of indicator can be outlined. The first type of indicator would measure change in the total volume of industrial production. The second type would measure change in the absolute amounts and relative proportions of environmentally significant inputs to industrial production. The first two types of indicator are, of course, common today (measures of GDP and raw material inputs, for example). Finally, the third type of indicator would measure changes in ecosystems resulting from the above two. This third type of indicator resembles such currently applied indicators as eco-efficiency (services produced per environmental input) and MIPS (material input per service unit), the quantification of which requires a life cycle analysis. The difference is that here the emphasis is on context-specific ecosystem implications of production, not some universal calculus of material and energy efficiency throughout the product lifecycle (Heiskanen 2000). Single-minded improvement of the eco-efficiency of all production units in a complex industrial network would be foolish, because the overall efficiency of the system might require 'inefficiencies' at the individual production unit level. Sensitivity to context is also necessary to overcome the key problem of eco-efficiency and MIPS, namely, their blindness to the micro- and macro-rebound effects resulting from efficiency gains (Hukkinen 2003). Indicators specific to the Baltic case would be likely to include not only the existing indicators on nutrient loads but also measurement of, say, oil output in Russia and maritime logistics in the Baltic Sea.

Second, indicators of what I have termed bounded carrying capacity are needed. The aim of this recommendation is to get around the problematic notion of carrying capacity, which is impossible to determine in absolute terms for an ecosystem without meaningful boundary conditions specified by the human-environmental interaction typical for that particular ecosystem (Arrow et al. 1995; Ayres 1997). The idea in bounded carrying capacity is to utilize alternative sustainability scenarios as a way of articulating ecosystem-specific limits that different interest groups in the society want to set for industrial production. Scenarios are here descriptions of possible pathways of development, aiming to facilitate informed decisions about the future (Bruun et al. 2002). Alternative scenarios provide an attractive way of anticipating the possible outcomes of the interaction of many decisions and actions, both past and present (Clark et al. 2001). Carrying capacity indicators would be developed for a set of alternative sustainability scenarios, constructed for the specific ecosystem in which the human-environmental interaction takes place and reflecting the variation in viewpoints about what is sustainable environmental management in that particular ecosystem (following the ideas presented in Norgaard 1995). A sustainability scenario would contain a bundle of indicators, the values of which would have 'bandwidths' (van Eeten and Roe 2002) specific to that particular scenario. Bandwidths determine the minimum and maximum values for the ecosystem functions and services that the managers of a particular ecosystem need to achieve.

The use of carrying capacity indicators bounded by alternative sustainability scenarios would have several benefits. First, alternative scenarios fix an adequate number of factors in human-environmental interaction so that the carrying capacity calculations can be performed. Second, scenarios have the capacity to illustrate the complexities of human-environmental interaction. Third, scenarios explicitly introduce multiple viewpoints and trajectories of development into the consideration of sustainability policies. Finally, alternative scenarios provide a socio-ecological context for the interpretation of carrying capacity indicators. The same value of a carrying capacity indicator means different things when interpreted in the context of alternative scenarios. In Fig. 1, for example, the dramatic shift in the value of the indicator at time t' would mean abandoning the sustainability range when sustainability is interpreted in the context of scenario B, but at the same time it would mean entering the sustainability trajectory as interpreted in scenario A (obviously, changes in other policies would be required to push the values of other indicators within the bandwidths specific to scenario A).

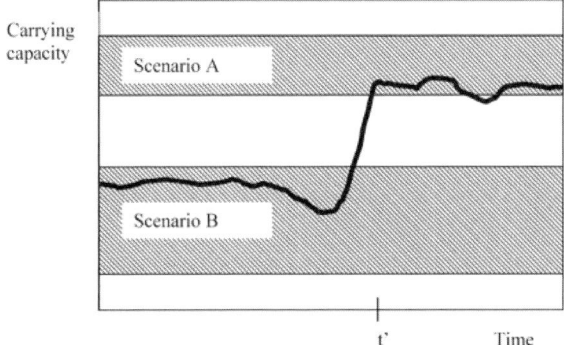

Fig. 1. Carrying capacity indicator bounded by alternative sustainability scenarios. The shift in the value of the indicator at time t' means abandoning the sustainability range when sustainability is interpreted in the context of scenario B, but at the same time entering the sustainable trajectory as interpreted in scenario A.

In the Baltic Sea, alternative sustainability scenarios could open up new perspectives and potential solutions to the blue green algae problem, which is currently dominated by the notion that the carrying capacity for nutrients in the Baltic has been exceeded and that nutrient inputs must therefore be reduced (Baltic 21 Secretariat 2000; Rosenström and Palosaari 2000). An alternative scenario might accept the fact that historical accumulation of nutrients in the bottom sediments has raised the nutrient level in the Baltic Sea for centuries to come. Consequently, the scenario would explore possibilities for developing aquaculture of algae-feeding organisms as an ecosystem-wide integrated management approach (as proposed in Jackson et al. 2001). In the new ecosystem management regime, the carrying capacity for nutrients would of course be much higher than today. Furthermore, should the algae-feeding organisms or some other organism at a higher trophic level become highly valued, the elevated nutrient levels would be seen less as a problem and more as a prerequisite for sustainable management.

The third challenge is to develop indicators of <u>congruence between ecosystems, institutions and production</u>. This is a reaction to the earlier mentioned lack of functional linkages between state indicators. The recommendation has three parts. Indicators are first of all needed to measure the agreement between the functions of a particular ecosystem and the institutions governing the management of that ecosystem. Indicators are also needed to measure the agreement between the combination of production inputs and the institutional environment. Finally, indicators need to measure the agreement between formal and informal institutions in environmental management (Ostrom 1994; North 1992; Hukkinen 1999). An example from reindeer management in Lapland, Finland, illustrates these needs. In Lapland, there is a mismatch between ecosystem functions and institutions governing the use of the ecosystem, because the Ministry of Agriculture and Forestry

(MoAF) has interpreted EU regulations to require the construction of expensive centralized slaughterhouses, although the cold ecosystem would function as a superior 'slaughterhouse' during winter. There is a mismatch between production inputs and institutions, because MoAF's interpretation of EU rules forces an economically unviable change from low-capital and labor-intensive slaughtering at the culling site to low-labor and capital-intensive slaughtering at the slaughterhouse. And there is a mismatch between formal and informal institutions, because the socially significant tradition of slaughtering at the culling site has been disrupted by formal MoAF and EU rules (Hukkinen et al. 2002).

Indicators of congruent ecosystem management should ideally be based on empirical historical evidence, as in the Lapland case. Since reindeer management practices there have been tested over centuries, it is relatively easy to identify mismatches between ecosystems, institutions and production. Where empirical evidence is lacking or there is no consensus over appropriate management, as is often the case in modern environmental management under rapidly changing social and ecological circumstances, knowledge about congruent ecosystem management will have to come from the trials and errors (and trials and successes) obtained in experiments (Ostrom et al. 1999; van Eeten and Roe 2002). In such cases, indicators of congruence between ecosystems, institutions and production would have to measure the management experiments themselves. Indicators might, for example, measure the congruence between the state of the ongoing experiment and its historical analogs, the congruence between different stakeholders' views on the management of the system, and the existence of institutions through which divergent views might be reconciled.

Finally, response indicators of <u>technological, institutional and ecological path dependence</u> are needed. The purpose of such indicators would be to inform us to what extent current decisions or developments may limit our options in the future. Indicators of path dependence should quantify or at least classify the concepts of path dependence that have been developed in fields of inquiry relevant to human-environmental interaction. In technology studies, such concepts are technological momentum (Hughes 1987) and technology life cycle (Grübler 1998), for example. In institutional theory and economic history, path dependence is described as a lock-in between institutional rules and the organizations that have evolved as a consequence of the incentive structure provided by the institutions (North 1981, 1992). Cognitive science offers us a closely related concept, cognitive lock-in, to describe individuals' tendency to stick to their mental models (Gentner et al. 2001; Hukkinen 1999). Finally, ecologists have described ecosystem evolution in terms of a path dependent renewal cycle, in which steady state climaxes and unexpected reorganizing events follow each other (Holling and Sanderson 1996; Gould 1990; Ehrlich 2000).

Assessing path dependence will require resorting to the same sustainability scenarios that were necessary for interpreting the bounded carrying capacity indicators. However, the use of scenarios is different in the two cases. In the case of bounded carrying capacity indicators, sustainability scenarios describe the socio-ecological context against which to assess the state of the system of human-environmental interaction at a particular point in time, as indicated by the carrying capacity indicators (thus helping us to answer the question: Where are we now?). In the case of path dependence indicators, each sustainability scenario represents a possible trajectory of development to which the decision taken now (i.e., response) is likely to commit us, and the indicators are used to assess the path dependence of each scenario (thus helping us to answer the question: Which direction should we be going?). The difference is illustrated in Fig. 2.

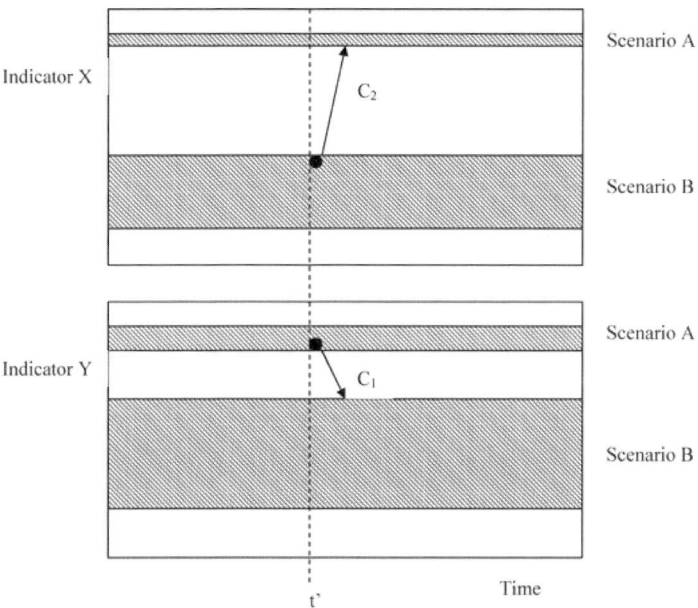

Fig. 2. Weighing short term constraints against long term path dependence. Choice C_1 immediately restrictive because it requires reductions in activities that increase the value of indicator Y, but less path dependent because it enhances scenario B with broad bandwidths. Choice C_2 is immediately relaxing because it permits increases in the value of indicator X, but more path dependent because it enhances scenario A with narrow bandwidths.

In Fig. 2, the simplified sustainability assessment system is made of just two sustainability scenarios (A and B) and two indicators of sustainability (X and Y). At time t', the measured value of the carrying capacity indicator X is within the recommended bandwidth for sustainability scenario B, whereas the value of carrying capacity indicator Y exceeds the value recommended for scenario B. The value of indicator Y does, however, fall within the recommended bandwidth of sustainability scenario A. In this case, decision makers have two policy options: begin a program to bring the measured value of indicator Y down to the bandwidth consistent with scenario B (choice C_1 in Fig. 2), or pursue sustainability scenario A and begin a program to push the measured value of indicator X to the bandwidth consistent with scenario A (choice C_2 in Fig. 2). Choice C_1 is more restrictive in terms of immediate measures (because it requires reductions in activities that tend to increase indicator value Y) but less restrictive in terms of path dependence (because it moves the trajectory of development into a scenario characterized by indicators with broad bandwidths). For reverse reasons, choice C_2 is less restrictive in terms of immediate measures but more restrictive in terms of path dependence.

The measurement or classification of path dependence would benefit from the concept of tight coupling developed in organizational reliability studies. What have there been termed tightly coupled systems come close to what is here meant by path dependent systems: they have little slack and permit only limited substitutions in resources and personnel, their buffers and redundancies are rare and deliberate, they have only one method to achieve a goal, they contain invariant sequences, and they permit no delays (Perrow 1999). The strongly path dependent scenario A in Fig. 2, for example, illustrates little slack (since the bandwidths are narrow for both indicators in scenario A), no substitutions (since both indicator X and indicator Y need to have the required value for the sustainability scenario A to be fulfilled), poor buffers (since the gaps between the bandwidths of the two scenarios are relatively wide) and no redundancy (since the indicator bandwidths for the two scenarios do not overlap each other).

When considered together with the earlier mentioned indicators of production impacts and management congruity, path dependence indicators can function as an early warning system of catastrophic shifts in human-environmental interactions. Earlier studies have indicated that the coincidence of tight coupling with complexity (defined as many interactions in an unexpected sequence) contributes to the vulnerability of technological systems to accidents (Perrow 1999; Rochlin et al. 1987). This is an important message to carry over from large scale technological systems to systems of human-environmental interaction more generally, since we are increasingly dealing with human-dominated ecosystems (Vitousek et al. 1997; van Eeten and Roe 2002). Path dependence indicators measure coupling. Complexity can in turn be measured with indicators of functional linkage, namely, ecosystem impacts of production and the congruence of ecosystem management. The early warning system would thus scan for the coincidence of high values in indicators of path dependence and functional linkage. In the Baltic Sea, the these indicators would assess the degree of complexity and coupling in the various components of the Baltic sustainability scenarios, such as oil output in Russia, nutrient loading, maritime logistics, maritime emergency preparedness and the state of the aquatic ecosystem.

Path dependence is not always catastrophic, however. The development of indicators of path dependence has a self-reinforcing aspect that can benefit environmental policy and management. Path dependence is a complex phenomenon the identification of which requires expertise in the particular technology, governance system, or ecosystem being assessed. Such expertise is most likely to be found among the individuals working within the systems, who should therefore be incorporated in the observation system that relies on the indicators. When the development and maintenance of an observation system of human-environmental interaction in a particular ecosystem is built upon the participation of those engaged in the interaction, a beneficial institutional and technological path dependence is created by virtue of the indicator system itself. The institution of an observation system grounded in local expertise reinforces itself by becoming an integral component of the local ecosystem management regime. In fact, this reinforcing process has been empirically observed in many systems of environmental management in which significant components of resource governance are in local hands (Ostrom 1994). In the Baltic Sea system, the local experts who should be involved in the observation system include fishermen, aquaculture facility managers, sea captains, recreational boaters and permanent and temporary inhabitants of coastal communities.

4 Conclusion: the new role of indicators

The critique of current PSR-based sustainability indicators and the recommendations for improving them lead us to rethink the purpose of indicators in general. Indicators can no longer be thought of as universal measures of sustainability but rather as measures of improvement in a particular ecosystem. Furthermore, indicators are integrators of pragmatic and scientific knowledge on a particular ecosystem, and an essential monitoring tool for high reliability ecosystem management. Finally, sustainability indicators should tell us not just about long term trends, but about sudden events with long term consequences. I will discuss each notion in further detail.

When at issue is the monitoring of ecological, economic and socio-cultural sustainability, we should drop the traditional notion of an indicator as a universal measure. It is useful to make a distinction between universal and general measures. Universal measures are based on a set of observations, which are then subsumed under a governing scientific theory that is applicable everywhere all of the time. General measures are also based on a set of observations, but they are understood as signs that are meaningful only within the particular context in which they were observed. The context therefore needs to be specified. Clifford Geertz likens general measures with clinical diagnosis, which looks for symptoms that deviate from theoretical predictions and then explains the theoretical peculiarities. Diagnostics is characterized by continuous testing of theoretical ideas that might be useful in explaining what is going on in the case under observation. If the ideas provide meaningful explanations of the observations, they are adopted in the explanatory frame; otherwise they are dropped (Geertz 1993). The notion of general indicators of sustainability is at the core of this article. Sustainability indicators can only help us assess the sustainability of a particular system of human-environmental interaction with respect to a particular trajectory of development, not its universal sustainability. Being case specific, however, also permits the design of indicators so that they provide detailed advice about the management of a particular system, which leads us to the second discussion point.

The scenario-based sustainability indicators proposed here can be understood as the knowledge management interface between pragmatic and scientific expertise on a particular ecosystem. The main purpose of scenario-based indicators is to provide relevant information on an ecosystem for the users of that ecosystem, without assuming broader comparability with other ecosystems. Broader comparability would not be feasible, because different ecosystems would require highly complex indicator systems with large amounts of often mutually incompatible indicators at the local level. Such complexity is not a problem, however, when the purpose of the indicator system is to link practical local knowledge about the intricacies of ecosystem management with science-based models and measurements of ecosystems. Integration of different types of expertise in real-time ecosystem management has already been successfully implemented (van Eeten and Roe 2002) and proposed for application elsewhere (Hukkinen and Roe 2002), which leads me to my next comment.

The proposed sustainability indicators would be a systematic tool for monitoring the performance of so-called high reliability ecosystem management regimes. High reliability ecosystem management is characterized by simultaneous demands for providing highly reliable ecosystem services, such as a reliable water supply from an estuary, and ensuring high quality ecosystem functions, such as a biologically diverse aquatic ecosystem in the same estuary. Experience has shown that such management will require intensive collaboration and communication among diverse experts to specify dynamic management goals, also known as bandwidths. To determine the bandwidths, the experts of both pragmatic and scientific knowledge on the ecosystem must be brought together in the same operating room to test their management models with respect to alternative scenarios (van Eeten and Roe 2002). Sustainability indicators are the monitoring data which inform the experts about the status of and options for the system of human-environmental interaction.

My final comment has to do with the objection raised by the officials against their critics in the hypothetical Baltic Sea example. Was it unfair to criticize sustainability indicators for the failure in emergency response? On the basis of the above, I think not. Sustainability and emergency response are two sides of the same coin. Current sustainability indicators, such as life expectancy, GDP growth, unemployment, emissions of pollutants and energy intensities, assume more or less stable trend-based development. But it would be against empirical evidence on the evolution of systems of human-environmental interaction to assume trends were the only relevant issue for sustainability. At least equally important are the contingent events that can radically alter and determine the path of development for decades or centuries to come. We therefore need to base our decisions and indicators on better models that take into account contingencies. 'Model' here refers not only to mathematical models describing abrupt shifts in systems of human-environmental interaction (as reviewed in Scheffer et al. 2001). More importantly the term includes gaming exercises between the diverse mathematical and cognitive models that different stakeholders in a management issue hold (as described in van Eeten and Roe 2002).

Now, could such a system of sustainability indicators have prevented the disasters envisioned for the Baltic region? Probably not. But the difference is that under the proposed system we would at least be discussing the specific deficiencies of management. With the existing indicator system, we can but wonder why the events that do happen are so often beyond all our expectations.

Acknowledgements

A first draft of the article was presented at the NATO Advanced Research Workshop 'Technological Choices for Sustainability' in Maribor, Slovenia on 13-17 October 2002. Some of the arguments of this article have been presented in a different form and context in Hukkinen (2003). I would like to thank two anonymous reviewers for helpful comments. The research was supported by Academy of Finland grant no. 47624 and European Commission 5th Framework Project no. QLRT-1999-30745.

References

Arrow K, Bolin B, Costanza R, Dasgupta P, Folke C, Holling CS, Jansson B-O, Levin S, Mäler K-G, Perrings C, Pimentel D (1995) Economic growth, carrying capacity, and the environment. Science 268:520–521
Ayres R (1997) Comments on Georgescu-Roegen. Ecol Econ 22:285–287
Baltic 21 Secretariat (2000) Development in the Baltic Sea region towards the Baltic 21 goals—an indicator based assessment. Baltic 21 Series No. 2/2000. Ministry of the Environment, Stockholm
Broecker WS, Hemming S (2001) Climate swings come into focus. Science 294:2308–9
Bruun H, Hukkinen J, Eklund E (2002) Scenarios for coping with contingency: the case of aquaculture in the Finnish Archipelago Sea. Technol Forecasting Soc Change 69:107–127
Clark JS, Carpenter SR, Barber M, Collins S, Dobson A, Foley JA, Lodge DM, Pascual M, Pielke Jr R, Pizer W, Pringle C, Reid WV, Rose KA, Sala O, Schlesinger WH, Wall DH, Wear D (2001) Ecological forecasts: an emerging imperative. Science 293:657–660
Commission of the European Communities. Proposal for a Decision of the European Parliament and of the Council Laying down the Community Environment Action Programme 2001-2010, Commission proposals, Document 501PC0031, Brussels, 2001 (http://europa.eu.int/eur-lex/en/com/pdf/2001/en_501PC0031.pdf)
Ehrlich P (2000) Human natures: genes, cultures, and the human prospect. Island Press, Washington, D.C.
Finnish Environment Institute (2002) WWW pages on the Baltic Sea (http://www.ymparisto.fi/tila/vesi/suomenl/suomenl.htm)
Flyvbjerg B (2001) Making social science matter: why social inquiry fails and how it can succeed again. Cambridge University Press, Cambridge
Geertz C (1993) The interpretation of cultures. Fontana Press, London
Gentner D, Holyoak KJ, Kokinov BN (eds) (2001) The analogical mind: perspectives from cognitive science. MIT Press, Cambridge, Mass.
Georgescu-Roegen N (1971) The entropy law and the economic process. Harvard University Press, Cambridge, Mass.
Gould SJ (1990) Wonderful life: The Burgess Shale and the nature of history. WW Norton, New York

Grassl H (2000) Status and improvements of coupled general circulation models. Science 128:1991–7
Grübler A (1998) Technology and global change. Cambridge University Press, Cambridge
Heiskanen E (2000) Translations of an environmental technique: institutionalization of the life cycle approach in business, policy and research networks. Acta Univ Oeconom Helsingiensis A-178, dissertation submitted in partial satisfaction of the requirements for the degree of Doctor of Economics. Helsinki School of Economics and Business Administration, Helsinki
Hinterberger F, Luks F, Schmidt-Bleek F (1997) Material flows vs. "natural capital"—what makes an economy sustainable? Ecol Econ 23:1–14
Holling CS, Sanderson S (1996) Dynamics of (dis)harmony in ecological and social systems. In: Hanna SS, Folke C, Mäler K-G (eds) Rights to nature. Island Press, Washington, D.C., pp 57–85
Holmberg J (1995) Socio-ecological principles and indicators for sustainability. PhD Thesis, Gothenburg, Chalmers University of Technology and University of Gothenburg, Institute of Physical Resource Theory
Hughes TP (1987) The evolution of large technological systems. In: Bijker WE, Hughes TP, Pinch T (eds) The social construction of technological systems. MIT Press, Cambridge, Mass., pp 51–82
Hukkinen J (1993) Institutional distortion of drainage modeling in Arkansas river basin. J Irrig Drain Eng ASCE 119:743–755
Hukkinen J (1999) Institutions in environmental management: constructing mental models and sustainability. Routledge, London
Hukkinen J (2001) Eco-efficiency as abandonment of nature. Ecol Econ 38:311–315
Hukkinen J (2003) From groundless universalism to grounded generalism: improving ecological economic indicators of human–environmental interaction. Ecol Econ 44:11–27
Hukkinen J, Jääskö O, Laakso A, Müller-Wille L, Raitio K (eds) (2002) Poromiehet puhuvat: Poronhoidon ongelmat, ratkaisumahdollisuudet ja tutkimustarpeet Suomen Lapissa poromiesten näkökulmasta (Reindeer herders speak: problems, potential solutions and research needs of reindeer herding in Finnish Lapland from the viewpoint of reindeer herders). Technol Soc Environ 1/2002, Espoo, Helsinki University of Technology Laboratory of Environmental Protection Publication
Hukkinen J, Roe E (2002) Satama ja Natura voivat liittyä yhteen (The harbor and Natura can join together) (in Finnish). Helsingin Sanomat, OpEd, 9 February, p A5
Jackson JBC, Kirby MX, Berger WH, Bjorndal KA, Botsford LW, Bourque BJ, Bradbury RH, Cooke R, Erlandson J, Estes JA, Hughes TP, Kidwell S, Lange CB, Lenihan HS, Pandolfi JM, Peterson CH, Steneck RS, Tegner MJ, Warner RR (2001) Historical overfishing and the recent collapse of coastal ecosystems. Science 293:629–638
Kurlansky M (1998) Cod: a biography of the fish that changed the World. Penguin Books, New York
Liebowitz SJ, Margolis SE (1990) The fable of the keys. J Law Econ 33:1–26
Longhurst A (1998) Cod: perhaps if we all stood back a bit? Fish Res 38:101–108
McNeill JR (2000) Something new under the sun: an environmental history of the twentieth-century World. WW Norton, New York
Norgaard RB (1995) Metaphors we might survive by. Ecol Econ 15:129–131
North DC (1981) Structure and change in economic history. WW Norton, New York
North DC (1992) Institutions, institutional change and economic performance. Cambridge University Press, Cambridge

OECD (1994) Environmental indicators, indicateurs d'environnement: OECD core set. OECD, Paris
OECD (1998) OECD policy brief. Sustainable development: a renewed effort by the OECD. No. 8 (http://www.oecd.org/publications/Pol_brief/1998/9808-eng.htm)
Ostrom E (1994) Governing the commons: the evolution of institutions for collective action. Cambridge University Press, Cambridge
Ostrom E, Burger J, Field CB, Norgaard RB, Policansky D (1999) Revisiting the commons: local lessons, global challenges. Science 284:278–282
Perrow C (1999) Normal accidents. Princeton University Press, Princeton, N.J.
Perttu J (2002) Suomenlahden kalusto ei riitä isossa öljytuhossa (Equipment in the Gulf of Finland not adequate to handle a big oil spill) (in Finnish). Helsingin Sanomat, 20 January, p A6
Pyykkönen A-L (2002) Vuosaaren sataman lähin naapuri on golfkenttä (Nearest neighbor to the Vuosaari harbor is golf course) (in Finnish). Helsingin Sanomat, 8 February, p B3
Redman CL (1999) Human impact on ancient environments. University of Arizona Press, Tucson, Az.
Rochlin GI, La Porte TR, Roberts KH (1987) The self-designing high-reliability organization: aircraft carrier flight operations at sea. Naval War College Rev, Autumn, pp 76–90
Rosenström U, M Palosaari (eds) (2000) Kestävyyden mitta—Suomen kestävän kehityksen indikaattorit 2000 (Measure of sustainability—Finland's sustainable development indicators 2000) (in Finnish). Suomen Ympäristö 404. Ympäristöministeriö, Helsinki
Scheffer M, Carpenter S, Foley JA, Folke C, Walker B (2001) Catastrophic shifts in ecosystems. Nature 413:591–596
Taivalkoski M, Kaurala H (2002) Venäläiset kaavailevat Uuraaseen uutta suurta öljysatamaa (Russians planning a new large oil port in Uuras) (in Finnish). Helsingin Sanomat, 18 June, p A7
The Economist (2002) When trade and security clash. 6 April, pp 65–67
van Eeten MJG, Roe E (2002) Ecology, engineering, and management: reconciling ecosystem rehabilitation and service reliability. Oxford University Press, Oxford
Vitousek PM, Mooney HA, Lubchenco J, Melillo JM (1997) Human domination of Earth's ecosystems. Science 277:494–499
World Business Council for Sustainable Development (2000) Eco-efficiency: creating more value with less impact. Conches-Geneva WBCSD (http://www.wbcsd.org/projects/pr_ecoefficiency.htm)
World Commission on Environment and Development (1990) Our common future. Oxford University Press, Oxford

Measuring Sustainability – Index of Balanced Sustainable Development

Janko Seljak[a], Damjan Krajnc[b], Peter Glavič[b]*

[a] University of Ljubljana, Faculty of Administration, Gosarjeva ulica 5, SI-1000 Ljubljana, Slovenia

[b] University of Maribor, Faculty of Chemistry and Chemical Engineering, Smetanova 17, P.O. Box 219, SI- 2000 Maribor, Slovenia

* Corresponding author. Tel.: ++386 2 229 44 51; Fax: ++386 2 252 77 74. E-mail address: glavic@uni-mb.si (P. Glavič).

Abstract

The need for aggregated sustainability information has become a concrete goal of nations committed to contributing to sustainable development and to the establishment of an integrated economic, environmental and social information system for the sustainability assessments.

The paper presents a method for aggregating a set of indicators into the Index of balanced sustainable development (I_{BSD}). The method constructs the Index from economic, environmental and social sub-indices, equally weighted together to describe total impact on sustainable development of nations. Thus, large number of indicators are aggregated in order to compress information into the index and provide problem-oriented information for decision-making.

On the basis of the methodology, index of balanced sustainable development for 24 countries (EU and candidate countries) was calculated for 1990, 1995 and 1998.

Keywords: sustainable development, sustainability indicators, sustainability assessment, index of balanced sustainable development

List of Abbreviations and Symbols

Abbreviations
ASG   Applied Statistics Group
CES   Center for Environmental Strategy
EC   European Community
HDI   Human Development Index
HDR   Human Development Report
IMAD   Institute of Macroeconomic Analysis and Development
JRC   Joint Research Center
NEF   New Economics Foundation
OECD   Organization for Economic Co-operation and Development
UNDP   United Nations Development Programme
WHO   World Health Organization
WRI   World Resources Institute

ISO 3166 Country Codes (2 characters)
AT   Austria
BE   Belgium
CH   Switzerland
CS   Czech Republic
DE   Germany
DK   Denmark
EE   Estonia
ES   Spain
FI   Finland
FR   France
GB   Great Britain
GR   Greece
HU   Hungary
IE   Ireland
IT   Italy
LT   Lithuania
LV   Latvia
NL   Netherlands
NO   Norway
PL   Poland
PT   Portugal
SE   Sweden
SK   Slovakia
SL   Slovenia

Symbols
I_A^+ indicator whose increasing value has positive impact (impact model A)

I_A^- indicator whose increasing value has negative impact (impact model A)
I_B^+ indicator whose increasing value has positive impact (impact model B)
I_B^- indicator whose increasing value has negative impact (impact model B)
I_{turn} turning value of indicator
I_{EC} economic sub-index
I_{EN} environmental sub-index
I_i actual value of indicator for observed country
I_{max} maximal value of indicator across all measured indicators
I_{min} minimal value of indicator across all measured indicators
I_{SO} social sub-index
f fraction of sustainability impact of indicator up to the turning value

1 Introduction

Sustainable development - to meet the needs of the present generation without compromising those of future generations - has become a fundamental objective of development planning that requires dealing with economic, social and environmental policies in a mutually reinforcing way. As the concept of sustainable development is becoming increasingly important, interest in monitoring development strategies in terms of sustainability has increased, resulting in a need to formulate special monitoring tools.

Nowadays, much work has been put into development and improvement of indicators that will enable a more statistically based follow-up of moves toward a sustainable development. Indicators can provide crucial guidance for decision-making in a variety of ways. Indicators are pieces of information that summarize the characteristics of a system or highlight what is happening in a system. They can help to measure and calibrate progress towards sustainable development goals. They can provide an early warning, sounding the alarm in time to prevent economic, social and environmental damage. They are also important tools to communicate ideas, thoughts and values.

Enterprises and Governments can use indicators to set targets and monitor consequent success. Indicators can also be used to compare development of countries. Many international organizations such as the UN, OECD, and the EU have developed indicators of their own. They have gathered information based on detailed instructions from member states and published them regularly.

Two different trends have become more and more obvious within the monitoring of sustainable development. Firstly, an increased understanding of the complexity of the ecosystems and the processes surrounding sustainable development has led to a trend of increased demand in data and other types of monitoring. Secondly, the need for a simplification of the information about the environment and sustainable development has increased – without easily understandable information it is difficult to get decision-makers, the public or others to act for a sustainable development.

The need for aggregated sustainability information has become a concrete goal of nations committed to contributing to sustainable development and to the establishment of an integrated economic, environmental and social information system for the sustainability assessments. In recent years, international development work has focused on the development of composite indicators. Composite indicators have already been successfully used at both national and international level in a number of different policy fields where it is necessary to summarize complex multidimensional phenomena. Table 1 comprises short selection of case studies on composite indicators in a wide variety of application fields such as environment, economy, society and sustainability.

Table 1. Short selection of developed composite indicators

Aspect	Name of composite indicator	Constructor
Environment	Pilot Environmental Performance Index	World Economic Forum
	Index of Environmental Friendliness	Statistics Finland
	Eco-indicator 99	Pré Consultants
Economy	Internal Market Index	ASG of JRC
	Economic Sentiment Indicator	European Commission
	Composite Leading Indicators	OECD
	Index of Sustainable and Economic Welfare	CES and NEF
Society	Human Development Index	UNDP
	Overall Health System Attainment	WHO
Sustainability	Dow Jones Sustainability Index	Dow Jones Indixes
	Index of Balanced Sustainable Development	IMAD

It has been foreseeable that further aggregation of indicators to sustainability indices could provide a chance for new policy guiding instruments and better integration of decision-making, as well as public participation in sustainability discussion. Although the common principle to aggregate indicators for problem assessments has gained international acceptance, it has also become evident that methods for the aggregation of indicators are either not sufficiently well-established yet, or are under development, or are not available with respect to all sustainability concerns. In addition, those that are available will impose, in the near future, considerable requirements for data production. As the credibility of aggregation methodologies is of crucial importance for the quality of new information categories, more research is needed on the aggregation methodologies and on the relevance of basic data for comprehensive assessments (Statistics Finland 2003).

2 Tools for monitoring and reporting of sustainable development

Commonly, the monitoring and reporting of sustainability include three independent areas: economic, social and environmental development. Each of them is measured using a special system of indicators. A measurement of sustainable development should take into account the connections between the economy, environment and society. Understanding the three parts and their links is a key to understanding and monitoring sustainability.

Data, indicators, headline indicators and indices are common terms in monitoring and reporting of sustainable development. The relationship between these tools, as well as their different contexts, can be depicted in a so-called information pyramid (Fig. 1). Data is the most basic component of indicator work. Most data cannot be used to interpret change in the state of the environment, economy or society. Indicators can be considered to be superior to data in several ways: they provide decision-makers and other target groups with enough knowledge to formulate responses and decisions and are more easily interpreted than complex data, which can simplify communication.

Indicators can be defined as parameters, or values, which build on data and commonly are the first, most basic tool to analyze changes, as well as to illustrate and communicate trends in the sustainable development. The types of few, well-selected, easily understandable and communicative indicators are commonly called headline indicators. If two or more headline indicators are combined an index is created, which is meant to function as a messenger.

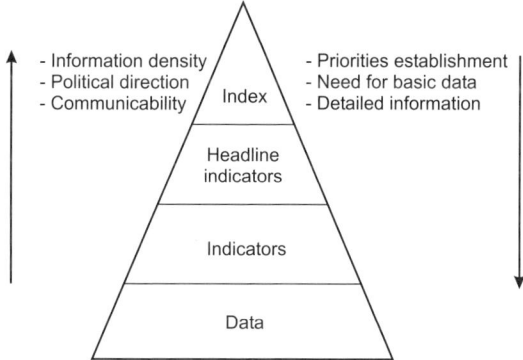

Fig. 1. The relationship between data, indicators and indices (SEI 2003)

3 Aggregating indicators of sustainable development

A mathematical combination (or aggregation as it is termed) of a set of indicators is most often called a composite indicator or an index (JRC 2002b). An index is often constructed from several sub-indices weighed together to describe total impact on a certain aspect.

The methodology of indicators aggregation into an index is crucial for a question of indices development. The methodology underlying indices development is valuable because it gives greater credibility to indices application (Malkina-Pykh 2000). In general, if the index is designed properly it can present a simplified picture of the state of the sustainability, but broader vision for decision-makers than the single indicator or even the set of indicators. In practice, indices are the result of a compromise between scientific accuracy, concise informativeness and usefulness for strategic decision-making (Lenz et al. 2000).

Indices have the advantage of providing a broader coverage of information than can be included in the current list of structural indicators and they also allow for a reduction in the number of indicators presented in the list. Indices are nevertheless useful to provide experts, stakeholders and decision-makers with:
- the direction of developments
- comparison across places (e.g. countries) and situations
- assessment of trend in relation to goals and targets
- early warning
- identification of areas for action
- anticipation of future conditions and trends
- communication channel for general public and decision-makers.

Because indices invite strong policy messages to be concluded they need to be robust and based on a sound methodology. They should be assessed on a case-by-case basis and should meet the following quality criteria (COM 2002):
- add value compared to the use of simpler indicators
- include only sub-indices which are relevant to the phenomenon to be measured
- be based on high quality data for all the sub-indices
- the inter-correlation between the sub-indices should be investigated
- the method for weighting the sub-indices should be transparent, simple and statistically sound
- the indices should be tested for robustness and sensitivity.

4 Index of Balanced Sustainable Development

The use of indices to assess progress towards sustainability is an emerging and pioneering field. Seljak (2001) has developed a method for combining indicators displaying them on a performance scale. The idea originates from the Human Development Index (HDI) developed by the United Nations Development Programme (UNDP) since 1990. However, the HDI is an aggregate of the three indicators, i.e. life expectancy, education and GDP per capita, while the method of Seljak aggregates many more indicators and could hardly be compared to the HDI.

Called "The Index of balanced sustainable development" (I_{BSD}), the method divides basic (standardized) indicators of each three aspects of sustainability into 27 sub-groups that represent crosscutting themes of the sustainable development. The method standardizes indicators to a scale from 0 to 1 and combines them into sub-indices in the way, which enables determination of positive and negative shifts. Finally, the method combines sub-indices to provide an overall index. Thus, large numbers of parameters are aggregated in order to compress information into a reduced number of parameters and provide problem-oriented information for decision-making.

On the basis of 154 indicators, index of sustainable development for 24 countries (EU and candidate countries) was calculated for 1990, 1995 and 1998. The procedure of calculating the I_{BSD} is divided into several parts, presented in Fig. 2 and briefly described in the rest of paper.

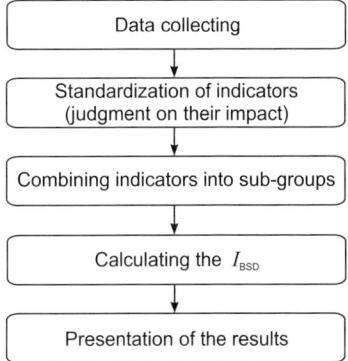

Fig. 2. The procedure of calculating the I_{BSD}

4.1 Data collecting

The availability of complete time series for all the assessed countries and component indicators is very important for the calculation of indices, since gaps in data are compounded when aggregating across many variables, countries and years. An important criterion for the selection of the component indicators (along with quality and comparability) is, therefore, the completeness of the datasets. However, for comparisons among countries we are limited to data widely available in international data sets.

As the model for the I_{BSD} is a comprehensive model for data aggregation, considerable demands for data improvements were found in the model evaluation.

Collecting more extensive and more reliable data is a challenge that the international community should face.

To calculate I_{BSD}, the data for 24 nations was collected. If possible data were obtained from the same source since data from different sources are often varying. When this was impossible the data were collected and combined from different sources. Since it is endeavored to use the most recent, reliable and internally consistent data the following order of data importance was considered:
1. Data of international world organizations (UN organizations and programs, World Bank, OECD, WRI).
2. Data of international regional organizations (EC, Baltic Union).
3. Data from national sources (national statistic offices, national institutes).

4.2 Standardization of indicators

All methods of calculating an index must transform indicators that are measured in different units into the same unit. Standardization is a more or less simple part of index calculation. There are several different methods of standardization, which range from the simplest to the most complex.

One of the methods is standardization of indicators using standardized scale. Central point of this method is determining the impact of each indicator in the aspect of sustainability and setting its upper and lower boundary. When countries are compared in the chosen time (year), the lowest value of all measured values of an indicator is determined as the lower boundary and the highest value of all the measured values of indicator is determined as the upper boundary. However, when sustainable development of countries is compared across the time, it is more preferable to use fixed (long-term) lower and upper boundaries.

In the case of calculating I_{BSD} the lower and upper boundaries (i.e. the highest and the lowest values of the indicator within the selected countries in the period of 1990 - 1998) were chosen for each underlying indicator. Various methods of aggregation and standardization of indicators could be applied. As the most appropriate of the index, standardization with the transformation of scale was selected. The basis of such standardization is the index which shows the place occupied by the country on the scale between 0 and 1 for an individual indicator. If a certain value of an indicator of a country is the lowest one within all the countries compared, the value zero is attributed, if it is the highest, the value one is attributed.

The standardization took into account the positive or negative impact of an indicator to sustainable development. Indicators whose increasing values have positive impact and indicators whose increasing values have negative impact in the perspective of sustainability were considered in order to assure appropriate calculation of the I_{BSD}.

The impact of indicator was determined according to two models (*model A* and *model B*) into which individual indicators were categorized. *Model A* (Fig. 3) has assumed that the positive or negative impact of an indicator is linear. For example, increased gross investment, fraction of active population, or life expectation in years clearly have a linear and positive impact (Fig. 3a), while increased consumption of alcohol and cigarettes or increased unemployment level are values with a linear and negative impact on the sustainable performance of the nation (Fig. 3b).

In the procedure of standardizing I_{BSD}, the relative values of indicators with positive impact were calculated by applying Eq. 1, while the relative value of indicators with negative impact applied Eq. 2.

$$I_A^+ = \frac{I_i - I_{min}}{I_{max} - I_{min}} \quad (1)$$

$$I_A^- = 1 - \left(\frac{I_i - I_{min}}{I_{max} - I_{min}}\right) \quad (2)$$

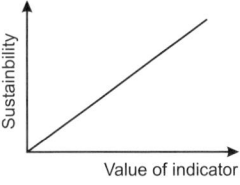
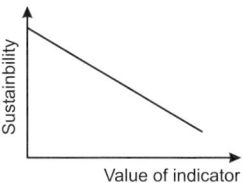

(a) Positive impact of indicator   (b) Negative impact of indicator

Fig. 3. Presentation of *Model A* for standardization of indicator considering its impact to sustainability

The second *model B* (Fig. 4) has assumed that the increasing value of indicator has linear and high impact to sustainable development up to a certain turning value I_{turn}, afterwards its impact diminishes (Fig. 4a). On the contrary, the indicator could have linear and low sustainability impact up to I_{turn}, afterwards its impact increased (Fig. 4b). Very high positive impact to certain turning value has for example gross domestic production (GDP) per capita. Afterwards, its impact diminishes (however, the positive impact still increases but with lower intensity) (Fig. 4a). On the other side, air and water pollution has negative but low impact to certain turning value (determined by the environmental health regulations), afterwards its negative impact increases with higher intensity (Fig. 4b).

In the calculation of I_{BSD}, *model B* included both types of indicators, those with positive and those with negative impact to sustainable development. The relative value of an indicator i with the positive impact was calculated using Eq. 3 and with the negative one using Eq. 4.

$$I_B^+ = \left(\frac{I_i - I_{min}}{I_{turn} - I_{min}}\right) \cdot f \qquad \text{for } I_i \leq I_{turn} \quad (3)$$

$$I_B^+ = f + \left(\frac{I_i - I_{turn}}{I_{max} - I_{turn}}\right) \cdot (1-f) \qquad \text{for } I_i \geq I_{turn} \quad (4)$$

$$I_B^- = 1 - \left(\frac{I_i - I_{min}}{I_{turn} - I_{min}}\right) \cdot f \qquad \text{for } I_i \leq I_{turn} \qquad (4)$$

$$I_B^- = 1 - f - \left(\frac{I_i - I_{turn}}{I_{max} - I_{turn}}\right) \cdot (1-f) \qquad \text{for } I_i \geq I_{turn}$$

For indicators, classified into *model B*, the determination of the value at which the slope of indicator line is changed was needed. Turning values are selected on individual basis for each indicator separately. For economic indicators they have been determined as the average values within countries compared (in 1998). An exception was made on the inflation indicator where I_{turn} was determined to be 0,1 (10 %). For environmental indicators turning values were determined according to the environmental health regulations.

Fraction (f) of sustainability impact of economic indicators up to their turning value was determined to be 0,8. To the most of environmental indicators, the fraction of sustainability impact up to the turning value, $f = 0,2$ has been attributed. None of social indicators has been classified into *model B*, therefore, the determination of fraction f was unnecessary.

The standardization method, represented above, was used at the 4th aggregation level. However, to assure comparable impact of composite indicators, standardization at the 2nd and the 3rd aggregation level has also been performed using the *model A*.

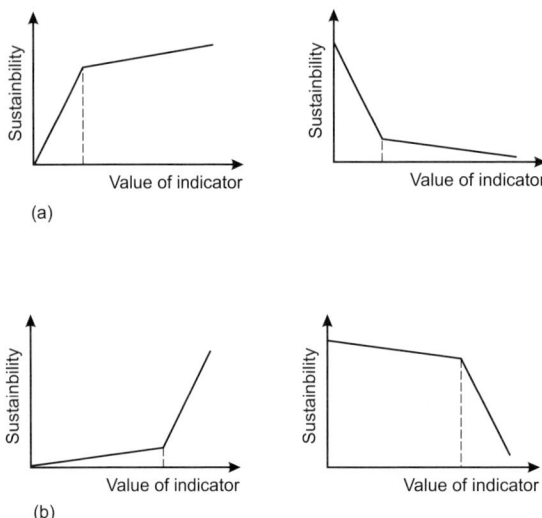

Fig. 4. Presentation of *Model B* for standardization of indicator considering its impact to sustainability

4.3 Combining indicators into the sub-groups

The procedure of combining indicators into sub-groups (economic, environmental and social sub-group) applied a multi-level approach, where the selection of 154 indicators was combined into three main sub-groups, which represent the main aspects of sustainable development. Clearness and intelligibleness of analysis increased this approach (Fig. 5). On the first level, indicators have been classified into three main sub-groups of economic, environmental, and social indicators according to the widely used three-dimensional model of sustainability.

The I_{BSD} was constructed from economic, environmental and social sub-indices. Each sub-index was represented with 9 indicators of sub-groups (second level). The main advantage of such a method is assuring the equal impact of each main group to the value of I_{BSD}. 27 indicators of the second level are presented in Table 2. The third level is introduced mainly because of representation of social and environmental indicators in several sub-fields. For economic indicators this level of aggregation was unnecessary. 47 third level indicators were combined into the second level indicators of the sub-groups. 154 individual fourth level indicators were aggregated into these 47 indicators.

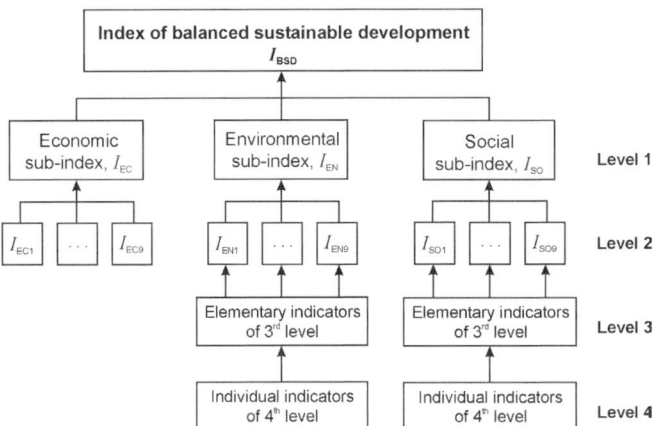

Fig. 5. Multilevel composition of the I_{BSD}

4.4 Calculating the index of balanced sustainable development

The aggregation of indicators into sub-indices I_{EC}, I_{EN} and I_{SO} was performed by averaging the relative values of the second level indicators, i.e. each of the indicators receives equal weight (Eqs. 5, 6 and 7). However, if indicators would be fully understood and there would be scientific consensus about the relative contributions of different factors to sustainability, the model of calculating the index of balanced sustainable development (I_{BSD}) would almost surely support an algorithm of unequal weighting.

$$I_{EC} = \sum_{i=1}^{9} I_{ECi} \Big/ 9 \qquad (5)$$

$$I_{EN} = \sum_{i=1}^{9} I_{ENi} \Big/ 9 \qquad (6)$$

$$I_{SO} = \sum_{i=1}^{9} I_{SOi} \Big/ 9 \qquad (7)$$

Finally, the I_{BSD} was constructed by combining the three sub-indices I_{EC}, I_{EN} and I_{SO} (Eq. 8). Equal weights (1/3), which should be combined with each sub-index, have been attributed to each sub-index to derive the I_{BSD}. Certainly, other methods of weighting the sub-indices of the I_{BSD} could be applied, e.g. by using public opinion polls or involving expert judgment. However, the sustainability concept is by definition giving equal weight to its three aspects, which makes equal weighting a sensible option.

$$I_{BSD} = (I_{EC} + I_{EN} + I_{SO})/3 \qquad (8)$$

Table 2. Presentation of second level indicators

Sub-group	Indicator notation	Description of second level indicator
Economic	I_{EC1}	Production
	I_{EC2}	Macro-economic stability and level of consumption
	I_{EC3}	Factors of economic growth - capital
	I_{EC4}	Factors of economic growth - human resources
	I_{EC5}	Factors of economic growth - technological resources
	I_{EC6}	Factors of economic growth - natural resources
	I_{EC7}	International trade
	I_{EC8}	Consumer habits of inhabitants
	I_{EC9}	Structure of production
Environmental	I_{EN1}	Air polluting
	I_{EN2}	Air pollution
	I_{EN3}	Water polluting
	I_{EN4}	Rivers pollution
	I_{EN5}	Soil and area
	I_{EN6}	Noise
	I_{EN7}	Nonrenewable resources
	I_{EN8}	Renewable resources
	I_{EN9}	Inflowing resources
Social	I_{SO1}	Number and social structure of inhabitants
	I_{SO2}	Communities, migrations and regional structure
	I_{SO3}	Economic inequality
	I_{SO4}	Gender inequality
	I_{SO5}	Life expectancy
	I_{SO6}	Illnesses, harmful habits and health care infrastructure
	I_{SO7}	Education
	I_{SO8}	Human rights, liberties and cooperation
	I_{SO9}	Safety

4.5 Presentation of results

Sub-indices I_{EC}, I_{EN}, I_{SO} and index of balanced sustainable development I_{BSD} show the position of the country concerned, compared with other countries: the country with a higher index is in a better position (the value 1 represents the highest level of development possible and the value 0 represents the lowest one). Following one particular index for several years, it shows how the country is progressing over time. If the index is higher in year $n+1$ than it was in year n, the performance of the country has improved over that period. The value of an index for a year N can be compared with its average value, too.

Tables 3 - 5 present sub-indices for EU and candidate countries compared in years 1990, 1995 and 1998.

Table 3. Relative values of economic sub-index of balanced sustainable development

Country	I_{EC} 1990	I_{EC} 1995	I_{EC} 1998	Growth of I_{EC} in 1990-1998
Norway	0,87	0,96	1,00	0,13
Finland	0,77	0,83	0,94	0,17
Sweden	0,75	0,81	0,90	0,14
Switzerland	0,83	0,85	0,89	0,06
Denmark	0,71	0,73	0,83	0,12
Ireland	0,60	0,73	0,82	0,22
Netherlands	0,74	0,74	0,82	0,08
Austria	0,74	0,74	0,81	0,08
Belgium	0,60	0,65	0,69	0,09
France	0,53	0,64	0,67	0,14
Great Britain	0,52	0,58	0,64	0,12
Germany	0,66	0,66	0,63	– 0,03
Italy	0,45	0,52	0,55	0,10
Slovenia	0,28	0,45	0,52	0,25
Spain	0,38	0,42	0,52	0,14
Estonia	0,34	0,32	0,52	0,18
Slovakia	0,20	0,40	0,48	0,28
Czech Republic	0,32	0,47	0,43	0,12
Portugal	0,29	0,36	0,41	0,12
Hungary	0,21	0,23	0,36	0,16
Lithuania	0,20	0,09	0,35	0,16
Latvia	0,38	0,11	0,35	– 0,02
Greece	0,09	0,22	0,33	0,24
Poland	0,00	0,19	0,25	0,25
Average	**0,48**	**0,53**	**0,61**	**0,14**

Results show that, in economic field (Table 3), Slovenia has ranked nineteenth in year 1990 and has climbed up to the fourteenth place in year 1998. In the 1990 it has had lower values in the most fields measured, especially in the field of economic growth (capital, technological resources), production, macro-economic stability and international trade. Until the year 1998 there was a great improvement in the field of production indicators and most factors of economic growth (capital, human, technological resources). The international trade and the structure of production have also improved. In the economic field only the consumer habits decreased. In 1998 Slovenia has ranked fourteenth in economic and thirteenth in social and environmental development. From the ranks it could be deducted that the development was relatively balanced. Nevertheless, it can be seen from the values of indicators in individual fields that Slovenia lags behind the other EU countries particularly in the social field and less in the environmental one.

Table 4. Relative values of environmental sub-index of balanced sustainable development

Country	1990	I_{EN} 1995	1998	Growth of I_{EN} in 1990-1998
Latvia	0,79	1,00	0,94	0,15
Austria	0,78	0,84	0,92	0,14
Sweden	0,81	0,84	0,89	0,09
Norway	0,70	0,76	0,87	0,17
Switzerland	0,82	0,82	0,81	– 0,01
Ireland	0,75	0,71	0,80	0,04
Lithuania	0,62	0,52	0,79	0,17
Finland	0,73	0,75	0,79	0,06
Estonia	0,55	0,71	0,73	0,18
Denmark	0,59	0,58	0,68	0,10
Slovakia	0,46	0,50	0,59	0,14
France	0,48	0,57	0,55	0,07
Slovenia	0,42	0,59	0,55	0,13
Portugal	0,56	0,57	0,55	– 0,01
Poland	0,24	0,37	0,51	0,26
Germany	0,34	0,43	0,47	0,13
Hungary	0,47	0,43	0,46	0,00
Spain	0,17	0,08	0,39	0,23
Greece	0,22	0,23	0,37	0,15
Czech Republic	0,00	0,34	0,37	0,37
Italy	0,11	0,13	0,36	0,25
Great Britain	0,05	0,26	0,33	0,29
Netherlands	0,25	0,31	0,24	– 0,01
Belgium	0,01	0,06	0,16	0,15
Average	**0,45**	**0,52**	**0,59**	**0,13**

For a selection of 24 countries the I_{BSD} has been calculated for 1990, 1995 and 1998 (Table 6). Similar ratings could be made for regions, cities or companies to analyze their performance and to design next development steps.

In 1998 (Fig. 6), the highest value of the I_{BSD} was reached by Norway, which was in all aspects the most sustainable country. The main reasons for this achievement were in the economic field, a relatively high quality and an improvement (in the last eight years) of factors of economic growth (human, technological and natural resources), favorable consumer habits of inhabitants, as well as a high level of the production activity (GDP and its growth). In the social field, Norway was placed relatively high in all spheres in 1998. Compared to 1990, the level of social development in the country improved notably in the fields of economic inequality, life expectancy, education, human rights, liberties, cooperation and safety.

Table 5. Relative values of social sub-index of balanced sustainable development

Country	1990	I_{SO} 1995	1998	Growth of I_{SO} in 1990-1998
Norway	0,87	0,97	1,00	0,13
Sweden	0,91	0,88	0,95	0,04
Denmark	0,60	0,70	0,86	0,26
Finland	0,84	0,69	0,78	– 0,06
Netherlands	0,70	0,73	0,75	0,05
Austria	0,60	0,70	0,72	0,12
Switzerland	0,54	0,58	0,52	– 0,03
Germany	0,51	0,50	0,51	0,01
Great Britain	0,31	0,51	0,51	0,19
Ireland	0,34	0,41	0,45	0,12
Slovakia	0,60	0,33	0,42	– 0,18
Czech Republic	0,49	0,52	0,42	– 0,07
Slovenia	0,53	0,32	0,41	– 0,12
Italy	0,40	0,45	0,40	0,00
France	0,35	0,41	0,39	0,04
Portugal	0,37	0,43	0,36	– 0,01
Belgium	0,39	0,36	0,31	– 0,09
Poland	0,17	0,18	0,30	0,13
Spain	0,33	0,31	0,20	– 0,12
Estonia	0,37	0,12	0,13	– 0,24
Hungary	0,18	0,16	0,10	– 0,09
Latvia	0,39	0,18	0,05	– 0,34
Greece	0,08	0,12	0,05	– 0,03
Lithuania	0,22	0,16	0,00	– 0,22
Average	**0,46**	**0,45**	**0,44**	**– 0,02**

Sweden, which was next to the top in all the three fields, was particularly good in the field of consumer habits of inhabitants, life expectancy, education and air pollution. Aggregated values of its development indicators also increased in all the three main fields. The loss of the first place from 1990 could principally be attributed to the even faster progress of Norway.

In the last place there was Greece, which had an extremely low value particularly in the social and partly also in the economic field. In the economic field the indicators measuring macro-economic stability, consumer habits of inhabitants and international trade achieved an extraordinary low level, but all the other indicators were relatively low, too. In the social field a low level was achieved by the indicators in the field of the number and social structure of inhabitants, economic and gender inequality, education, human rights, liberties, cooperation and safety. Hungary can be found in the last but one place. It was not unsuccessful in any of the included areas, but the ranks between 17 and 21 caused it to occupy the last but one place in the total. The country was less successful in the field of macro-economic stability, consumer habits, illnesses, harmful habits of individuals and health care infrastructure. In the economic field the conditions substantially improved in the last eight years, but slept in the social and the environmental fields (loss of four ranks).

Table 6. Values of index of balanced sustainable development (I_{BSD}) for compared countries

Country	1990	1995	1998	Growth of I_{BSD} in 1990-1998
Norway	0,82	0,92	1,00	0,18
Sweden	0,84	0,86	0,94	0,10
Finland	0,78	0,76	0,85	0,07
Austria	0,69	0,76	0,83	0,14
Denmark	0,61	0,65	0,80	0,19
Switzerland	0,72	0,75	0,73	0,01
Ireland	0,52	0,59	0,67	0,15
Netherlands	0,52	0,56	0,57	0,05
France	0,39	0,49	0,49	0,10
Germany	0,45	0,48	0,49	0,04
Great Britain	0,20	0,39	0,44	0,24
Slovakia	0,35	0,34	0,44	0,09
Slovenia	0,34	0,39	0,44	0,10
Estonia	0,35	0,31	0,40	0,05
Latvia	0,47	0,36	0,38	– 0,09
Italy	0,23	0,29	0,37	0,14
Portugal	0,33	0,39	0,37	0,04
Czech Republic	0,17	0,38	0,33	0,16
Belgium	0,24	0,27	0,31	0,07
Lithuania	0,26	0,15	0,30	0,04
Spain	0,20	0,17	0,29	0,09
Poland	0,01	0,14	0,27	0,26
Hungary	0,19	0,17	0,21	0,02
Greece	0,00	0,07	0,14	0,14
Average	**0,40**	**0,44**	**0,50**	**0,10**

One of the tasks of I_{BSD} is also to measure the balance of development (proportionate development in the economic, social and environmental fields). The most balanced sustainable development level in 1998 was achieved by Sweden and Czech republic where the values of indicators in all the three included fields differed only a little. But Sweden achieved this balance level at a much higher level of development. The most unbalanced level of development can be found in the Baltic States where the most problematic is the status in the social field.

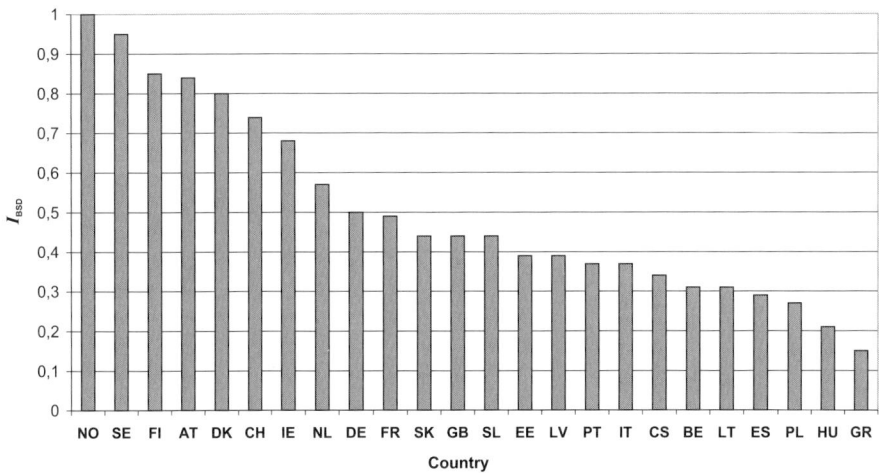

Fig. 6. Index of balanced sustainable development (I_{BSD}) for countries compared in 1998

5 Conclusions

The proposed IMAD method to calculate I_{BSD} seems to meet the general requirements for a good index, as it captures the concept of sustainability as a whole, includes its different components and is relatively insensitive to minor changes in its composition, as missing data can be left out without considerably affecting the method to calculate an average score. The sub-indices produced by the model can be used separately from each other or as a part of the I_{BSD}. Consequently, the model provides effective means for processing information into integrated assessments of sustainability.

As regards the quality of the information provided with the model, this is naturally very much dependent on the quality of basic data and methods used in studies on the valuation of sustainability concerns. The sub-indices produced by the model are of good information quality and they are suitable for political decision-making.

Although the time concept is not directly included in the calculation model of the I_{BSD}, it can easily be made visible by showing the development of the index in time. Therefore, the I_{BSD} reflects the performance of countries compared in time, which is an important criterion of sustainability: providing future generations with better prospects than were inherited by the present generation.

The I_{BSD} enables comparisons across places since the boundaries can be determined for a certain region (e.g. EU) or for the entire world. The I_{BSD} permits national comparisons of sustainability in a systematic and quantitative fashion. It assists the move towards analytically more rigorous, data driven approach to decision-making. In particular, the I_{BSD} enables identification of issues where national performance is above or below expectations. It allows priority-setting of policy areas within countries and regions and quantitative assessment of the success of policies and programs. The I_{BSD} also enables investigation of interactions between economic, environmental and social performance.

Formation and structure of the I_{BSD} enables theoretical determination of the impact of an individual index on the sustainability level by defining the form and direction of a relation, marginal values, upper and lower boundaries. Also, upgrading with the system of weighted measures is possible for individual fields, which would be constructed on the basis of opinion polls of inhabitants (Seljak 2001).

Summarized, the I_{BSD} has the following advantages:
- Easier interpretation than trying to find a trend with many separate indicators. It facilitates the task of ranking countries on complex issues.
- Reducing the size of a list of indicators or including more information within the existing size limit.
- Summarizing the complex or multi-dimensional issues, in a view of supporting decision-makers.
- Helping to attract public interest by providing a summary figure with which to compare the performance within countries and their progress over time.

One could argue that the individual indicators can present more transparent and comprehensive picture than any composed measure, which would inevitably be subject to criticism about the choice of indicators and weights used. However, individual and composed indicators may seem incompatible, but in reality they should not be combined at all. Instead, different target groups, with partly different needs for the information, should be provided with the type of information they need to be able to continue the efforts towards a sustainable development. Hence, if the needs of different target groups are to be fulfilled, the choice of reporting tool (index, headline indicators, set of indicators, data etc.) should be considered (SEI 2003).

Future reports on sustainability might proceed in a different, more analytical direction. This would involve a more explicit analysis of dependencies among indicators as well as different schemes for weighting them. However, the overall evaluation of the I_{BSD} model reveals that the selected approach and methodology are feasible with respect to current information availability. The model meets the problem-orientation of widely used aggregation methodologies, although there are still problem fields such as the choice of indicators, and the way in which they are weighted together. Therefore, further development of the model itself would be welcomed with respect to better assessment of sustainable development.

Acknowledgements

The authors gratefully acknowledge the financial support from the Ministry of Education, Science and Sport of Slovenia, Research Grant No. 3311-02-831226. Permission to publish this paper by the IMAD is also appreciated.

References

COM (2002) 551 final. Commission of the European Communities. Structural indicators - Communication from the Commission, Brussels, 16.10.2002. http://europa.eu.int/eur-lex/en/com/cnc/2002/com2002_0551en01.pdf

JRC - Joint Research Center (2002a) Internal Market Index 2002: Technical details of the methodology. Institute for the Protection and Security of the Citizen, Technological and Economic Risk Management, Applied Statistics Group. http://www.jrc.cec.eu.int/

JRC - Joint Research Center (2002b) State-of-the-art Report on Current Methodologies and Practices for Composite Indicator Development. Institute for the Protection and Security of the Citizen, Technological and Economic Risk Management, Applied Statistics Group. http://www.jrc.cec.eu.int/

Lenz R, Malkina-Pykh GI, Pykh Y (2000) Introduction and overview. Ecological Modelling 130: 1–11

Malkina-Pykh IG (2000) From data and theory to environmental models and indices formation. Ecological Modelling 130: 67–77

Pré Consultants (2001) The Eco-indicator 99 - A damage oriented method for Life cycle Impact Assessment. Methodology Report. http://www.pre.nl/

SEI - Stockholm Environment Institute (2003) Indicators of environment and sustainable development. http://www.sei.se/policy/INDIC.pdf

Seljak J (2001) Sustainable Development indicators (in Slovene). Institute of Macroeconomic Analysis and Development, Ljubljana

Statistics Finland (2003) Index of Environmental Friendliness. http://www.stat.fi/tk/yr/ye22_en.html

UNDP - United Nations Development Programme, various years (1990 – 2003) Human Development Report. Oxford University Press, New York.

World Economic Forum (2002) An Initiative of the Global Leaders of Tomorrow Environment Task Force, Annual Meeting 2002. Environmental Sustainability Index. http://www.ciesin.columbia.edu/indicators/ESI/EPI2002_11FEB02.pdf

Evaluating the Environmental Friendliness, Economics and Energy Efficiency of Chemical Processes: Heat Integration

Teresa M. Mata[a,†], Raymond L. Smith[b,††], Douglas M. Young[b,†††] and Carlos A. V. Costa[a,††††]

[a] Laboratory of Processes, Environment and Energy Engineering, Faculty of Engineering, University of Porto, *Rua Dr. Roberto Frias, 4200-465 Porto, Portugal*

[b] National Risk Management Research Laboratory, Office of Research and Development, U.S. Environmental Protection Agency, 26 W. Martin Luther King Drive, Cincinnati, OH 45268 USA

[†] tmata@fe.up.pt
[††] smith.raymond@epa.gov
[†††] young.douglas@epa.gov
[††††] ccosta@fe.up.pt

Abstract

The design and improvement of chemical processes can be very challenging. The earlier energy conservation, process economics and environmental aspects are incorporated into the process development, the easier and less expensive it is to alter the process design. In this work different process design alternatives with increasing levels of energy integration are considered in combination with evaluations of the process economics and potential environmental impacts. The example studied is the hydrodealkylation (HDA) of toluene to produce benzene. This study examines the possible fugitive and open emissions from the HDA process, evaluates the potential environmental impacts and the process economics considering different process design alternatives. Results of this work show that there are tradeoffs in the evaluation of potential environmental impacts. As the level of energy integration increases process fugitive emissions increase while energy generation impacts decrease. Similar tradeoffs occur for economic evaluations, where the capital and operating costs associated with heat integration could be optimised. From the example designs considered here, an intermediate amount of energy integration produces the most economically beneficial and environmentally friendly process.

1 Introduction

Heat integration is an important aspect of chemical process design, not only for economic, but also for environmental reasons. The concept of heat integration was first introduced via heat exchange networks (HEN) in the 1970s (Linnhoff and Flower 1978). Linnhoff and Hindmarsh (1983) indicate that typically 20-30% energy savings, coupled with capital savings, can be realised in state-of-the-art flowsheets by improved HEN design. The optimisation of HEN design consists of finding feasible sequences of heat exchangers in which pairs of streams are matched, with the purpose of minimising the total costs, i.e., capital and operating costs. Normally a designer chooses between decreasing the total heat transfer area to reduce capital costs and increasing the heat transfer area to reduce operating costs. One simple method to estimate the sensitivity of the total costs consists of devising several heat integration alternatives. (If several process alternatives exhibit similar economics, additional criteria can be considered, as for example, the controllability or the simplicity of the process.) With this approach, it is not guaranteed that the optimum heat-exchanger network is found, but we can gain a better understanding of the economic importance of energy integration.

Reductions in energy usage and waste treatment generally improve the economics of a process. However, a design with less waste does not necessarily have lower impacts on the environment, since the wastes may have a higher contribution to the potential environmental impacts than another design with a larger amount of waste but a lower environmental effect (Young et al. 2000). For this reason when designing a chemical process it is important to evaluate the potential environmental impacts (PEIs) of the different design alternatives. A possible way to accomplish this task is to use the waste reduction (WAR) algorithm (Young and Cabezas 1999). The original version of the WAR algorithm, developed by Hilaly and Sikdar (1994), introduced the concept of a pollution balance, which was strictly mass based. Cabezas et al. (1999) introduced the generalised WAR algorithm with a PEI balance, which assigned environmental impact values to different pollutants, as an improvement upon the original WAR algorithm. Young and Cabezas (1999) extended the PEI balance to include the consumption of energy by the process into the environmental evaluation.

Before evaluating a process one needs to know stream flowrates and energy flows. One method to create such a detailed process flowsheet is to use a process simulator. Examples of process simulators are ASPEN PLUS™ by Aspen Technology Inc., CHEMCAD™ by ChemStations, Inc., HYSYS™ by Hyprotech Ltd. and PRO/II® by Simulation Sciences Inc. These software programs can be very helpful to do rigorous analyses. They help to analyse alternative processes, predict the performance of a process, locate malfunctions and solve specific problems that arise from the primitive design problem. Also, they calculate heats of reaction, heat added to or removed from a stream, power requirements for pumps and compressors, performance of a flash separator at various temperatures and pressures, bubble and dew point temperatures associated with distillates and bottoms products, among many other quantities (Seider et al. 1998). The results of such an analysis can more easily include detailed reaction kinetics, separations with sloppier splits, rigorous recycle streams, etc. While this further level of detailed analysis takes longer to set up and calculate, the resulting flowsheets can increase the confidence in a design (Smith et al. 2001).

This paper presents an illustrative case study, the HDA process with energy integration, which demonstrates the use of the WAR algorithm in conjunction with economic evaluation, with the purpose of distinguishing which of the process design alternatives is more environmentally friendly and economically beneficial. The case study presents four design alternatives with increasing levels of energy integration: an original design without any energy integration and three modified designs where the level of energy integration increases. Although in this paper only four energy integration alternatives for the HDA process are studied, Terril and Douglas (1987) developed other HEN design alternatives, which are also viable for the HDA process. The four HDA process design alternatives evaluated here were simulated using the process simulator PRO/II® of Simulation Sciences Inc. The economics of the HDA process and the heat-exchanger network calculations are described by Douglas (1988), while the kinetics for the simulations are from Luyben (2000). The theory of the PEI calculations or the WAR algorithm, including the incorporation of energy, is described by Young et al. (2000), and an abbreviated description of WAR is given by Smith (2002). Results of this work show the tradeoffs between process economics, potential environmental impacts and energy integration, revealing where attention should be focused when designing a chemical process.

2 Case study – the hydrodealkylation process

The HDA process was used following World War II, when it became favourable to convert large quantities of toluene (which was no longer needed to make the explosive TNT) to benzene for use in the manufacture of cyclohexane, a precursor of nylon. Nowadays, this process may follow a reformer in a refinery, which turns cyclic compounds into benzene, toluene, and xylene (among other species), depending on the demand for such chemicals, where the hydrodealkylation of toluene is one process for altering the supply distribution. The main reaction of the HDA process involves the conversion of toluene to benzene as follows

$C_7H_8 + H_2 \rightarrow C_6H_6 + CH_4$

This main reaction is accompanied by the unavoidable side reaction that produces biphenyl

$2C_6H_6 \rightarrow H_2 + C_{12}H_{10}$

In this study the HDA plant capacity is based on the conversion of 1.650×10^2 kgmol/hr of toluene, or approximately 1.386×10^6 kgmol/yr, assuming operation 350 days per year. An excess of hydrogen gas is used to prevent carbon deposition. The benzene product obtained has a purity of 99.6% (by mass).

The toluene and hydrogen raw material streams are heated and combined with recycled toluene and hydrogen streams before they are fed to the reactor. The hydrodealkylation reaction takes place at temperatures in the range of 622-688°C and a pressure of about 34atm. Approximately 75.4% (by mass) of toluene is converted to benzene and approximately 1.5% (by mass) of the benzene produced in the hydrodealkylation reaction is converted to biphenyl. Large quantities of heat are needed to raise the temperature of the feed chemicals to 622°C, which is accomplished using a small furnace. Similarly, large quantities of heat must be removed to partially condense the reactor effluent. These enthalpy changes (heat duties) were calculated using PRO/II®. With the process simulator the heat duty of a stream for a specified heat exchanger inlet and outlet conditions are most easily obtained, especially for streams that are multi-component mixtures undergoing phase change. These calculations are relatively complex because effects of temperature, pressure and composition on enthalpy are taken into account and the phase condition is established by a phase equilibrium calculation.

The product stream leaving the reactor contains hydrogen, methane, benzene, toluene and the unwanted biphenyl. The separation section in this process involves a flash separator and three distillation towers. The flash separator works at a temperature of 38°C and a pressure of 32atm. Most of the hydrogen and methane are separated from the other components by using a partial condenser of the aromatics and then the light gases are flashed away. The flash gas stream is composed essentially of methane and hydrogen, which is partially recycled, being first compressed and then mixed with the reactor's feed stream. A purge is required to prevent the accumulation of methane in the gas-recycle loop. A purge/recycle ratio of 0.20 is maintained. The purge stream leaving the process contains mainly methane with an unavoidable amount of hydrogen. Not all the hydrogen and methane are separated from the aromatics in the flash drum. Most of the remaining hydrogen and methane in the flash liquid stream are removed in a first distillation column, in order to prevent the contamination of the benzene product. The benzene is recovered in a second distillation column, and finally the recycled toluene is separated from the unwanted biphenyl in a third distillation column.

Figure 1 shows four HDA process design alternatives, a, b, c and d, with increasing levels of energy integration. Alternative a indicates only the need for heating and cooling and alternatives b, c and d are energy-integrated flowsheets.

In alternative a the furnace is used to heat stream 1 to 622°C, which is then fed to the reactor. The reactor effluent, stream 2, is cooled to 38°C in the heat exchanger HX1. Another heat exchanger, HX2, is used to cool stream 3 to 25°C. Water is used as a coolant fluid at 15°C. Reboilers use saturated steam at 180°C and 10atm, except the reboiler of the third column, which uses saturated steam at 240°C and 34atm. At these conditions of pressure and temperature the steam condenses in the reboilers (Smith and Van Ness 1987).

In alternative b, the feed-effluent heat exchanger, HX3, is introduced to match streams 1 and 2. The heat available in stream 2 is used to pre-heat stream 1, which simultaneously cools stream 2. In this alternative, the furnace (used mainly for start-up and process control) is also included. Otherwise, the process comparisons would not be realistic, since without a furnace the process could only operate at steady state conditions.

In alternative *c* a stream leaving the first plate of tower 2 is pre-heated in the condenser of tower 3. Similarly, in alternative *d* a stream leaving the first plate of tower 1 is heated in the condenser of tower 2. The condensers of towers 2 and 3 and the reboilers of towers 1 and 2 are included in the designs, otherwise the comparisons would not be operationally realistic. Note that these results were obtained with the process simulator for steady state conditions.

For the process design alternatives *a*, *b*, *c* and *d*, process simulator PRO/II® calculates data such as the streams' flowrate, temperatures and pressures and also the heat available in the hot streams and the heat required for the cold streams.

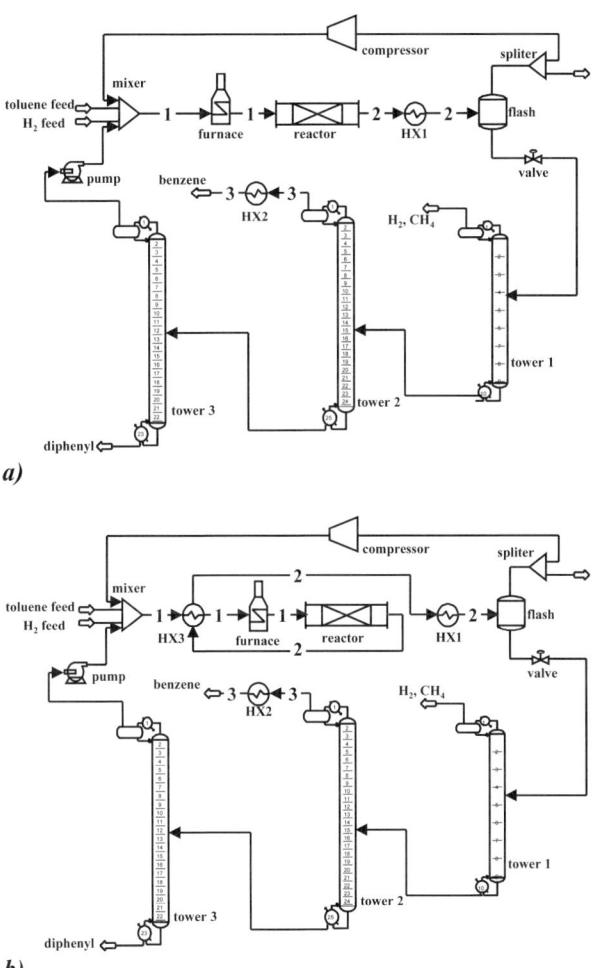

a)

b)

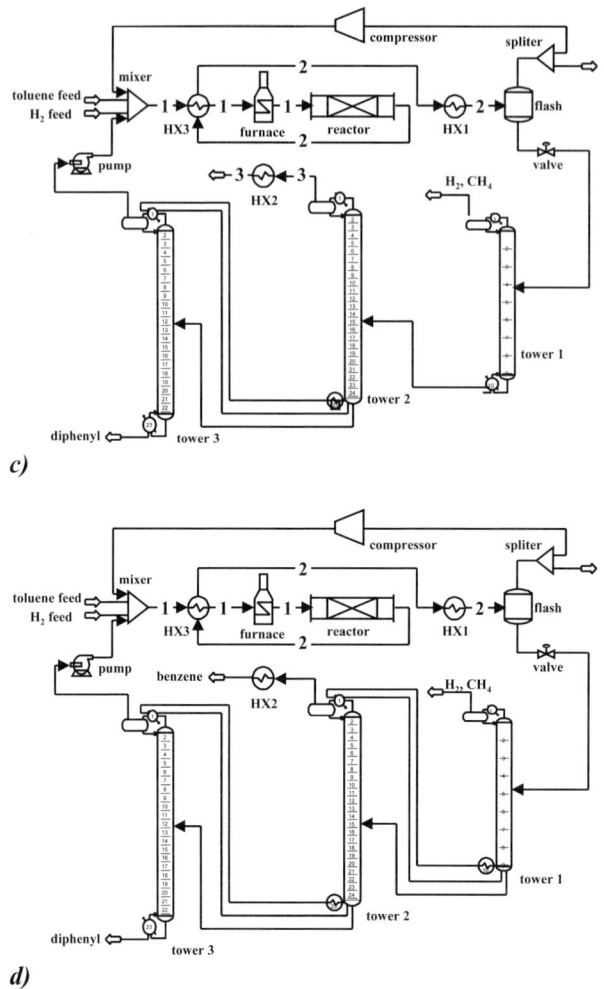

Fig. 1. HDA process alternative designs *a*, *b*, *c* and *d* for increasing levels of energy integration

3 Results and discussion

In process design, heating and cooling of liquids and vapours should be treated with regard to the source or sink of thermal energy transferred to or from the stream, the rate at which the energy is transferred and the type and size of the heat exchanger needed. Heat is transferred to or from process streams using other process streams or heat transfer media. Heat transfer media are classified as coolants (heat sinks) when heat is transferred to them from process streams and as heat sources when heat is transferred from them to process streams. In a final process design, efforts are made to exchange heat between process streams to decrease the use of utilities. Inevitably, however, some use of cooling water, steam and the products of combustion is necessary. The associated heat exchangers are called utility exchangers.

Table 1 shows the enthalpy change (Q) in the cold and hot process streams, the supply (T_s) and target temperatures (T_t) for the process streams and utilities used. Table 2 shows the enthalpy change, column reboiler and condenser's temperature and supply and target temperatures for the utilities used. Negative values mean enthalpy surplus and positive values mean enthalpy deficit. All the values of Tables 1 and 2 were obtained from the simulations of the process design alternatives a, b, c and d using the process simulator PRO/II®.

Table 1. Enthalpy change for the cold and hot process streams. Supply and target temperatures for the specified heat-transfer equipment and utilities for the alternatives a, b, c and d

Stream no. and type	Equipment description	T_s (°C)	T_t (°C)	Q (kJ/hr)
Alternative a				
1 (Cold)	Furnace	48	622	97006000
2 (Hot)	HX1	665	38	-104280000
3 (Hot)	HX2	101	25	-6284000
Alternatives b, c, d				
1 (Cold)	HX3	48	622	97042000*
2 (Hot)	HX3	665	81	-97042000*
2 (Hot)	HX1	81	38	-7274000
3 (Hot)	HX2	101	25	-6284000
Utilities				
Fuel (Hot)	Furnace	630	630	
Water (Cold)	HX1 and HX2	15	40	

*heat exchanged between stream 1 and 2

The Q values of Tables 1 and 2 are calculated using the process simulator. They depend on the stream characteristics, i.e. flowrate, supply and target temperatures, and heat capacities. When phase change occurs or a stream is a mixture of liquid and vapour the calculation of Q must consider the stream composition and the different heat capacities of the liquid and vapour fractions (as for example streams 1 and 2). Also, the simulation results of different process topologies can present small differences in the stream characteristics and consequently in the Q values. This is the case when HX3 is added to the process flowsheets. The flowrate of streams 1 and 2 vary from 47624kg/hr in alternative a to 47677kg/hr in alternatives b, c and d, together with small variations in the temperatures, liquid and vapour fractions and the heat capacities of the streams. Stream 1 is the sum of two feed streams (which have fixed flowrates) and two recycle streams that vary depending on the reactor conversion and on the separation efficiency of the flash and distillation column, which also depend on the stream characteristics. Therefore, slightly different Q values were calculated by the process simulator PRO/II® as shown in Table 1 (see Q values for furnace and HX3).

Table 2. Enthalpy change, column condenser and reboiler temperatures, and supply and target temperatures of utilities for the alternatives a, b, c and d

Stream type	Equipment Description	T_s (°C)	T_t (°C)	Q (kJ/hr)		
				Alternatives a, b	Alternative c	Alternative d
Hot	Condenser 1		33	-8300	-8300	-8300
Cold	Reboiler 1		85	1484600	1484600	0
Hot	Condenser 2		101	-5673700	-5673700	-4189100
Cold	Reboiler 2		133	11193700	9523000	9523000
Hot	Condenser 3		160	-1670700	0	0
Cold	Reboiler 3		229	2012700	2012700	2012700
Utilities						
Water (Cold)	Condenser 1	15	20			
Water (Cold)	Condensers 2 and 3	15	40			
Steam (Hot)	Reboilers 1 and 2	180	180			
Steam (Hot)	Reboiler 3	240	240			

3.1 Environmental Analysis

The potential environmental impacts (PEI) of the processes have been evaluated using the waste reduction (WAR) algorithm (Young and Cabezas 1999). The WAR algorithm uses a database of potential environmental impacts for more than 1600 chemicals to evaluate streams that cross the system boundaries. The mass flowrates of these streams are multiplied by potential environmental impact scores, which are given on a mass basis for the potential environmental impact categories. These environmental impact scores reflect the potential environmental harm of each chemical in eight potential environmental impact categories: human toxicity by ingestion and by dermal/inhalation routes, terrestrial toxicity, aquatic toxicity, photochemical oxidation, acidification, global warming and ozone depletion. The PEI evaluation of these categories is discussed in abbreviated form by Smith (2002) and in detail by Young and Cabezas (1999). Smith (2002) also presents the potential environmental impact scores used in this study for the HDA process' components (hydrogen, methane, benzene, toluene and biphenyl). For the evaluation of the potential environmental impacts associated with energy generation, process values from Young (2002) were used in this study.

Weighting factors are used to combine PEI categories in order to obtain a total PEI index. The weighting factors represent the relative or site-specific concerns of the user, according to their social, economic or ecological values. Since this paper discusses an illustrative case study with no specific site in mind, the weighting factors for all the categories have been assigned equivalent values of unity. Any setting of environmental priorities reflects value judgements and preferences, which introduces subjectivity in a study. To handle the subjectivity introduced one could do a sensitivity analysis on the weighting factors for the environmental impact categories.

In this study, to evaluate the potential environmental impacts of the HDA process, the work of Young and Cabezas (1999) on the WAR algorithm is extended to include both open and fugitive emissions. Open emissions are those associated with certain exit streams from the process known as waste streams. Fugitive emissions are those from equipment leaks, which occur in the form of gases or liquids that escape to the atmosphere through many types of connection points (e.g., flanges, fittings, seals, connectors, etc.) or through moving parts (e.g., valves, pumps, compressors, pressure relief devices and certain types of process equipment). In this study it is assumed that 0.1% of each stream in the flowsheet is lost as a fugitive emission. This assumption can be refined at a later time through emission factors (e.g., U. S. EPA 1995). This study doesn't consider the fugitive emissions of the heat transfer fluids (nor fuel, cooling water and steam) for the evaluation of the potential environmental impacts associated with the process. However, the consumption of fuel, electricity and saturated steam is accounted for in the evaluation of the potential environmental impacts of the energy generation. Figure 2 distinguishes the potential environmental impact associated with the process emissions (open and fugitive emissions) and the potential environmental impact associated with the energy generation process (Young 2002), for the four HDA process design alternatives a, b, c and d.

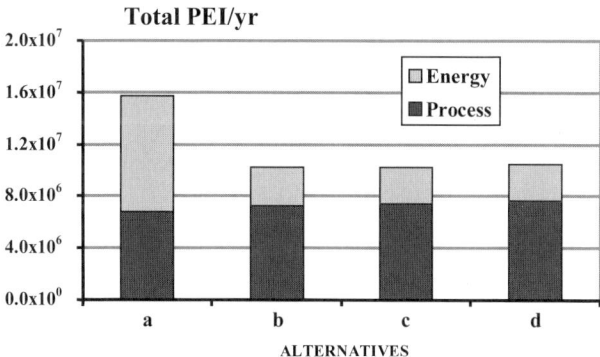

Fig. 2. Total PEI per year of the HDA process design alternatives a, b, c and d. Shown are the contributions to the total PEI attributable to process (open and fugitive) emissions and to energy generation

Figure 2 shows that the contributions to the total PEI attributable to the process emissions increase with increasing levels of energy integration, from alternative a to b, c and d. This is explained as fugitive emissions increasing when more pieces of equipment and streams are added to the process. As shown in Figure 1 the more energy-integrated the process is the more complicated it is, i.e., there are many more interconnections and streams. Note that from alternative *a* to *b*, *c* and *d*, that one more heat exchanger, HX3, and process streams were added without replacing others. The furnace and condensers of towers 2 and 3 and the reboilers of towers 1 and 2 were maintained in the designs. Thus, fugitive emissions increase. However, the contribution to the potential environmental impacts due to energy generation decreases as the level of energy integration increases because energy consumption decreases. Considering the combined potential environmental impact of the process and of the energy generation, the process design alternatives *b, c* and *d* are superior to alternative *a*. (Note that this is the result of our uniform weighting. If a different set of weighting factors had been chosen, these results could have possibly been different.) This evaluation of the potential environmental impacts indicates that the addition of the feed-effluent heat exchanger (HX3) makes a large environmental difference, whereas integrating the distillation columns (alternatives *c* and *d*) is not environmentally beneficial due to relatively small energy savings and increased fugitive emissions.

3.2 Economic Evaluation

Before an industrial plant can be put into operation it is necessary to purchase and install the necessary machinery and equipment. Also, land and service facilities must be obtained and the plant must be erected complete with all piping, controls and services. In addition, it is necessary to have capital available for the payment of expenses involved in the plant operation. The capital needed to supply the necessary manufacturing and plant facilities is called *the fixed-capital investment*, while the capital necessary for the operation of the plant is termed the *working capital*. The sum of the fixed-capital investment and the working capital is known as the *total capital investment*. Generally, the working capital represents 10 to 20% of the total capital investment (Peters and Timmerhaus 1991). An estimate of the capital investment for a process may vary from a pre-design to a detailed estimate by increasing the accuracy depending upon the stage of development of the project. Various methods can be employed for estimating fixed-capital investments depending on the information available and desired accuracy. Peters and Timmerhaus (1991) have outlined seven such methods. The method used in this study calculates the fixed-capital investment by percentage of the total purchased equipment cost (E_c), including the *direct costs* (D_c) and the *indirect costs* (I_c). Direct costs are calculated by summing the total purchased equipment cost (E_c), with the equipment installation (39% E_c), instrumentation (28% E_c), piping (31% E_c), electrical (10% E_c), buildings (22% E_c), yard improvements (10% E_c), service facilities (55% E_c) and land (6% E_c). The indirect costs are calculated by summing the engineering and supervision (32% E_c) with construction expenses (34% E_c). Finally, to calculate the fixed-capital investment D_c and I_c are summed together with the contractor's fee (5%(D_c+I_c)) and contingency (10%(D_c+I_c)).

In this work the purchase costs for the process equipment such as the reactor, condenser, flash vessel, columns, pump, valves, pipes, etc., have been calculated as indicated by Douglas (1988), put on an annualised basis (using a capital charge factor), and actualised using the Marshal & Swift (M&S) index equipment cost index published by *Chemical Engineering* (2000). The M&S index takes into consideration the cost of machinery and major equipment plus costs for installation, fixtures, tools, office furniture and other minor equipment. The average purchase costs for the heat exchanger equipment have been evaluated using the graphs of capacity versus cost presented by Peters and Timmerhaus (1991).

Table 3 shows the surface area of the heat-transfer equipment (calculated as $A = Q/U\Delta T_m$) and the log mean temperature difference (ΔT_m) across the heat-transfer equipment. Since the heat transfer surface area calculated for some heat exchangers is very small, a higher surface area was assumed for the calculation of purchase costs. The values of the overall heat transfer coefficient, U, were obtained from the literature (Holman 1987; Krajnc and Glavic 1995).

Determination of the necessary capital investment is only part of a complete economic estimate. Also important is the estimation of the costs for operating a plant and of the revenues from selling the products.

In the chemical industry one of the major costs in a production operation is for the purchase of raw materials involved in the process. Market prices for toluene and benzene were obtained from *Chemical Market Reporter* (2002) and the prices of hydrogen, methane and biphenyl were scaled up from Douglas (1988). The biphenyl is used as fuel as are the chemicals leaving the process in the purge, which are mainly hydrogen and methane but also some aromatics. The fuel value of these chemicals was calculated from Douglas (1988). Therefore the prices used in this study for the raw materials are 0.0272€/mol for toluene and 0.0001€/mol for hydrogen. The price of the benzene product is 0.0322€/mol and the fuel values of biphenyl and methane in monetary terms are 0.0337€/mol and 0.0048€/mol, respectively.

For this study the costs of utilities were estimated based on information available in Douglas (1988) and Peters and Timmerhaus (1991): 0.0105€/kg for saturated steam at 10atm and 180°C, 0.0127€/kg for saturated steam at 34atm and 240°C, 0.0559€/kwh for electricity, 0.2955€/m^3 for cooling water and 0.0053€/MJ for fuel (oil or gas). The utilities are included in the *operating costs* estimation.

Table 4 shows the results of the economic analysis of the HDA process for the alternative designs *a*, *b*, *c* and *d*. Bold values represent sums of non-bold values directly above. The economic potential is determined by subtracting the annualised total capital investment from the net operating profit.

Table 4 shows that the total capital investment increases when the process is more heat integrated, from alternative *a* to *d*, due to the additional piping system and heat exchangers. Table 4 also shows that the operating costs decrease because the consumption of utilities is reduced when the process is heat integrated.

Table 3. Surface area needed for the heat-transfer equipment, log mean temperature difference, overall heat transfer coefficient, and initial and final temperatures of the hot and cold streams

Stream no. and type		T_1 / °C	T_2 / °C	ΔT_m / °C	U / (kJ/(hr.m². °C))	Area / m²
Alternative a						
1 (Cold)	Furnace	48	622	134	360	2006
2 (Hot)	HX1	665	38	182	4320	133
3 (Hot)	HX2	101	25	28	3960	56
Alternatives b, c, d						
1 (Cold)	HX3	48	622	-	-	-
2 (Hot)	HX3	665	81	38	1260	2032
2 (Hot)	HX1	81	38	31	4320	54
3 (Hot)	HX2	101	25	28	3960	56
Hot	Condenser 1	33	33	16	4680	0
Cold	Reboiler 1	85	85	95	7560	2
Hot	Condenser 2	101	101	73	4680	17
Cold	Reboiler 2	133	133	48	7560	31
Hot	Condenser 3	160	160	132	5760	2
Cold	Reboiler 3	229	229	11	7560	25
Utilities						
Fuel (Hot)	Furnace	630	630		360	
Water (Cold)	HX1 and HX2	15	40		3960	-
Water (Cold)	Condenser 1	15	20		3960	-
Water (Cold)	Condensers 2 and 3	15	40		3960	
Steam (Hot)	Reboilers 1 and 2	180	180		-	
Steam (Hot)	Reboiler 3	240	240		-	

Table 4. Economic analysis of the HDA process for alternative designs *a*, *b*, *c* and *d*

Values in million €/year	Alternatives			
	a	b	c	d
Revenue from benzene product	42.56	42.56	42.56	42.56
Revenue from other streams	8.63	8.63	8.63	8.63
Total operating revenue	51.19	51.19	51.19	51.19
Cost of feed stocks	37.94	37.94	37.94	37.94
Operating costs	8.21	3.63	3.52	3.42
Total operating expense	46.15	41.57	41.46	41.36
Net operating profit	5.04	9.62	9.73	9.83
Total capital investment	6.61	6.87	7.15	7.42
Economic potential	-1.57	2.75	2.59	2.41

From Table 4 one can see that design alternative *b* has the largest economic potential and alternative *a* has the lowest economic potential. There is a considerable improvement from alternative *a* to *b*, where the economic potential increases 4.32 million €/yr. This is explained by the operating costs decreasing 4.58 million €/yr while the capital costs increase 0.26 million €/yr. Alternative *a* is the one with the largest operating costs because it is not heat integrated and it has a high consumption of utilities. In design alternative *b* the hot stream leaving the reactor is used to heat the reactor feed stream. This operation considerably reduces the costs of utilities (fuel and cooling water).

From alternatives *b* to *c* and *d* the economic potential decreases slightly as the process becomes more energy integrated. From alternatives *a* to *d* the operating costs decrease from 8.21 to 3.42 million €/yr, since the last alternative is more heat integrated. However the capital costs increase from 6.61 to 7.42 million €/yr. Therefore, from an economic perspective, heat integration using a feed-effluent heat exchanger (HX3) is advantageous, however, integrating the distillation columns does not decrease operating costs enough for the investment to be worthwhile. From alternative a to b the total capital investment increases 3.81% (or 0.26 million €/yr), which corresponds to the installation of HX3 and piping. The purchased cost of HX3 represents 2.74% of the total capital investment in alternative b. From alternatives b to c and c to d the total capital investment increases 3.85% and 3.70% respectively (or 0.27 million €/yr), which corresponds to a column/reboiler integration in both cases.

In considering both the economic and environmental effects, the process modifications in changing from alternative *a* to *b* are much more significant than changes between alternatives *b*, *c* and *d*. This result, that adding HX3 in changing from alternative *a* to *b* is much more significant than changes in *b* to *c* to *d*, is not surprising because the heat loads integrated in alternatives *c* and *d* are considerably smaller (see the changes in Q in Tables 1 and 2). Similarly, the effect of integration on process (fugitive) emissions is relatively small between these alternatives (see Figure 2). Therefore, considering the economic evaluation in combination with the environmental evaluation, alternative *b* is superior to the others. However, it is important to note that this result comparing the various alternatives may be affected if a different set of weighting factors had been chosen or if costs of utilities and/or equipment change. For example, if the market costs of utilities increase alternative *d* may be superior to the others because the operating costs will have a higher importance in the economic potential than the capital costs. On the other hand if the market capital costs increase it will emphasize the difference between the economic potential of the various design alternatives; however, the alternative *b* will continue to be superior to the others.

4 Conclusions

In chemical process design it is important to perform environmental impact analysis in concert with the traditional economic analysis. This point is demonstrated in this study using the HDA process. Four HDA process design alternatives with increasing levels of energy integration were evaluated from an environmental and economic viewpoint. The design option which exchanges a large amount of heat between the reactor effluent and feed is superior to other designs in both analyses because of large energy savings that reduce operating costs and potential environmental impacts.

There were a number of tradeoffs found in this study. The process fugitive emissions increase as the level of energy integration increases, and subsequently the potential environmental impacts increase. However, the potential environmental impacts due to energy generation decrease as the level of energy integration increases. As for the economics, the operating costs decrease as the level of energy integration increases. However, the capital costs increase with an increase in energy integration, so that with these trade-offs an intermediate amount of energy integration produces the most economically beneficial process for the example designs considered.

References

Cabezas, H, Bare JC, Mallick SK (1999) Pollution Prevention with chemical process simulators: the generalized waste reduction (WAR) algorithm – full version. Comput Chem Eng 23:623-634
Chemical Engineering, mid-August (2000):410
Chemical Market Reporter, (2002) October 7, 262(12)
Douglas JM (1988) Conceptual Design of Chemical Processes. McGraw-Hill, New York, pp.267-280
Hilaly AK, Sikdar SK (1994) Pollution balance: a new method for minimizing waste production in manufacturing processes. J Air Waste Manag Assoc 44:1303-1308
Holman JP (1987) Heat Transfer. McGraw-Hill, New York, p.527
Krajnc M, Glavic P (1995) The Influence of Different Temperature Contributions on Heat Integrated Process Structure. Trans IchemE Part A 73:880-888
Linnhoff B, Flower JR (1978) Synthesis of Heat Exchanger Networks. AIChE J 24(4):633-54
Linnhoff B, Hindmarsh E (1983) The Pinch Design Method for Heat Exchanger Networks. Chem Eng Sci 38(5):745-763
Luyben WL (2000) Effect of Kinetic, Design, and Operating Parameters on Reactor Gain. Ind Eng Chem Res 39:2384-2391
Peters MS, Timmerhaus KD (1991) Plant Design and Economics for Chemical Engineers, 4th ed. McGraw-Hill International Editions, Chemical and Petroleum Engineering Series
Seider DW, Seader JD, Lewin DR (1998) Process Design Principles. Synthesis, Analysis and Evaluation. John Wiley & Sons, New York

Smith JM, Van Ness HC (1987) Introduction to Chemical Engineering Thermodynamics, 4th ed. McGraw-Hill International Editions, pp 574-597

Smith RL, Mata TM, Young DM, Cabezas H, Costa CAV (2001) Designing Efficient, Economic and Environmentally Friendly Chemical Processes. In Computer-Aided Chemical Engineering 9, edited by Jorgensen S and Gani R. Elsevier Science B. V., Proceedings of the 11th European Symposium on Computer Aided Process Engineering, Kolding, Denmark, 27-30 May, 1165-1170

Smith RL (2002) Evaluating the Economics and Environmental Friendliness of New and Retrofitted Chemical Processes. Clean Techn Environ Policy, 3:383-391

Terril DL, Douglas JM (1987) Heat- Exchanger Network Analysis. 1. Optimization. Ind Eng Chem Res 26:685-691

Terril DL, Douglas JM (1987) Heat- Exchanger Network Analysis. 2. Steady-State Operability Evaluation. Ind Eng Chem Res 26:691-696

U. S. EPA (1995) Protocol for Equipment Leak Emission Estimates. EPA-453/R-95-017. U.S. EPA, Office of Air and Radiation, Office of Air Quality Planning and Standards, Res. Triang. Park

Young DM, Cabezas H (1999) Designing Sustainable Processes with Simulation: The Waste Reduction (WAR) Algorithm. Comput Chem Eng 23:1477-1491

Young DM, Scharp R, Cabezas H (2000) The waste reduction (WAR) algorithm: environmental impacts, energy consumption and engineering economics. Waste Manage 20:605-615

Young DM (2002) Personal communication

Eco-efficiency reporting exemplified by case studies

Annik Magerholm Fet

The Norwegian University of Science and Technology (NTNU), Department of Industrial Economics and Technology Management

Abstract

This paper presents current and future trends and requirements for environmental, eco-efficiency and sustainability reporting. Further it defines the concept of eco-efficiency, and describes ways of developing eco-efficiency indicators for production sites and for product chains. Eco-efficiency measures give indications both on economic and on environmental performance. These indicators are then exemplified by results from case studies within Norwegian and European industrial companies. Some of these projects have laid the foundation for environmental accounting and reporting systems in local communities. Eco-efficiency as a tool for measuring the performance along product value chains is demonstrated in the paper. Product oriented eco-efficiency indicators are seen in the context of the international efforts on standardisation of environmental product declarations (EPDs) which are ways to report the environmental performance of products. This is exemplified with cases from furniture production value chains. The presentation focuses further on the concept of corporate social responsibility and on the challenges of how to incorporate this in future sustainability reporting.

1 Introduction

There are different types of reporting systems on environmental performance and eco-efficiency. Eco-efficiency covers economic performance in addition to the environmental performance, while sustainability reporting encompasses social, economic and environmental aspects: the "triple bottom line". Today, the trend has shifted from traditional environmental reporting to eco-efficiency reporting and sustainability reporting.

Indicators are frequently used to report environmental performance. Fig. 1 shows the three pillars in sustainable development as the corners of a triangle, and indicates reporting at different levels.

Organisations such as the United Nations' Environment Program (UNEP), the World Business Council for Sustainable Development (WBCSD 2000) and the Organisation for Economic Co-operation and Development (OECD) have a strong influence on the requirements for such reporting (OECD 2001). One of the initiatives by UNEP is the Global Reporting Initiative (GRI). GRI was established in 1997 with the mission of developing globally applicable guidelines for reporting on economic, environmental, and social performance. The GRI's Sustainable Reporting Guidelines (UNEP 2002) represents the first global framework for comprehensive sustainability reporting. The latest version came in September 2002, which gives guidance to reporters on selecting generally applicable and organisation specific indicators, as well as integrated sustainability indicators. In order to compare different systems, it is necessary to follow standardised ways of reporting by means of a set of understood indicators.

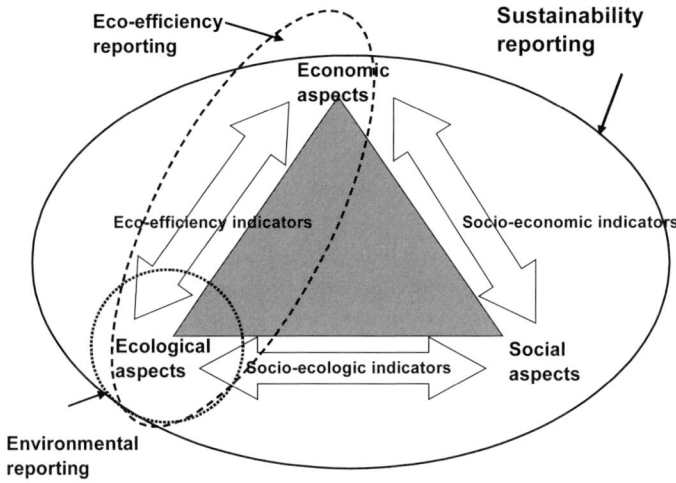

Fig. 1. Sustainability reporting

2 Performance indicators

The GRI Indicator Framework organises the performance indicators in accordance with the following hierarchy, see Fig. 3:
- Category: The broad areas, or groupings, of economic, environmental, or social issues of concern to stakeholders (e.g., human rights, direct economic impacts).

- Aspect: The general subsets of indicators that are related to a specific category. A given category may have several aspects, which may be defined in terms of issues, impacts, or affected stakeholder groups.
- Indicator: The specific measurements of an individual aspect that can be used to track and demonstrate performance. These are often, but not always, quantitative. A given aspect (e.g. water) may have several indicators (e.g., total water use, rate of water recycling, discharges to water bodies). A pillar of the GRI framework is that aspects and indicators derive from an extensive, multi-stakeholder consultative process.

GRI performance indicators are classified along the following lines:
- Core indicators (or general applicable indicators) are those relevant to most reporters; and of interest to most stakeholders.
- Additional indicators (or business specific indicators) are viewed as leading practice in economic, environmental, or social measurements, and in providing information of interest to stakeholders who are particularly important to the reporting entity.

In addition to the sustainability indicators on economic, social and environmental aspects, a fourth dimension of information is necessary: *integrated performance*. GRI has not identified a standardised set of integrated performance indicators, but integrated measures are categorised as:
- Systemic indicators that relate the activity of an organisation to the larger economic, environmental, and social systems of which it is a part. For example, an organisation could describe its own performance in relation to the overall system.
- Cross-cutting indicators that directly relate two or more dimensions of economic, environmental, and social performance as a ratio. Eco-efficiency measures are the best-known examples (see Fig. 1 and Fig. 2 for other cross-cutting indicators).

The economic performance indicators used in eco-efficiency primarily focus on the profitability of an organisation for the purpose of informing its management and shareholders. The focus of economic performance measurement in sustainability reporting is on how the economic status of the stakeholder changes as a consequence of the organisation's activities (direct impact) rather than on changes in the financial condition of the organisation itself (indirect impact). Indirect impacts include externalities that create impacts on communities, e.g. costs or benefits arising from a transaction that are not fully reflected in the monetary amount of the transaction. A community can be considered as anything from a neighbourhood, to a country, or even a community of interest such as a minority group within a society. See Table 1 for the aspects under each category suggested by GRI. Under each aspect GRI suggested a set of core indicators and additional indicators.

Table 1. Categories and aspects for economic, environmental and social performance indicators (UNEP 2002)

	Category	Aspect
Economy	Direct Economic Impact	Customer
		Suppliers
		Employees
		Providers of capital
		Public Sector
Environmental	Environmental	Materials
		Energy
		Water
		Biodiversity
		Emissions, effluents and waste
		Suppliers
		Products and services
		Compliance
		Transport
		Overall
Social	Labour Practices and Decent Work	Employment
		Labour / management relations
		Health and Safety
		Training and education
		Diversity and opportunity
	Human Rights	Strategy and management
		Non-discrimination
		Freedom of association and collective bargaining
		Child labour
		Forced and compulsory labour
		Disciplinary practices
		Indigenous rights
	Society	Community
		Bribery and corruption
		Political contributions
		Competition and pricing
	Product Responsibility	Customer health and safety
		Products and services
		Advertising
		Respect for privacy

3 Eco-efficiency

WBCSD defines eco-efficiency as "*the delivery of competitively priced goods and services that satisfy human needs and bring quality of life, while progressively reducing ecological impact and resource intensity throughout the life cycle, to a level at least in line with the earth's estimated carrying capacity*" (DeSimone and Popoff 1997).

To develop eco-efficiency measures, information on both the economic and the environmental performance is needed. The eco-efficiency can be calculated using the following formula:

eco-efficiency = product or service value per environmental influence (1)

For a production site the value can be yearly production volume, total sale or turn over. The environmental influence can be the environmental impact within one aspect, or an aggregated value, which requires weighting between the aspects. So far measures of eco-efficiency have mainly focused on specific production sites. To evaluate the eco-efficiency of a product, information concerning its entire life cycle is required to allow an evaluation of its environmental and economic performance.

The performance can be measured/calculated by economic performance indicators and environmental performance indicators (EPIs). Equation (1) can be transformed to:

$$Eco\text{-}efficiency\ indicator = \frac{economic\ performance\ indicator}{environmental\ performance\ indicator} \quad (2)$$

The WBCSD's framework of indicators is shown in Fig. 2.

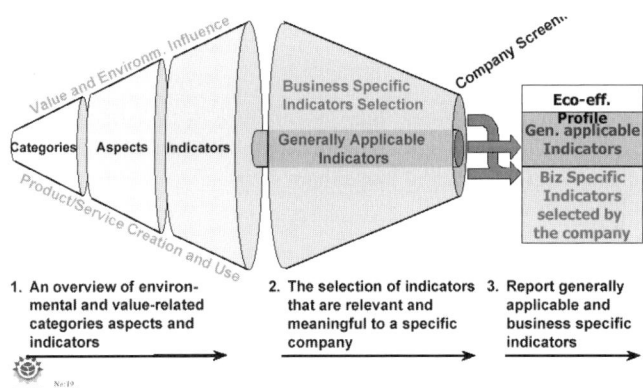

Fig. 2. WBCSD Framework of indicators (WBCSD 2000)

4 Methodology

The GRI framework focuses on indicators that are most relevant to the stakeholder. However, to decide upon the most significant environmental aspects, relevant data is needed. One way of collecting the data and selecting the most relevant indicators is to use the methodology described in the ISO 14031 code on "Environmental Performance Evaluation" (ISO 1998). The methodology is illustrated in Fig. 3.

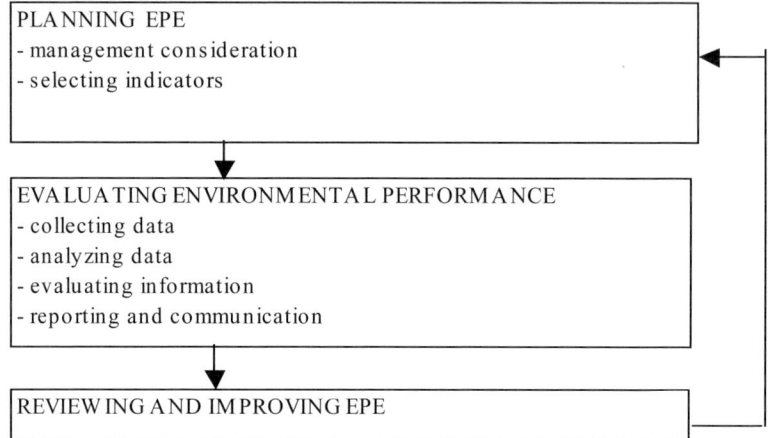

Fig. 3. ISO 14031 methodology

To analyse the data and evaluate the information, as in assessing the environmental impact, the methodology recommended by ISO 14040 can be used (ISO 1996). The methodology consists of these main steps:
1. Goal and scope definition
2. Inventory
3. Impact assessment
4. Interpretation

For impact assessment the following procedures are used:
3a: Classification; the step where the different substances are classified under the impact category they contribute to.
3b: Characterisation; the step in which the relative contribution of each substance within each category is calculated.
3c: Normalisation and evaluation; the step where the total contribution within each category is evaluated against the mean values in e.g. a society. Very often different weightings are used to compare the impact categories against each other.

The choice of environmental impact categories can be e.g. global warming, acidification, eutrophication, biodiversity, etc., or the aspects recommended by GRI. A similar methodology should be established for the selection of economic performance indicators. However, that is not a part of this presentation. An appro-

priate methodology for deciding the most relevant eco-efficiency indicators used for reporting can be described by:
1. Define the system, subsystems and system boundaries, and the functional unit if the system is a product.
2. Establish economic and environmental performance indicators.Define eco-efficiency indicators for the system (by means of established methods, see above).Collect and evaluate data for quantification of value creation and environmental impact.Test the eco-efficiency indicators among the most important stakeholders.Use the eco-efficiency indicators in reporting.

5 Case studies

The use of indicators is demonstrated by a few case studies. Case 1 demonstrates the use of site-specific EPIs and eco-efficiency indicators, while case 2 demonstrates a few systemic indicators used by companies and their related municipalities. Eco-efficiency indicators are not yet included. Case 3 demonstrates how eco-efficiency indicators can be used for the purpose of comparing alternative fuel types for recreational boats. The last case, case 4, shows an attempt to use eco-efficiency of products.

5.1 Case 1: Site-specific indicators

These results are from the company A/S Olivin in Norway (Olivin 2001). It is the world's largest supplier of olivine based products. During the year 2001, 2.08 million tons of bulk sand, 50 600 tons of packaged sand and 22 530 tons of refractory products were produced. The total sales in 2001 were 365 million NOK, and the company had 194 employees. Manufactured products are mainly transported by ship from the company's own harbour. The company received the award for the best environmental report among Norwegian production industry and the international prize for the best report among small and medium sized companies last year. In their environmental policy they focus on the most important environmental aspects (dust, noise, waste and the efficiency of material and energy utilisation). They use EPIs in their reporting, e.g. emissions of climatic and acidifying gases. The amount of emissions from internal transport and combustion of oil is estimated from motor characteristics and oven specifications. The company uses this for eco-efficiency measures. The created value is expressed by yearly sale, see Fig. 4. This shows sales in proportions to emissions of climatic gases (here, CO_2) and acidic components (NO_x and SO_2). The eco-efficiency indicator shows a positive development over the past years. This does not include last year.

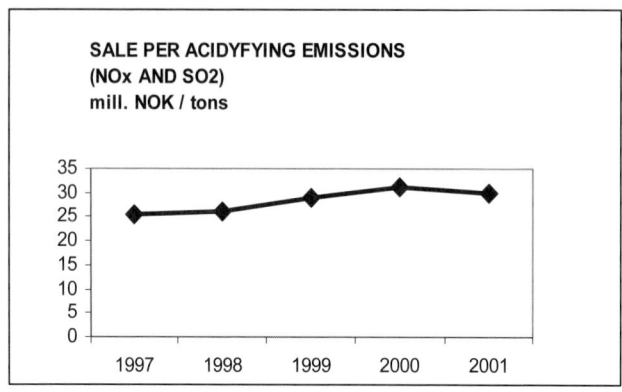

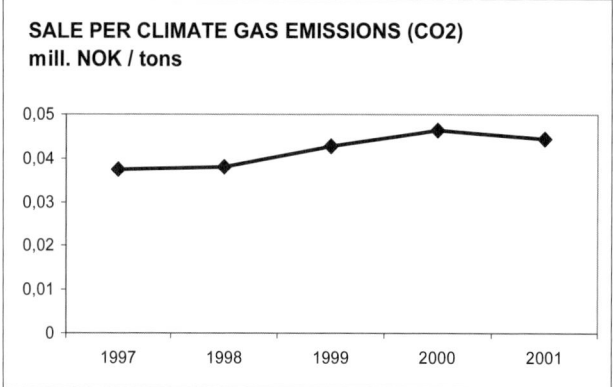

Fig. 4. Eco-efficiency indicators expressed as sale per climate gas emissions and sale per acidifying emissions

Other important EPIs are related to waste and waste treatment. By measuring the proportion of waste against the production volume, the efficiency of resource consumption can be evaluated. The waste indicators also give information regarding hazardous waste, waste for recycling and disposable waste. In addition, an indicator on costs per delivered ton of waste together with indicators on environmental investments are measured. These are further developed to eco-efficiency indicators, see Fig. 5.

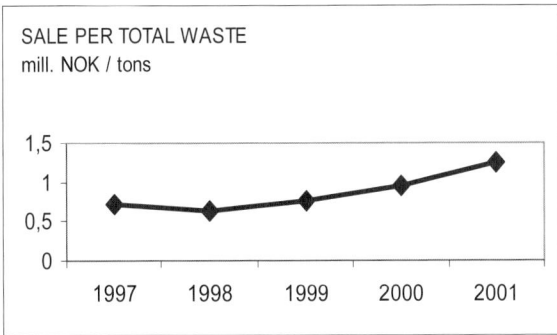
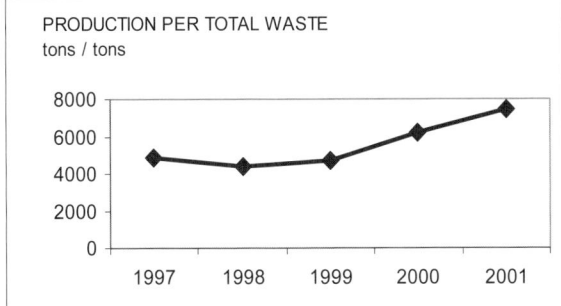

Fig. 5. Eco-efficiency indicators for waste

5.2 Case 2: Reporting in local communities by means of systemic indicators

As part of the Local Agenda 21 project, the following goals were set:
- Establish comparable environmental accounting systems for a group of industrial companies in a community and for the community itself.
- Select a set of appropriate EPIs to meet the standards requested by interested parties.
- Establish reporting systems and the use of indicators in a local society.

The companies in this project wanted to use EPIs as shown in Table 2, and the waste indicators were of special interest to the community because the community wanted to secure a safe environmental treatment of the waste. A system for the reporting of waste streams and amounts of these from the companies to the community was established. Indicators such as "Amount of each waste fraction delivered to the waste treatment plant" and "Amount of waste delivered from the treatment plant for reusage, recycling, incineration and landfill" were further established.

The community also developed a reporting system concerning their own activities, including the administration of the community, seven schools and a nursery home. Examples of the use of EPIs for energy use and waste treatment are shown in Fig. 6 and Fig. 7. The most obvious systemic indicators are for energy use and waste. They are of interest both for each participant and for the community, and they are of great value for further planning and improvements.

Table 2. EPIs for small companies

Group	EPI
Purchase	Proportion of products with environmental declaration (%)
	Number of suppliers with an environmental management system.
Energy usage	Electrical energy use per year (kWh)
	Energy use based on fossil fuel (kWh)
	Total energy use per area (kWh/m2)
	Total energy use per turn-over (kWh / NOK)
Waste:	Yearly amount in total (ton)
	Yearly amount per production volume (ton / ton)
	Yearly amount per turnover (ton / NOK)
	Yearly amount to recycling per total waste (%)
	Yearly amount to disposal (ton)

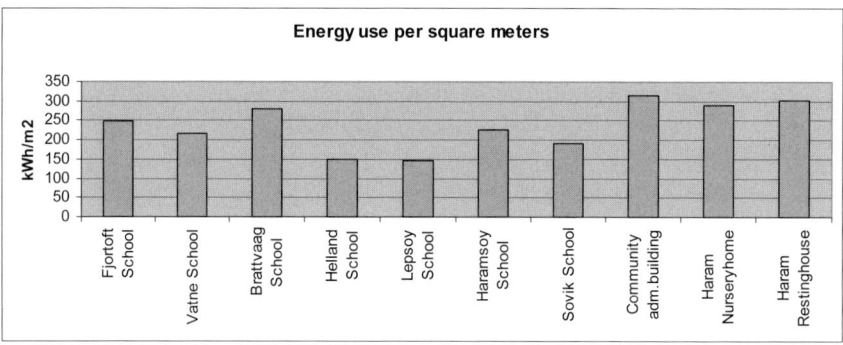

Fig. 6. Yearly energy use per m^2 in the building owned by the community

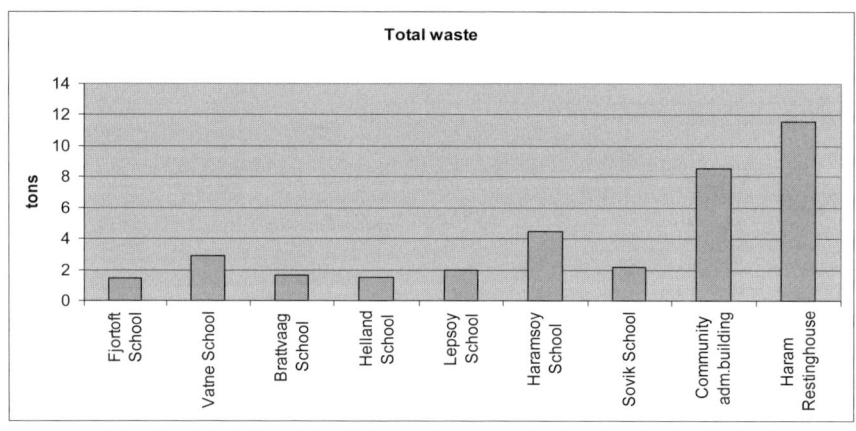

Fig. 7. Total waste delivered from schools and other activities in the community

5.3 Case 3: Eco-efficiency indicators for the purpose of comparing alternative fuel types

This study was conducted focusing on biodiesel (rapeseed oil) production and on the recreational boat market in the UK (Fet 2000). Biodiesel has been considered as an environmentally-friendly fuel which offers advantages of cleaner emissions, less pollution, less toxicity, biodegradability, less odour, ease of handling, high lubricity, smoother operation and complete combustion. These facts make biodiesel an attractive fuel for recreational boats where a clean environment is desired. It was concluded in the study that the biodiesel production rate in the UK would be about 430,000 tons per year if all UK set aside land were used for growing rapeseed corps. For fuelling all sailing yachts in UK, only 4.5 % of this is needed. Table 3 presents the emissions from the sailing yachts in the UK, fuelled by either fossil diesel or biodiesel (BABFO 2002; Statistics UK 2002). Column 5 shows the characterisation values (Fet et al. 2000). It indicates that the contribution of NO_x emissions to acidification is only 70% of that of SO_2. Column 6 shows the normalisation factors based on total emissions per year in UK. The yearly emissions are divided among these factors. This makes the figures comparable (see last two columns in Table 3). Figure 8 shows the characterised and normalised EPIs from sailing yachts in the U.K.

Table 3. Life cycle emissions of fuel from sailing yachts operation in UK (Zhou et al 2002)

Impact category	Substances	Fuelled by Fos-D* (kg/year)	Fuelled by Bio-D (kg/year)	Characterisation	Normalisation factors	Normalised values fossil diesel	Normalised values biodiesel
Climate change	CO_2	$1.4 \cdot 10^8$	$3.4 \cdot 10^7$	1	$4.16 \cdot 10^{11}$	$3.46 \cdot 10^{-4}$	$8.17 \cdot 10^{-5}$
Acidification	SO_X	$1.5 \cdot 10^5$	$2.9 \cdot 10^4$	1,0	$1.62 \cdot 10^9$		
	NO_X	$6.2 \cdot 10^5$	$8.2 \cdot 10^5$	0,7	$1.75 \cdot 10^9$	$3.40 \cdot 10^{-4}$	$3.45 \cdot 10^{-4}$
Local air pollution	Particulars / soot	$5.2 \cdot 10^5$	$3.1 \cdot 10^5$	1	$0.44 \cdot 10^9$		
						$1.28 \cdot 10^{-3}$	$8.26 \cdot 10^{-4}$
	CO	$4.8 \cdot 10^5$	$5.8 \cdot 10^5$	1	$4.76 \cdot 10^9$		
Photo oxidant formation	NMVOC	$2.9 \cdot 10^5$	$1.5 \cdot 10^5$	1	$1.96 \cdot 10^9$	$1.48 \cdot 10^{-4}$	$7.66 \cdot 10^{-5}$
Eutrophication	NO_X	$6.2 \cdot 10^5$	$8.2 \cdot 10^5$	1	$1.75 \cdot 10^9$	$3.54 \cdot 10^{-4}$	$4.68 \cdot 10^{-4}$

*Fos-D represents fossil diesel fuel; Bio-D represents biodiesel.

In order to calculate the eco-efficiency, it is necessary to evaluate the economic values created by sailing activity (see Eq. 1). However, it is difficult to obtain values/incomes on recreational activities, although one solution is to measure the value as the inverse of costs based on the fuel prices and yearly costs shown in Table 4 (BASFO 2002).

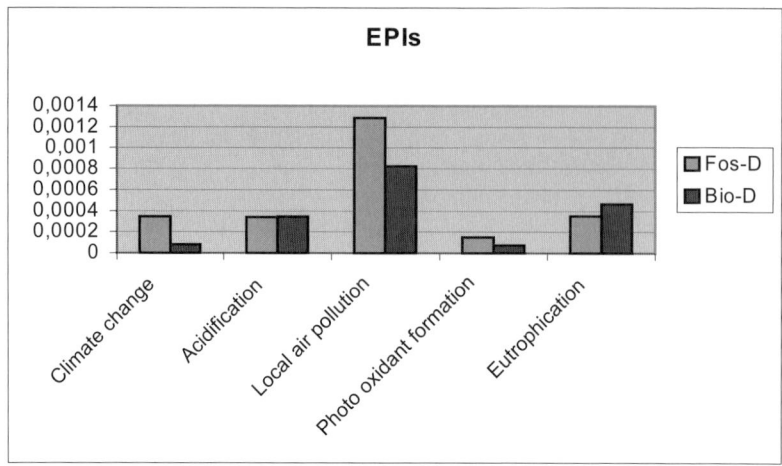

Fig. 8. Characterised and normalised inventory results presented by EPIs

Table 4. Fuel cost per year by sail yacht. Sensitivity study on fuel costs

	A*	B	C	D
Bio-D	40632	90.2	874	32032156
Red Diesel	40632	34.79	852	12043764
Fos-D	40632	77,9	852	26967783
Bio-D, no tax	40632	50,56	874	17955053

*A = fuel consumption (t/y), B = average fuel cost pence per litre, C= density of fuel (kg/m^3), D = total fuel cost per year (GBP/y)

The environmental influence can be expressed by the impact categories as shown in Table 3. The eco-efficiency of sailing yachts in the UK on a yearly basis could then be calculated by:

$$Eco\text{-}efficiency = (1/\ yearly\ costs)\ /\ environmental\ impact \qquad (3)$$

The values in Table 4 are used to calculate eco-efficiency. The results from the different scenarios are presented in Fig.9. They show that non-taxed biodiesel has the best eco-efficiency for climate change and photo oxidant formation. For acidification, normal priced biodiesel has the worst eco-efficiency. This is due to NO_x-emissions. However, for other impact categories, fossil diesel shows a good eco-efficiency. For red diesel the environmental performance of fossil diesel is used, which is partly incorrect. By comparing and analysing Fig. 8 and Fig. 9, the influence of costs on the eco-efficiency can be derived. However, the use of cost-factors is just for the purpose of demonstration. Other pricing mechanisms and taxation systems could be included here. This is a subject for further studies.

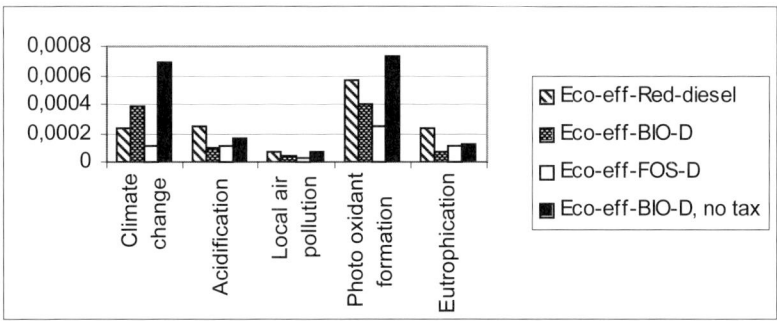

Fig. 9. Eco-efficiency for different alternatives, see Table 4 (Zhou et al. 2002)

5.4 Case 4: Eco-efficiency of products

To determine the eco-efficiency of products, the total value chain has to be evaluated. Appropriate EPIs can be found in LCA data. Different measures of value creation can be used even though monetary terms are found most useful. Eco-efficiency indicators can thus be used to track changes in eco-efficiency over time in different parts of the value chain (Michelsen and Fet 2002). An example is shown in Fig. 11. Here both an eco-efficiency measure (net sale in NOK per mega Joule energy consumed) and the total environmental pressure (total energy consumed) are shown.

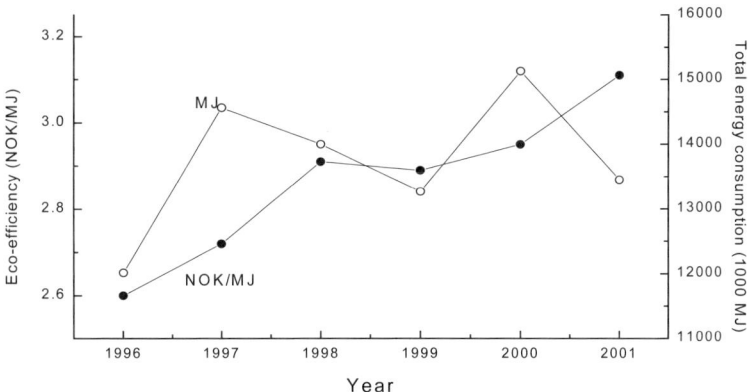

Fig. 10. Eco-efficiency for value chain calculated by energy use (Michelsen et al. 2003)

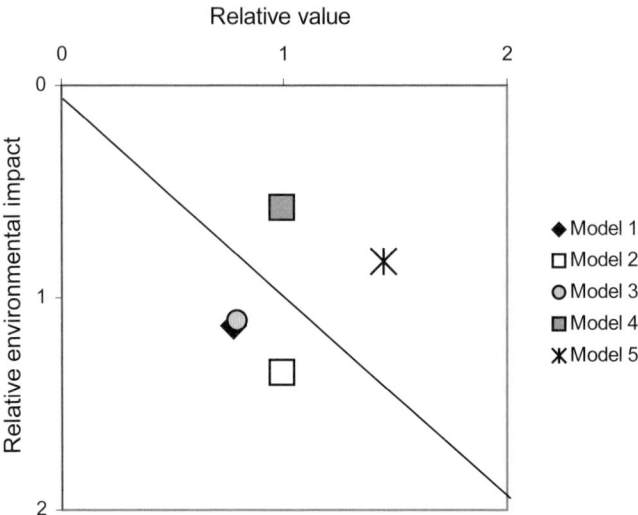

Fig. 11. Eco-efficiency for product value chains presented for different models of a product

Another set of eco-efficiency indicators is visualised graphically, see Fig. 11. They are found to be useful for comparing different models of a product. The values are given a relative value based on arithmetic mean values. The measure of environmental impact can either be a single EPI (i.e. energy consumption) or aggregated values where different impact categories are weighted and added. Fig. 11 shows the value as the sale value for a product and the environmental impact is the normalised value for aggregated impact categories. This is based upon the method used by BASF (Saling et al. 2002), and the values represent five different models of an office chair. It is possible to start with only a few important environmental aspects, i.e. energy and material consumption, and includes more aspects when more data are available. This makes it easier for small and medium sized enterprises to evaluate eco-efficiency for their products when resources to identify all environmental aspects, e.g. to do a complete LCA, are limited. It is also possible to use different measures of value creation. In this example the monetary terms are used, but function fulfilment can also be used (Michelsen et al. 2003).

The method for the establishment of eco-efficiency indicators presented earlier in this paper can be expanded to yield products by including the two next steps:
1. Develop a weighting model for the environmental aspects so the environmental performance can be aggregated to one single indicator (this step is controversial).
2. Implement the eco-efficiency indicators in the value chain to help decision-makers track the performance and changes in performance of products.

Hopefully in the near future eco-efficiency indicators will also be of great use for environmental product declarations (EPD).

6 Corporate Social Responsibility (CSR)

In line with the introduction in this paper, the term Corporate Social Responsibility (CSR) is often used to identify the business role in this context. However, CSR has no clear definition, and to discuss what it exactly embraces is a point in it itself. Growing public awareness, and demands for greater transparency, add a new requirement to communication and reporting. WBCSD has given the following description of CSR: "Corporate social responsibility is the commitment of business to contribute to sustainable economic development, working with employees, their families, the local community and society at large to improve quality of life". This means that the balance between a company's social responsibility and the corresponding responsibility for the efficient utilisation of key resources, including labour and capital, must be addressed. CSR requires open dialogue and constructive partnerships with the authorities at various levels, inter-governmental and non-governmental organisations and other elements of civil society, in particular, local communities; CSR is sometimes explained as "corporate citizenship". In addition to the economic and environmental performance indicators already mentioned, GRI also provides a list of social performance indicators. See Table 1 for the grouping and aspects of social performance indicators.

The Norwegian Research Foundation has established a research group on CSR. The mandate for this group is to plan the need for research within CSR seen from the industry's point of view as well as from the governmental and researchers/universities point of view. Given a clear statement of the need for further research, the planning group will come up with a program plan for future research programs. One area of concern will most likely be the development of appropriate social performance indicators (SPI) and cross-cutting indicators, see Fig. 1.

7 Conclusion

The function of GRI's performance indicators is to provide information about the economic, environmental, and social impacts of the reporting organisation in a manner that enhances comparability between reports and reporting organisations. In the case of GRI, the indicators are designed to inform both the reporting organisation and any stakeholders seeking to assess the organisation's performance. To achieve these goals, performance must not only be defined in terms of internal management targets and intentions, but must also reflect the broader external context within which the reporting organisation operates. In the end, it speaks to how an organisation contributes to sustainable development by virtue of its economic, environmental, and social interactions with its diverse stakeholders. The combination of better methods (both for selecting indicators and for analysing them), and rising stakeholder demands for richer disclosure is likely to continue this movement toward a new generation of performance reporting. Full integration in the form of single reports that communicate performance along the three dimensions is already practised by a handful of leading companies. National and local efforts should focus on the development of appropriate indicators and testing of reporting systems. This presentation has shown that EPIs and economic indicators can be used on a small scale. However, it does not include social aspects or cross cutting indicators in the areas of socio-economic and socio-environmental. This is an area for further studies.

References

BABFO (2002) British Association for Bio Fuels and Oils, http://www.biodiesel.co.uk, May

DeSimone LD, Popoff F (1997) Eco-efficiency: the business link to sustainable development. MIT Press, Cambridge, Mass.

Fet K (2000) The use of biodiesel in recreational boats, Final Year Project for BEng in Marine Technology, the University of Newcastle upon Tyne, UK, May

Fet AM, Michelsen O, Johnsen T, Sørgård E (2000), Environmental performance of transport –A comparative study. Norwegian University of Science and Technology (NTNU), Norway

ISO (1996) Life Cycle Analysis – general framework. ISO 14040, The International Standardisation Organisation

ISO (1998) Environmental Performance Evaluation. ISO 14031, The International Standardisation Organisation

Michelsen O, Fet AM (2002) Eco-efficiency along value chains – towards a methodology. Conference Paper, International Society of Industrial Ecology, Barcelona, Spain, December

Michelsen O, Fet AM and Dahlsrud A (2003) Eco-efficient value chains – status for research questions. Norwegian University of Science and Technology (NTNU), Norway, in preparation

OECD (2001) Policies to enhance sustainable development, OECD

Olivin A/S (2001) Environmental report

Statistics UK (2000) Digest of United Kingdom Energy Statistics. Department of Trade and Industry, London.

Saling P, Kicherer A, Dittrich-Krämer B, Wittlinger R, Zombik W, Schmidt I, Schrott W and Schmidt S (2002) Eco-efficiency analysis by BASF: the method. Int. J. LCA 7(4): 203-218

UNEP (2002) Sustainable reporting guidelines on economic, environmental, and social performance, August, UNEP's Global Reporting Initiative (GRI), see www.globalreporting.org

WBCSD (2000) Verfaille HA, Bidwell R (eds) Measuring eco-efficiency, a guide to reporting company performance, June

Zhou PL, Fet AM, Michelsen O, Fet K (2002) A feasibility study of using biodiesel in recreational boats in the UK, submitted to J Eng Mar Environ, May

Interpolation for creating hydrogeological models

A. Spalvins, J. Slangens, R. Janbickis, I. Lace

Environment Modelling Centre, Riga Technical University, 1 Meza Street, Riga.LV-1048, Latvia

Abstract

Credibility and sustainability of a hydrogeological model (HM) depends not only on worthy initial data, but also upon interpolation technologies applied to create HM. An interpolation system for HM has been developed.

1 Introduction

The algebraic equation system (Eq. 1) describing HM is specified on the *xyz*-grid built of ($h \times h \times m$)-sized blocks, h and m are a constant plane step and a variable height of blocks, respectively:

$$A \varphi = \beta - G \psi \qquad (1)$$

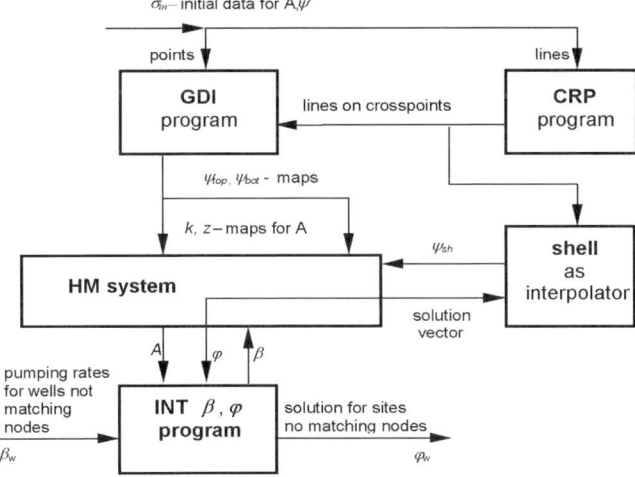

Fig. 1. Scheme of interpolation for creating of (Eq. 1)

In (Eq. 1), φ is the solution vector at nodes of the HM grid; it may be necessary to interpolate $\varphi \rightarrow \varphi_w$, at sites not matching the nodes.

Primary elements A, β, ψ of (Eq. 1) are obtained by interpolation (Fig. 1):

A – the symmetric matrix of the geological environment represented by a xy-layer system of aquifers and interjacent aquitards; to obtain A, permeability k and elevation z-maps for each layer should be created;

ψ - the boundary head vector; ψ_{top}, ψ_{bot}, and ψ_{sh}-maps should be prepared for HM top, bottom and shell (four vertical sides of HM) surfaces, correspondingly;

G – the diagonal matrix (part of A) assembled of elements linking the nodes where φ must be found with the ones where ψ is given;

β - vector of water pumping rates; to obtain β, interpolation $\beta_w \rightarrow \beta$ is needed.

For these interpolations, special tools have been developed:
- geological data interpolation (GDI) program (Spalvins and Slangens 1994) for creating the k, z, ψ_{top}, ψ_{bot}-maps (called σ-maps) and the program CRP (CRoss Point) for serving GDI and the shell (Slangens and Spalvins 2000);
- interpolation program INT β, φ for performing $\beta_w \rightarrow \beta$ and $\varphi \rightarrow \varphi_w$ (Lace et al. 1995);
- to provide ψ_{sh}, the HM shell is acting as an interpolation device (Spalvins and Slangens 1994).

2 CRP program

The CRP program (Slangens and Spalvins 2000) turns a raw data line l_{in} into standard l_σ and forms $l_\sigma \rightarrow l_c$ for the GDI program and the shell.

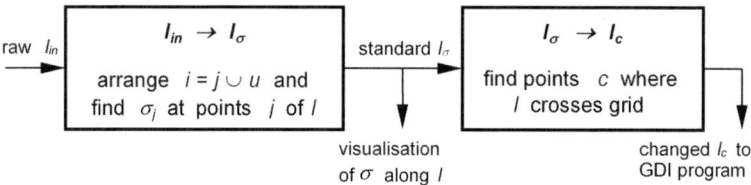

Fig. 2. Scheme of CPR program

These lines are based on vectorized line l passing master points j where its direction changes. The points $j, j+1$ are linked by the directed straight segment $l_{j,j+1}$, and l is the series:

$$l = \{\theta_i\} = \{x_i, y_i\}, \quad j = 1, 2, \ldots, J$$

$$d_{i,j+1} = \sqrt{(x_j - x_{j+1})^2 + (y_j - y_{j+1})^2} \qquad (2)$$

$$d_{1,j} = \sum_{j=1}^{j-1} d_{j,j+1}$$

where θ_j is coordinates of the j-th point; $d_{j,j+1}$ and $d_{l,J}$ are lengths of $l_{j,j+1}$ and l.
Raw l_{in} = { l, σ_{in} }, σ_{in} = { θ_u, σ_u }, u = 1, 2, ..., U; u and j may not coincide.

Standard l_σ = { θ_i, σ_i }, $i = j \cup u$ = 1, 2, ..., and interpolation on $\sigma_{in} \rightarrow \sigma_j$.

Changed l_c = { θ_c, σ_c }, c = 1, 2, ..., C, $d_{l,C} \rightarrow d_{l,J}$ if $h \rightarrow 0$, is based on points c where l crosses the grid and σ_c are treated by the GDI program as an irregular part of (Eq. 1).

Data lines including j and c-points are shown in Fig. 3. taken from (Spalvins et al. 2002).

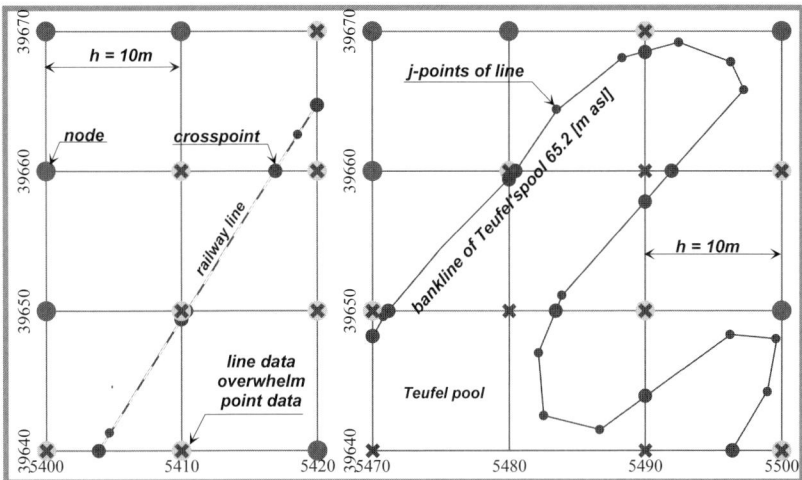

Fig. 3. Illustration of local interpolation for line σ_{in}

3   GDI program

The GDI program (Spalvins and Slangens 1994; Spalvins and Slangens 1995) provides a σ-map by applying (Eq. 3), (Eq. 4), (Eq. 5).

$$div\ (\rho\ grad\ \sigma) = 0 \qquad (3)$$

$$V\sigma = f_\psi - G_\sigma\ \sigma_\psi \qquad (4)$$

$$\rho = (\ n - d\ /\ h\ )\ !\geq 1 \qquad (5)$$

According to the scheme of Fig. 4, on the chosen xy-plane of (Eq. 1), the algebraic equation system (Eq. 4) approximates (Eq. 3) where ρ is given by the factorial function (Eq. 5) for positive rational numbers; V is the symmetric matrix of links $v_{xy} = \rho$; G_σ is the diagonal matrix (part of V) of elements connecting the nodes, where σ must be found, with the σ_ψ-nodes of the boundary condition σ_ψ interpolated from σ_{in}.

The shape of σ is changed by ρ specified for σ_ψ-nodes. In Eq. (5), n controls the radius $n \times h$ of the $\rho > 1$ area; d is the distance from the σ_ψ-node to the one where $\rho > 1$ must be specified. When $n = 1$, $\rho = 1.0$, and then peaks of σ may appear at the σ_ψ-nodes. These peaks may be turned dome-shaped when $n \geq 4$. If necessary, ρ may be controlled for each σ_ψ-node, or along any line chosen. The solution of (Eq. 4) upholds maximum/minimum and minimal energy principles. Then root sets of σ_{in} are minimal.

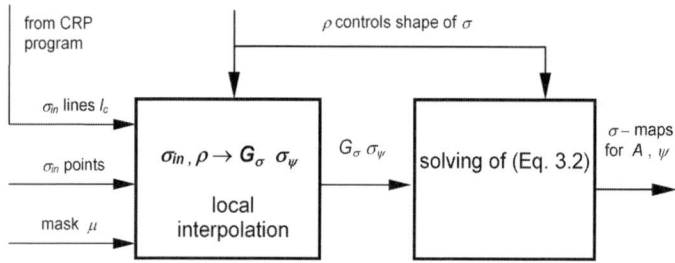

Fig. 4. Scheme of GDI program

A round of GDI starts with local interpolation $\sigma_{in} \to \sigma_\psi$ that involves pointwise and line data (e and c-data) (Spalvins and Slangens 1995). Hence the e-data has the lowest rank, they are processed first:

$$\sigma_0 = \sum_{i=1}^{t} C_i \sigma_i \qquad (6)$$

$$\sum_{i=1}^{t} C_i = 1.0$$

$$C_i = c_{i0} / \sum_{i=1}^{t} c_{j0}$$

$$c_{i0} = (1 - |\xi_i|/h) \cdot (1 - |\eta_i|/h)$$

$$c_{i0} = 0 \quad \text{if} \quad c_{i0} < 0.042$$

The index 0 runs through $p = 1, 2, ..., N$ nodes of (Eq. 4); σ_0 is found for the 0-th node, as the centre of the search area L_σ bounded by hyperbolic arcs; C_i and c_{i0} are total and partial weights of a source σ_i; $\xi_i = x_i - x_0$, $\eta_i = y_i - y_0$ are local coordinates of σ_i. For an $h \times h$ block, (Eqs. 6) are illustrated by Fig. 5 and Fig. 6.

The grid of (Eq. 4) is controlled by the mask μ: if for the p-th node, $\mu_p = 1$ or 0 then σ_{in} are allowed or blocked here, accordingly.

Commonly, c-data are carried by l_σ. For GDI, the CPR program finds $\{\theta_c, \sigma_c\}$, $c = 1, 2,...$ at the points where l_σ crosses the grid lines. These crosspoints are irregular nodes of (Eq. 4). Local c-interpolation eliminates them, thus providing f_ψ and annihilating the result of (Eq. 6) there, because l_c has higher rank upon e-data. Local conflicts of l_σ by different ranks may also cause serious errors. These ranks can be accounted for by repeating several rounds of applying (Eq. 4). This way is convenient for detecting possible errors, and much simpler σ_{in} may be applied than if one tries to obtain σ as the first solution of (Eq. 4). Unlike most of interpolation methods, the surface created by GDI may include sharp edges and rows specified by l_σ.

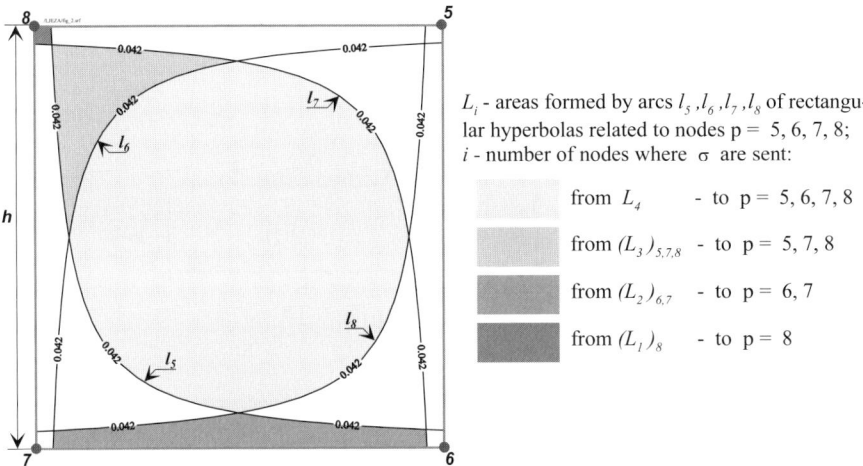

Fig. 5. Areas for point data search within an elementary $h \times h$ block if the optimal value $\lambda = 0.042$ (for the parameter of the hyperbolas-borderlines l_5, l_6, l_7, l_8) is applied

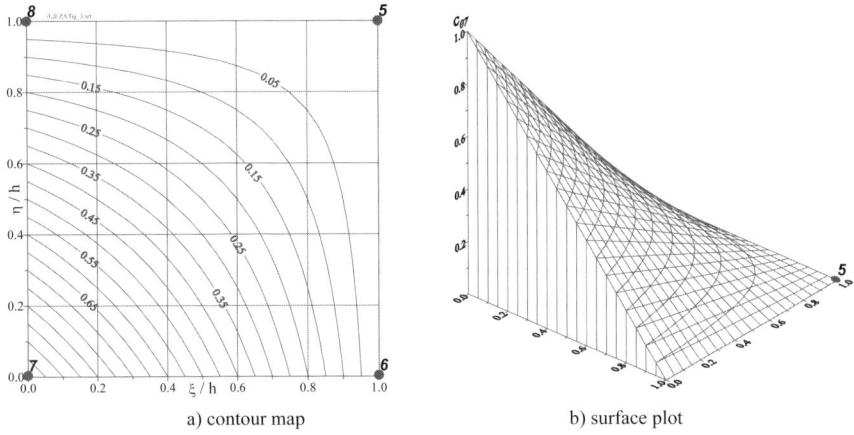

a) contour map   b) surface plot

Fig. 6. Rectangular hyperbolas as the contours (isolines) of c_{07} on the grid block of Fig. 5

4 INT β, φ program

When A and ψ of (Eq. 1) are ready, $\beta_w \rightarrow \beta$ ($\beta_0 \rightarrow \beta_{op}$) is interpolated $\beta_{op} = c_{op}\,\beta_0$ to nodes $p = 5, 6, 7, 8$ of the grid block (Fig. 5) by the weights c_{op}.

The back-interpolation $\varphi \rightarrow \varphi_w$, at the 0-th site, applies the formula:

$$\varphi_0 = \sum_{p=5}^{8} c_{op}\varphi_p - \Delta_0, \quad \Delta_0 = \sum_{j=1}^{t} \lambda_{0j}\beta_j \tag{7}$$

where Δ_0 is the local depression caused by β_j sources via the weights λ_{0j}. Obtaining of c_{op} and λ_{0j} are explained in (Spalvins et al. 2002). The principal element of λ_{0j} is the source function τ_0. Its contour map is shown in Fig. 8. The shape of τ_0 is close to a circle; in nodes, $\tau_0 = 0$.

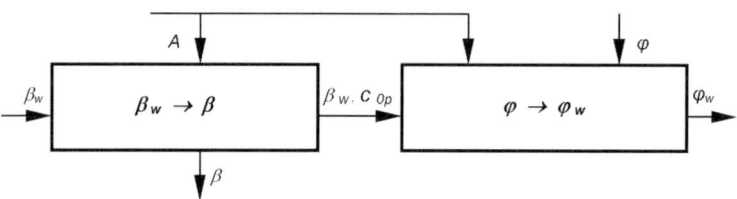

Fig. 7. Scheme of INT β, φ program

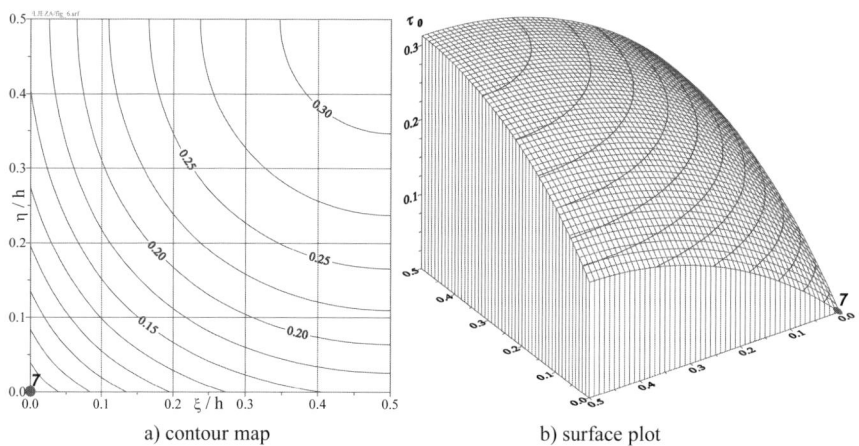

a) contour map b) surface plot

Fig. 8. . Contours of τ_0 on the quarter of the elementary grid block

5 The shell of HM as an interpolator

In many cases, conventional software or modeller is helpless to provide the right ψ_{sh}-distribution. The problem has been solved by converting the shell into an interpolation device by enlarging (10^3 - 10^5)-fold values of its links (Spalvins et al.2000). The shell then acts like an almost ideally conducting shield computing missing values of ψ_{sh}, as a special portion of φ where no ψ components are fixed. This method may be applied in all kinds of modelling programs developed for HM.

6 Conclusions

The following ideas have been implemented in software for creating HM:
- The method of local interpolation for pointwise data.
- Applying crosspoints of data lines as a part of the HM system.
- Using numerical solutions of boundary field problems for creating σ-maps by controlling the heterogeneousity parametre ρ.
- Complex σ–maps are obtained gradually by repeating interpolation rounds.
- Back-interpolation ($\varphi \rightarrow \varphi_w$) for irregular points of the HM body.
- Using the HM shell as an interpolation device.

The research has been financed by the Latvian Council of Science.

References

Lace I, Spalvins A, Slangens J (1995) Algorithms for accounting groundwater discharge in the regional programme. In: Proc. of Intern. Seminar on "Environment modelling". Boundary field problems and computers, Riga – Copenhagen, 36(1): 201-216

Slangens J, Spalvins A (2000) Creating reliable program for preparing line data of hydrogeological models. In: Scient. Proc. of Riga Technical University in series "Computer Science". Boundary field problems and computer simulation. Riga, 4(42): 35-40

Spalvins A, Janbickis R, Slangens J (2000) Boundary Shells of Hydrogeological Models as Interpolation Devices. In: Scient. Proc. of Riga Technical University in series "Computer Science". Boundary field problems and computer simulation. Riga, 4(42): 32-34

Spalvins A, Janbickis R, Slangens J, Lace I (2002) Modelling of remediation tools for the contaminated Bernau place, Germany. In: Scient. Proc. of Riga Technical University in series "Computer Science". Boundary field problems and computer simulation. Riga, 12(44): 20-28

Spalvins A, Slangens J (1994) Numerical interpolation of geological environment data. In Proc. of Latvian - Danish seminar on "Groundwater and geothermal energy". Boundary field problems and computers. Riga – Copenhagen, 35(2): 181-196

Spalvins A, Slangens J (1995) Local interpolation of geological environment data. In: Proc. of Intern. seminar on "Environment modelling". Boundary field problems and computers. Riga – Copenhagen, 36(1): 159-174

Spalvins A, Slangens J (1995) Updating of geological data interpolation of hydrogeological model and interpolation of simulation results at observation wells. In: Proc. of Intern. seminar on "Environment modelling". Boundary field problems and computers. Riga – Copenhagen, 36(1): 175-192

Spalvins A, Slangens J, Janbickis R, Lace I (2002) Interpolation as an important tool for creating credible hydrogeological models. In: Proc. of the 4th Intern. conf. on calibration and reliability In Groundwater modelling, Model CARE'2002 . Prague, pp. 84-87

Indicators of Sustainable Production

Damjan Krajnc, Peter Glavič[*]

University of Maribor, Faculty of Chemistry and Chemical Engineering, Smetanova 17, SI- 2000 Maribor, Slovenia

[*] Corresponding author. Tel.: ++386 2 229 44 51; fax: ++386 2 252 77 74. E-mail address: glavic@uni-mb.si (P. Glavič).

Abstract

The main cause of environmental damage is unsustainable production and consumption, especially in industrialized countries. Achieving sustainable development will require changes in industrial processes, in the type and quantity of resources used, in the treatment of waste, in the control of emissions, and in the products produced. One of the difficulties in measuring the company's level of sustainability is to determine which directions of change are leading towards sustainability. Hence, it is necessary to apply appropriate metrics that will enable these assessments. This paper presents indicators for assessing and promoting business sustainability - indicators of sustainable production. It first introduces the main concepts of such production and a set of necessary conditions that firms must fulfill in order to be sustainable. It identifies major functions of indicators and it proceeds to presenting the role of indicators.

The paper focuses on sustainable production, proposing indicators of sustainable production, which could be used as strategic metrics for assessing the sustainability level of a company and for identifying more sustainable options for the future. They enable a large amount of information to be compressed into a format easier to manipulate, compare and understand. The proposed indicators focus on the environmental, economic and social aspects of sustainable production. Most of the indicators included can be applied across industry. However, they are not aimed at being uniformly applicable to all sectors. According to the flows in the manufacture they are divided into input and output indicators and they are based on commonly measured environmental aspects of sustainable production (energy use, materials use, water consumption, products, wastes, and air emissions) covering key global issues. The paper represents a new approach to the systematization of indicators and their symbols and units.

Keywords: sustainability indicators, sustainable production

List of Abbreviations

CWRT	Center for Waste Reduction Technologies
EU	European Union
GRI	Global Reporting Initiative
ISO	International Organization for Standardization
PO	Production Output
SPI	Sustainable process index
UP	Unit of Production (e.g. mass in t or kg, volume in m^3, number, monetary value in EUR, etc.)
WBCSD	World Business Council for Sustainable Development

1 Introduction

In the present age there is extensive pressure on the ecosystems and biodiversity of the world. During the last few decades, humans have remodeled physical, chemical and biological systems in new ways and at faster rates than ever recorded on Earth. We have to solve problems associated with climate change, ozone depletion, global warming, acid rain, bioaccumulation of toxic substances, species loss, deforestation, depletion of natural resources, population growth, etc. The human race must confront the accumulated facts of a potential collapse of critical ecosystems and we are forced into quick changes over the entire planet. The environmental movement needs to address the global problems. However, environmental activity at local, regional, state and international levels seems to be an appropriate and necessary (but not sufficient) step towards sustainability.

At the moment the survival of humans and the planet as a whole depends upon guiding human development in positive directions that are healthy, diverse, and sustainable. According to the report of the Brundtland Commission (1987), sustainable development is defined as development that meets the needs of the present without compromising the ability of future generations to meet their own needs. Although the definition reaches the heart of the problem, it enables different interpretations, since people have different goals and sensitivities. Most endeavors on sustainable development are based on supporting cultural, social, economical, industrial, and technological development while preserving the natural environment. Achieving sustainable development is no easy task. Not only perceptible changes in decision-making at the highest levels, but also progress in production and consumption will be needed. Sustainability will be the driving force for 21st century industry much as automation was for 20th century industry, and steam was for 19th century industry (O'Brien 1999).

The industry sector is responsible for most material flows within human society as well as for the exchange of material and energy with the environment. Present industrial systems are not sustainable in the long term because of their excessive demands upon natural non-renewable resources. The achievement of sustainability requires a radical re-think of many of industry's practices.

Nowadays, the companies recognize the importance of coordinated sustainable management actions and to know whether they are meeting the sustainable development, they need to be able to measure their progress. One way to measure performance is through the use of sustainability indicators compressing large amounts of information from different sources into a format easier to understand, compare and act upon. However, efficient tools to harmonize sustainability metrics are still lacking.

In past, most companies have been using standard financial indicators to track their business effectiveness. Nowadays, due to demands from various parties and accounting more fully for their actions, the economic, environmental, and social reporting is becoming more and more important. The companies are starting to develop and implement different methods for their sustainability assessment moving from traditional environmental reporting schemes towards eco-efficiency reports and sustainability reporting. They use different indicator frameworks, however, the reports demonstrate that most indicator frameworks are still under development and no one is applicable as a whole to evaluate sustainable production.

2 Sustainable production

The concept of sustainable production is a key component of sustainable development, which balances three principal requirements: environmental, social and economic development of the company (Fig. 1).

The Lowell Center for Sustainable Production (LCSP) defines sustainable production as the creation of goods and services using processes and systems that are non-polluting, conserving energy and natural resources, economically viable, safe and healthful for employees, communities and consumers, and socially and creatively rewarding for all working people (Lowell Center 1998).

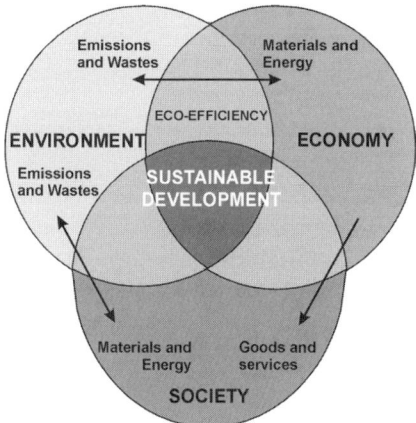

Fig. 1. A model of sustainable development (Azapagic 2000)

2.1 Achievement of Sustainable production

A number of characteristics must be satisfied in order to ensure that production processes and the use of products and materials operate within environmental limits. If the objectives of sustainable production are to be achieved, then companies must minimize all kinds of wastes as well as the use of natural resources, raw materials and energy. They must design, produce, distribute, and dispose of or recycle products in such a way that the associated environmental impacts and resource usage levels are at least in line with the Earth's estimated carrying capacity. This goal requires a fundamental re-think in the design of a product to take account of all stages of a product's life cycle, and a shift in manufacturing processes from cleaning technologies to clean technologies, which reduce the actual level of emissions produced as well as the energy and other resources used during processing (O'Brien 1999). A set of necessary conditions that firms must fulfill in order to be sustainable includes:

- Reducing the use of materials and energy in products and their production
- Closing of material loop systems, to conserve resources and prevent waste
- Minimization or avoidance of waste
- Reuse and recycling products
- Disposing of non-recyclable products or production waste in an environmentally acceptable way
- Planning of products which are easy to repair, adaptable, durable and with longer lifetime
- Minimization of transportation needs
- Cleaner production technologies and procedures throughout the product life cycle
- Improving a process technology
- Research and development in environmentally sound technologies
- Consideration of the social role played

3 Indicators of Sustainable Production

"Measure what can be measured, and make measurable what can not be measured." - Galileo Galilei

Many companies and sectors are involved in the development of sustainability indicators, either for internal use or as a tool for communication with stakeholders Sustainability reports are emerging as a new trend in corporate reporting, integrating the elements of financial, environmental, and social facets of the company into the sustainability report (GRI 2002). Recently have many leading companies begun to measure their sustainability performance. They have already developed some key indicators for their businesses and are tracking and reporting them, but the results are not readily comparable, because these indicators have been developed internally within business sectors (NRTEE 2001).

Having designed a comprehensive set of sustainable production indicators and put them to use, it should be possible to report on performance and to compare the performance of different companies by benchmarking. However, since there are no benchmarks that would indicate from which level firms could be declared sustainable, the focus is on comparing the firms between themselves. Some are more efficient, or inefficient, in all respects. Most often, however, they might be efficient in specific respects but inefficient in others (Callens and Tyteca 1999).

Developing a globally accepted reporting framework is a long-term endeavor. Numerous organizations are presently trying to develop a set of indicators to state the progress of a company towards sustainability. Veleva and Ellenbecker (2001) have analyzed four of the best-known indicator frameworks:

- International Organization for Standardization (ISO 14031)
- Global Reporting Initiative (GRI)
- World Business Council for Sustainable Development (WBCSD)
- Center for Waste Reduction Technologies (CWRT)

Results demonstrate that most indicator frameworks are still under development and none is applicable as a whole to evaluate sustainable production. Material use and environmental protection are best covered in all reviewed frameworks. This is particularly the case of ISO 14031, which lists about 100 environmental indicators. Unlike environmental indicators, social issues receive the least attention in existing indicator frameworks. All reviewed frameworks, with the exception of the GRI, use only quantitative indicators to measure the performance of the companies. There is a trend toward using a manageable number of indicators (between ten and twenty) that are simple and easy to apply. Exceptions are ISO 14031 and GRI, each of which includes about a hundred indicators. There is clear trend toward developing standardized indicators, which would be applicable to any organization. GRI, WBCSD, and CWRT have all suggested common measures for evaluation of sustainability performance. Most indicator frameworks attempt to address key global issues, yet these are typically environmental.

Indicators can be used alone or in thematic sets, which are useful for demonstrating the links between issues and for analyzing the reasons behind trend (SECRU 2001). The multiplicity of indicators and metrics being developed in this fast growing field shows the importance of the conceptual and methodological work in this area (Voinov 1997).

In spite of evidence that it is not possible to set up sustainability indicators that are applicable to any company or organization, thus far a number of different approaches to standardization have been proposed. The problem, however, is to introduce a quantitative measure of sustainable production, since some aspects of sustainability (especially the social aspect) cannot be quantitatively expressed. With some issues such as energy use and water use there are no difficulties since they are common for all companies. However, more specific indicators have to be defined separately, dependent on the sector.

Recently, some experts are trying to introduce fuzzy set theory and to develop fuzzy mathematical models to assess sustainable development. A potential problem in the practical application of the fuzzy model applying approximate reasoning concerns the combinatorial nature of the fuzzy rules. For example, the assessment of the contribution of n sustainability indicators to sustainable development using two linguistic values (e.g. acceptable and unacceptable) results in a fuzzy rule base of 2^n rules. Therefore, we have 1,024 rules for only ten indicators. Because of the exponentially increasing number of fuzzy rules, the fuzzy rule base soon becomes non-transparent and difficult to apply (Cornelissen et al. 2001).

Based on a set of criteria for sustainability and on conventional mass and energy balances, the concept of a sustainable process index (SPI) was introduced (Krotschak and Narodoslawsky 1996). The SPI measures the potential impact (pressure) of processes on the ecosphere and compares mass and energy flows induced by human activities with natural flows. As natural flows are always linked to area (examples are the growth of biomass, precipitation and, most importantly, solar radiation) the basic unit of the SPI is area. The lower the requirement of area for a given activity is, the less is the impact of this activity on the environment.

3.1 The role of indicators

Indicators have numerous applications. They compress large amounts of information from different sources into a format easier to understand, compare and manipulate. Companies can use indicators to set targets and monitor consequent success. These targets help the decision-maker visualize what actions will need to be emphasized in future (Finnish Environment 2000).

In the literature we can trace numerous definitions of indicators. However, it is more useful to state the primary role of indicators. Gallopín (1997) identifies the following major functions of indicators:
- Assessing conditions and trends in relation to goals and targets
- Reflecting the status of a system
- Providing early warning information
- Anticipating future conditions and trends
- Comparing across place and situations
- Highlighting what is happening in a large system (Changing views on change 1998)

The role of sustainable production indicators is to help measure a company's economic, environmental and social performance and to provide information on how it contributes to sustainable development.

However, in order to cope up with the complexity of sustainability-related issues for different systems, the indicators have to reflect the wholeness of the system as well as the interaction of its subsystems. Their purpose is to show how well the system is working and they are strongly dependent on the type of the system they monitor (Afgan et al. 2000).

The indicators of sustainable production enable identification of more sustainable options through (Azapagic and Perdan 2000):
- Comparison of similar products made by different companies
- Comparison of different processes producing the same product

- Benchmarking of units within corporations
- Rating of a company against other companies in the sector
- Assessing progress towards sustainable development of a sector

3.2 Proposed Indicators of sustainable production

"Everything should be as simple as possible, but not simpler." - Albert Einstein

The indicators framework presented here focuses on the main aspects of sustainable production, relating to the social, environmental and economic aspects of sustainability. The criteria for the sustainability assessment of the company have to reflect six aspects:
- Resource use aspect
- Product aspect
- Environment aspect
- Economic aspect
- Quality aspect
- Social aspect

The framework is based on examples of indicators that could be applied at a company. They are simple and easy to use, based on available data and commonly measured aspects of production (e.g. materials use, energy use, water consumption, products, waste, air emissions, etc.) (Fig. 2).

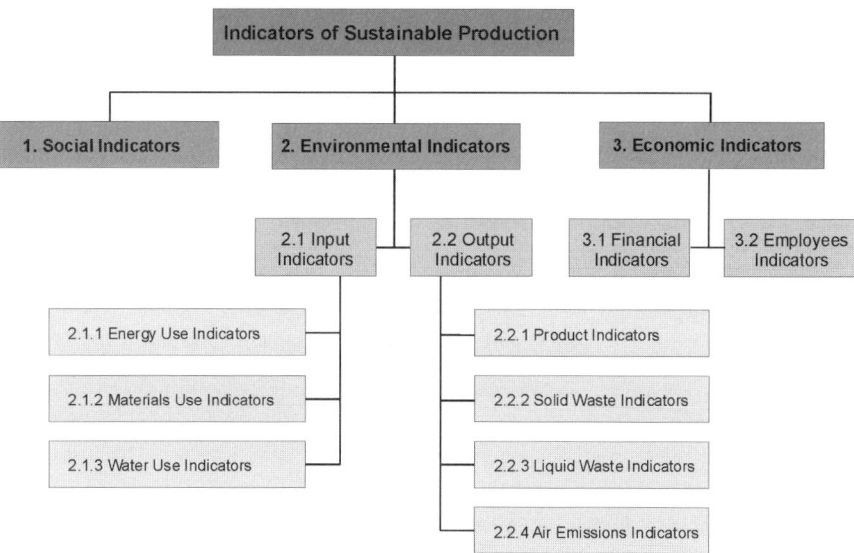

Fig. 2. Organizational chart of indicators of sustainable production

The environmental indicators are divided into input and output indicators based on flows in the manufacture (Fig. 3). The economic indicators are divided into financial and employees' indicators. The paper suggests examples for each indicator, yielding the first step toward sustainable production. Results from their plot testing could demonstrate which ones need to be modified and which ones are working well for most companies. Examples of indicators of sustainable production are presented in Table 1.

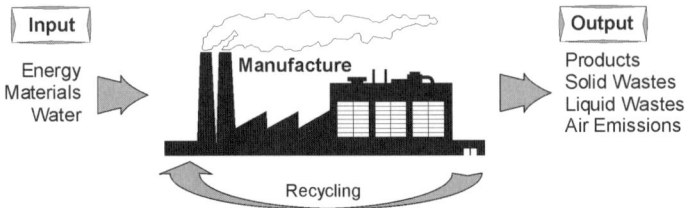

Fig. 3. Flows in manufacture

Proposed indicators of sustainable production aim to (Geiz and Kutzmark 1998; Veleva and Ellenbecker 2001):
- Reflect sustainability concept
- Provide a set of indicators applicable to most companies
- Suggest simple and easy way to implement indicators
- Avoid the use of too many indicators
- Suggest indicators that cover key global issues (e.g. global warming, ozone depletion, acidification, nutrification, etc.)
- Provide companies with a new metrics and guidance on how to measure their achievements toward sustainable production

To incorporate indicators of sustainable production the company must identify a period for tracking and calculating an indicator (e.g. fiscal year, calendar year, 6 months, quarter, month) and define units of measurement. It must also identify a type of measurement (absolute or adjusted) and boundaries, which determine how far a company wishes to go in measuring the indicator (Fig. 4).

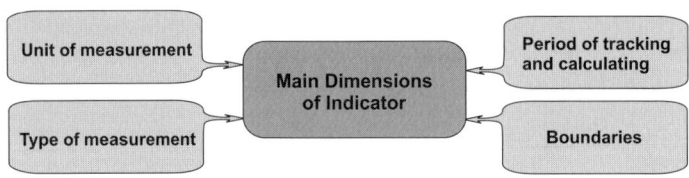

Fig. 4. Main dimensions of an indicator

It is recommendable that a company first begins with simple, easy to implement measures of compliance and resource efficiency and then moves toward more complex indicators, addressing supply-chain, social effects and life-cycle impacts. Using indicators of sustainable production is one part of a process of continuous improvement, where the goal is to move the organization from adopting primarily low-level measures to using all levels of indicators of sustainable production (Veleva and Ellenbecker 2001).

Table 1. Indicators of sustainable production

1. Social Indicators:			
Indicator	**Quantity**	**Symbol**	**Unit**
1. Specific employee number	Number of employees / Unit of production	$N_{employee}$	1 / UP
2. Employee turnover[e]	Number of employees who have resigned or been made redundant / total number employed	$X_{employee}$	$\frac{1}{1} = 1$
3. Payment ratio	$\dfrac{\text{The salary of the upper 10 \% of employees}}{\text{The salary of the lower 10 \% of employees}}$	$R_{payment}$	$\dfrac{EUR}{EUR} = 1$
4. Fraction of workers satisfied with their work	$\dfrac{\text{Number of employees satisfied with their work}}{\text{Total number of employees}}$	$X_{satisf.}$	$\dfrac{1}{1} = 1$
5. Promotion rate[e]	$\dfrac{\text{Number of promotions}}{\text{Total number employed}}$	$R_{promot.}$	$\dfrac{1}{1} = 1$
6. Time of employee illness	Time of lost workdays because of injuries and illnesses	$t_{illness}$	days, d
7. Fraction of charitable contributions	$\dfrac{\text{Charitable contributions to the community}}{\text{Total revenues}}$	$X_{contrib.}$	$\dfrac{EUR}{EUR} = 1$
8. Number of community projects	Number of projects of the company with its community	$N_{cooper.}$	1
9. Mass fraction of local consumption	$\dfrac{\text{Mass of locally consumed products}}{\text{Total output mass of products}}$	$w_{loc. cons.}$	$\dfrac{kg}{kg} = 1$
10. Index of community population growth	Population growth in the community in %	$r_{com. pop.}$	1
2. Environmental Indicators ▶ 2.1 Input Indicators ▶ 2.1.1 Energy Use Indicators:			
Indicator	**Quantity**	**Symbol**	**Unit**
1. Total energy consumption[d]	Total energy consumed	$E_{tot.}$	J
2. Specific energy consumption	$\dfrac{\text{Total energy consumed}}{\text{Production output}}$	$E_{spec.}$	$\dfrac{J}{UP}$
3. (Source of energy) fraction	$\dfrac{\text{Consumption per source of energy}}{\text{Total energy consumption}}$	E_{source}	$\dfrac{J}{J} = 1$
4. (Renewable energy) fraction	$\dfrac{\text{Renewable energy consumption}}{\text{Total energy consumption}}$	$E_{renew.}$	$\dfrac{J}{J} = 1$
5. Energy for recycling	Energy used for recycling	$E_{for recycl.}$	J
6. Energy intensity[a]	$\dfrac{\text{Total energy consumed}}{\text{Value of product sold or Value added}}$	$E_{intensity}$	$\dfrac{J}{EUR}$
7. Total energy costs[d]	Absolute	$C_{E, tot.}$	EUR
8. (Energy costs) fraction	$\dfrac{\text{Total energy costs}}{\text{Total production costs}}$	$C_{E, spec.}$	$\dfrac{EUR}{EUR} = 1$
9. Average costs of energy source	$\dfrac{\text{Costs per source of energy}}{\text{Consumption per source of energy}}$	$C_{E, source}$	$\dfrac{EUR}{J}$

Table 1: (cont.)

2. Environmental Indicators ▶ 2.1 Input Indicators ▶ 2.1.2 Materials Use Indicators:			
Indicator	Quantity	Symbol	Unit
1. Total material consumption[b]	Absolute mass	$m_{mat., tot.}$	kg
2. Specific material consumption	$\dfrac{\text{(Total material input) mass}}{\text{Production output}}$	$m_{mat., spec.}$	$\dfrac{kg}{UP}$
3. Fraction of renewable raw materials[d]	$\dfrac{\text{(Renewable raw material input) mass}}{\text{(Total material input) mass}}$	$w_{renw. mat.}$	$\dfrac{kg}{kg}=1$
4. Raw materials efficiency[d]	$\dfrac{\text{(Production output) mass}}{\text{(Raw materials input) mass}}$	$\eta_{raw\ mat.}$	$\dfrac{kg}{kg}=1$
5. Recycled material fraction[c]	$\dfrac{\text{(Recycled material input) mass}}{\text{(Total material input) mass}}$	$w_{recycl. mat.}$	$\dfrac{kg}{kg}=1$
6. Variety of hazardous materials[d]	Number	$N_{haz. mat.}$	1
7. Hazardous materials input mass[d]	Absolute mass	$m_{haz. mat.}$	kg
8. Total material costs[d]	Absolute value	$C_{mat., tot.}$	EUR
9. Material intensity[a]	$\dfrac{\text{(Total material input) mass}}{\text{Value of product sold or Value added}}$	$I_{mat.}$	$\dfrac{kg}{EUR}$

2. Environmental Indicators ▶ 2.1 Input Indicators ▶ 2.1.3 Water Use Indicators:			
Indicator	Quantity	Symbol	Unit
1. Total water consumption[d]	Absolute volume	$V_{water, tot.}$	m^3
2. Specific water consumption	$\dfrac{\text{Water consumption volume}}{\text{Production output}}$	$V_{water, spec.}$	$\dfrac{m^3}{UP}$
3. Volumic fraction of water type[d]	$\dfrac{\text{Consumption volume per type of water}}{\text{Total consumption volume}}$	$\varphi_{water\ type}$	$\dfrac{m^3}{m^3}=1$
4. Total water costs[d]	Absolute value	$C_{water, tot.}$	EUR
5. Water cost fraction[d]	$\dfrac{\text{Water costs}}{\text{Total production costs}}$	$C_{water, spec.}$	$\dfrac{EUR}{EUR}=1$
6. Volumic water type cost	$\dfrac{\text{Costs per type of water}}{\text{Consumption volume per type of water}}$	$C_{water\ type}$	$\dfrac{EUR}{m^3}$

2. Environmental Indicators ▶ 2.2 Output Indicators ▶ 2.2.1 Product Indicators:			
Indicator	Quantity	Symbol	Unit
1. Mass fraction of products with an environmental label[d]	$\dfrac{\text{Mass of products with environmental labels}}{\text{Total mass of products}}$	$w_{EL\ prod.}$	$\dfrac{kg}{kg}=1$
2. Mass fraction of products from recyclable materials[d]	$\dfrac{\text{Mass of products from recyclable materials}}{\text{Total mass of products}}$	$w_{recycl. prod.}$	$\dfrac{kg}{kg}=1$
3. Mass fraction of products designed for disassembly, reuse or recycling[b]	$\dfrac{\text{Mass of products designed for recovery}}{\text{Total mass of products}}$	$w_{recov. prod.}$	$\dfrac{kg}{kg}=1$
4. Product durability[c]	Time of durability	$t_{durability}$	d or a
5. Revenues from eco products[d]	Absolute value	REV	EUR
6. Revenue fraction of eco products[d]	$\dfrac{\text{Revenues from ecoproducts}}{\text{Total revenue}}$	$REV_{eco\ pr.}$	$\dfrac{EUR}{EUR}=1$

Table 1: (cont.)

2. Environmental Indicators ▸ 2.2 Output Indicators ▸ 2.2.1 Product Indicators:			
Indicator	Quantity	Symbol	Unit
7. Total packaging mass[d]	Absolute mass	$m_{pack.}$	kg
8. Packaging mass fraction of the product[d]	$\dfrac{\text{Packaging mass}}{\text{Total mass of products}}$	$w_{pack.}$	$\dfrac{kg}{kg}=1$
9. Mass fraction of reusable packaging[d]	$\dfrac{\text{Reusable packaging mass}}{\text{Total packaging mass}}$	$w_{reus.\ pack.}$	$\dfrac{kg}{kg}=1$
10. Packaging costs[d]	Absolute value	$C_{pack.}$	EUR
11. Specific packaging costs[d]	$\dfrac{\text{Packaging costs}}{\text{Production output}}$	$C_{pack.,\ spec.}$	$\dfrac{EUR}{UP}$

2. Environmental Indicators ▸ 2.2 Output Indicators ▸ 2.2.2 Solid Waste Indicators:			
Indicator	Quantity	Symbol	Unit
1. Total solid waste mass[d]	Absolute mass	$m_{s,\ tot.}$	kg
2. Specific solid waste mass	$\dfrac{\text{Mass of specific type of solid waste}}{\text{Production output}}$	$m_{s,\ spec.}$	$\dfrac{kg}{UP}$
3. (Solid waste mass) for recovery[d]	Recovered solid waste mass absolute	$m_{s,\ recov.}$	kg
4. (Solid waste mass) for disposal[d]	Non-recovered solid waste mass absolute	$m_{s,\ disp.}$	kg
5. Recycling mass fraction	$\dfrac{\text{Recycled solid waste mass}}{\text{Total mass of solid waste}}$	$w_{s,\ recycl.}$	$\dfrac{kg}{kg}=1$
6. Disposal mass fraction[d]	$\dfrac{\text{Mass of non-recovered solid waste}}{\text{Total mass of solid waste}}$	$w_{s,\ non-recy.}$	$\dfrac{kg}{kg}=1$
7. (Hazardous solid waste) mass fraction[d]	$\dfrac{\text{Mass of hazardous solid waste}}{\text{Total mass of solid waste}}$	$w_{s,\ haz.}$	$\dfrac{kg}{kg}=1$
8. (Hazardous solid waste) mass[d]	Mass of hazardous solid waste released into the environment	$m_{s,\ haz.}$	kg
9. Total solid waste costs[d]	Absolute value	$C_{s,\ tot.}$	EUR
10. Solid waste cost fraction[d]	$\dfrac{\text{Total solid waste costs}}{\text{Total production costs}}$	$C_{s,\ spec.}$	$\dfrac{EUR}{EUR}=1$

2. Environmental Indicators ▸ 2.2 Output Indicators ▸ 2.2.3 Liquid Waste Indicators:			
Indicator	Quantity	Symbol	Unit
1. Total volume of liquid waste[d]	Absolute volume	$V_{l,\ tot.}$	m³
2. Specific liquid waste volume	$\dfrac{\text{Total volume of liquid waste}}{\text{Production output}}$	$V_{l,\ spec.}$	$\dfrac{m^3}{UP}$
3. Non-polluted liquid waste volume[d]	Absolute volume	$V_{l,\ non-poll.}$	m³
4. Polluted liquid waste volume[d]	Absolute volume	$V_{l,\ poll.}$	m³
5. Specific pollution mass ratio[d]	$\dfrac{\text{Pollution load mass (P, N, AOX)}}{\text{Production output}}$	$R_{poll.,\ spec.}$	$\dfrac{kg}{UP}$
6. Pollution mass concentration in liquid waste[d]	$\dfrac{\text{Mass of pollutants}}{\text{Liquid waste volume}}$	$c_{l,\ poll.}$	$\dfrac{kg}{m^3}$
7. Total liquid waste costs[d]	Absolute value	$C_{l,\ tot.}$	EUR
8. (Liquid waste) cost fraction[d]	$\dfrac{\text{Total liquid waste costs}}{\text{Total production costs}}$	$C_{l,\ spec.}$	$\dfrac{EUR}{EUR}=1$

Table 1: (cont.)

2. Environmental Indicators ► 2.2 Output Indicators ► 2.2.4 Air Emissions Indicators:			
Indicator	**Quantity**	**Symbol**	**Unit**
1. Mass fraction of Greenhouse Gases[a]	$\dfrac{\text{Total mass of CO}_2 \text{ equivalents}}{\text{Total mass of products}}$	$w_{CO_2 \text{ equiv.}}$	$\dfrac{kg}{kg} = 1$
2. Greenhouse Gases intensity	$\dfrac{\text{Total mass of CO}_2 \text{ equivalents}}{\text{Value of product sold or Value added}}$	I_{GHGs}	$\dfrac{kg}{EUR}$
3. Acidification mass fraction[b]	$\dfrac{\text{Total mass of SO}_2 \text{ equivalents}}{\text{Total mass of products}}$	$w_{SO_2 \text{ equiv.}}$	$\dfrac{kg}{kg} = 1$
4. Acidification mass intensity	$\dfrac{\text{Total mass of SO}_2 \text{ equivalents}}{\text{Value of product sold or Value added}}$	$I_{acidif.}$	$\dfrac{kg}{EUR}$
5. Photochemical ozone creating potential mass fraction[a]	$\dfrac{\text{Total mass of ethylene equivalents}}{\text{Total mass of products}}$	$w_{C_2H_4 \text{ equiv.}}$	$\dfrac{kg}{kg} = 1$
6. Photochemical ozone creating potential mass intensity[a]	$\dfrac{\text{Total mass of ethylene equivalents}}{\text{Value of product sold or Value added}}$	I_{POCP}	$\dfrac{kg}{EUR}$
7. Eutrophication mass fraction	$\dfrac{\text{Total mass of phosphate equivalents}}{\text{Total mass of products}}$	$w_{PO_4^{3-} \text{ equiv.}}$	$\dfrac{kg}{kg} = 1$
8. Eutrophication mass intensity	$\dfrac{\text{Total mass of phosphate equivalents}}{\text{Value of product sold or Value added}}$	$I_{eutroph.}$	$\dfrac{kg}{EUR}$
9. Costs of purifying air[d]	Absolute value	$C_{pur.\ air}$	EUR
10. Costs fraction of purifying air[d]	$\dfrac{\text{Absolute purifying cost}}{\text{Total production costs}}$	$C_{pur.\ air,\ fract.}$	$\dfrac{EUR}{EUR} = 1$

3. Economic Indicators ► 3.1 Financial indicators			
Indicator	**Quantity**	**Symbol**	**Unit**
1. Fraction of Value Added in GDP	Value added / GDP	$GDP_{ctrb.}$	$\dfrac{EUR}{EUR} = 1$
2. Value of investments in sustainable development	Investments in sustainable R&D as fraction of the expenses of the company	I_{SD}	EUR
3. Value of investments in environmental protection	Investments of company in environmental protection	I_{env}	EUR
4. Environmental responsibility costs	Costs in case of environmental damage responsibility	$C_{env.\ resp.}$	EUR
5. Specific number of complaints of customers	$\dfrac{\text{Number of complaints}}{\text{Mass of products sold}}$	n_{compl}	$\dfrac{1}{kg}$
6. Value fraction of investments in ethical activity	A profit invested in ethical business activities	$f_{ethic.\ act.}$	$\dfrac{EUR}{EUR} = 1$
7. Number of sustainable environmental reports	Yearly number of positive/negative paper reports on environmental and social activity of the company	$N_{reports}$	1
8. Number fraction of suppliers	Fraction of suppliers without environmental, health and safety violations	$R_{sup.,\ unprobl.}$	1
9. Number of contact breaks	Number of contract breaks with suppliers because of disagreement with environmental, health and safety standards	$N_{contr.\ breaks}$	1

Table 1: (cont.)

3. Economic Indicators ▶ 3.2 Employees Indicators:			
Indicator	Quantity	Symbol	Unit
1. Cost of employee	Cost of employee per production output	$C_{employee}$	EUR/UP
2. Employee labor service duration	Average period of employee labor service	t_{labor}	a
3. Costs of health protection of employee	Total costs of health protection of employee	C_{health}	EUR
4. Noise level	Level of sound pressure at the working stations	L_{noise}	dB
5. Investments in employee development	Investments in employee's education and professional/personal development	$I_{educ.}$	EUR
6. Time of employee education	Average time of education per employee	$t_{educ.}$	h
7. Number of suggested improvements by employee	Number of suggested improvements in quality, social, environmental, health and safety aspect of production per employee	$N_{sug.\ imp.}$	1

[a](AIChE 2002)
[b](Veleva and Ellenbecker 2001)
[c](Azapagic and Perdan 2000)
[d](FEM and FEA 1997)
[e](IChemE 2003)

4 Indicators Objectives

4.1 Social Indicators Objective

The social dimension of sustainability concerns impacts of the company on the social systems within which it operates. It is very difficult to incorporate the social dimension of the sustainable development. Therefore, there are still very few social indicators developed and measured. Social performance can be gauged through an analysis of the impacts on stakeholders at the local, national, and global levels. In some cases, social indicators influence the intangible assets of the company, such as its human capital and reputation. Social performance measurement enjoys less of a consensus than environmental performance measurement (GRI 2002).

4.2 Energy Use Indicators Objective

The manufacturing sector is a major consumer of energy. Energy consumption and production have resulted in major pressures on the environment, from both a resource use and a pollution point of view (UN Sustainable Development 2001). Especially important are the effects of consumption of fossil fuels, of which the most significant are emissions of greenhouse gases. Improving energy efficiency in order to reduce fossil fuel consumption, greenhouse gas emissions and related air pollution emissions, and increasing the fraction of renewable energy sources are essential for sustainable production.

Energy use is usually measured at the point of consumption, i.e. the plant or establishment. To be able to add or compare the data determined, megajoules (MJ) should be used. Since natural gas usage is usually calculated in cubic meters (m^3), fuel oil in liters (L) and electricity in kWh these measurements must be converted. Table 2 illustrates the most important conversion coefficients of mass (in kg) or volume (in L or m^3) for the input of energy sources using their energy value (MJ). Another important coefficient is the mass of CO_2 emitted per energy value for fuel type. These coefficients range between about 100 kg/MJ for coal to about 50 kg/MJ for natural gas. Specific CO_2 emission coefficients for some fuel types are presented in Table 3.

Table 2. Energy contents of energy sources (Statistic Canada 1989)

	Fuel Type	Volumic energy
Petroleum products:	Heavy fuel oil	41,73 MJ/L
	Light fuel oil	38,68 MJ/L
	Diesel	38,68 MJ/L
	Kerosene	37,68 MJ/L
	Gasoline	34,66 MJ/L
	Petroleum coke	42,38 MJ/L
Natural gas:	Natural gas	37,78 MJ/m^3
	Propane	25,53 MJ/L
	Butane	28,62 MJ/L
	Fuel Type	**Massic energy**
Coal:	Anthracite	27,70 MJ/kg
	Bituminous	29,00 MJ/kg
	Sub-bituminous	18,30 MJ/kg
	Lignite	15,00 MJ/kg
	Coke	28,83 MJ/kg
Biomass:	Wood	18,00 MJ/kg
	Hog fuel	18,00 MJ/kg
	Black liquor	14,00 MJ/kg
	Fuel Type	**Conversion factor**
	Electricity	3,60 MJ/(kW·h)

Table 3. The conversion coefficients for calculating the mass of CO_2 emissions of combustion processes (FEM and FEA 1997)

Energy source	CO_2 production (g/MJ)
Brown coal	111
Brown coal coke	104
Hard coal	93
Mineral oil (crude)	80
Fuel oil: light	72
Fuel oil: heavy	78
Diesel	74
Turbine fuels	74
Natural gas	56
Petroleum gas	58
External supply of electricity	137

4.3 Materials Use Indicators Objective

The Earth's resources are not inexhaustible and some raw materials will eventually become limited in supply. The necessary reduction in the demand for virgin raw materials and non-renewable resources will only be achieved by developing disassembly technologies, recycling and remanufacturing capabilities on a commercial scale and by designing products with these concepts in mind. Recovery and re-use of materials can extend their useful life several times before eventual disposal to the environment (O'Brien 1999). Because of these combined problems, recycling technology is becoming a growing priority for society (CEFIC 2002).

Material use indicators lead to the replacement of problematic materials with environmentally safer alternatives (e.g. renewable raw materials, recyclable raw materials, solvent-free paints and varnishes). They report on the politics of the main raw, auxiliary and ancillary materials of the company. Since the company has to deal with a huge variety of materials, preparing an input–output balance sheet can assist in determining the structure. In order to be able to compare input quantities, they should be recorded in standard units of mass (in kilograms or tonnes).

4.4 Water Use Indicators Objective

Oil refineries, petrochemical plants, special chemical producers, pulp and paper industry, electric utilities, food and beverage industry, mining, etc. are large users of water. In industry, water is used as a cooling/heating medium, a cleaning agent, a reaction solvent, etc. Water use indicators track the water consumption of the company. They are intended to stimulate a reduction of water consumption by wastewater reuse, regeneration, recycling or process changes.

4.5 Product Indicators Objective

The exponential growth in world population and the recognition that materials, fuels and other resources are finite force us to change our widespread culture of disposal. The sustainable principle demands durable products that do not consume large quantities of resources during their production, use, maintenance, and repair. This requires new, flexible products with a long useful life. From the view of the company, product design, internal production, products, parts, packaging and material recovery have to be considered in relation to the usage of materials and the quality of products, processes and systems.

Product indicators measure improvements in the environmental impact of individual products or the complete range of products. They also indicate relative advantages or disadvantages in comparison to other products and/or competitors. Product indicators can refer to the environmental aspects of the internal manufacturing process of one company only or the entire life cycle of the product (e.g. including its use, preliminary and intermediate production, transportation and disposal).

4.6 Solid Waste Indicators Objective

Waste reflects inefficiencies in the production process and represents a failure in designing both the process and the product. The protection of the environment needs new techniques for treatment and disposal of wastes (CEFIC 2002). Several universities, governments, business and other organizations are working to develop, promote and apply a zero waste strategy as the ultimate goal of sustainability. It strives for (Zero Waste Alliance 2001):
- 100 % resource efficiency
- Zero solid and hazardous waste
- Zero emissions - to air, water or soil
- Zero waste in production
- Zero waste in product life
- Zero toxics

The company can achieve sustainability only by waste elimination, which leads to reducing extraction from nature, eliminating waste to nature, improving economic efficiency and making more resources available to all. To become sustainable the companies must follow examples in nature, which are cyclical and have no waste. That is not an easy task, but the quantities of waste taken to landfill sites must be reduced as much as possible. Waste should in the first case be minimized, subsequently recycled, recovered for raw materials, energy extracted from it by incineration and as a last alternative deposited on a landfill site. The waste indicators track the company's success in waste reduction.

4.7 Liquid Waste Indicators Objective

The once-through use of industrial water is becoming both uneconomical and environmentally unacceptable. The recovery and recycling of industrial wastewater is a more attractive option. However, before the recycling of wastewater, it is necessary to minimize the quantity of wastewater that appears during the process. There are a variety of reasons driving manufacturers to pursue wastewater minimization (Goldblatt et al. 1993):
- Reduced availability of fresh water
- Discharge permit compliance
- Legislation banning priority
- Economics (taxes)
- "Good neighbor" policy

Liquid waste indicators provide companies with metrics to measure their achievements in a reduction of water and other liquids consumption.

4.8 Air Emissions Indicators Objective

Air emissions have a particular significance due to their diverse environmental impacts (acid precipitation, stratospheric ozone depletion, greenhouse effect with climate change, etc.). The emissions of different substances can be used as basic air emissions indicators. Due to the variety of air pollutant types they can be limited to the most relevant substances (CO_2, CH_4, SO_2, NO_x, particulate matter, CFCs, VOCs, etc.).

Due to the high cost involved, small and medium-sized companies do not usually take a direct measurement of the mass of air emissions into consideration. In the context of climate change, it has become increasingly desirable to convert energy consumption to carbon emissions per unit of production. The fuels consumed can be converted to carbon emissions using the coefficients in Table 3. Carbon emissions will therefore change both with changes in energy efficiency and with changes in fuel type (UN Sustainable Development 2001).

4.9 Economic Indicators Objective

The economic dimension of sustainability concerns the impacts of the company on the economic well-being of its stakeholders and on economic systems at the local, national and global levels. Economic performance encompasses all aspects of economic interactions, including the traditional measures used in financial accounting, as well as intangible assets that do not systematically appear in financial statements (GRI 2002). The economic indicators of the case company are presented in Table 1.

5 Conclusions

Sustainable development is becoming increasingly important for industry. Recently some companies have begun introducing a sustainability assessment to be able to supervise the actual status of products and operations with respect to sustainable production. This paper focused on sustainable production and summarizing indicators, which could be used as strategic metrics for assessing sustainability level of the company and for identifying better options for the future. They compress a large amount of information into a format that is easier to manipulate, compare and understand. The proposed indicators are focused on environmental aspects of sustainable production. However, to achieve the sustainable production, a company should incorporate social and economic indicators as well. Therefore, social and economic indicators are also included in the indicator framework.

Most of the indicators included can be applied across industry, but they are not aimed at being uniformly applicable to all sectors. According to the flows in the manufacture, the environmental indicators are divided into input and output indicators and they are based on commonly measured environmental aspects of sustainable production (energy use, materials use, water consumption, products, wastes, and air emissions) covering key global issues.

One of the possible weaknesses of the developed framework could be in the subjectivity related to the choice of decision-makers in what to measure. However, any set of indicators is going to be affected by the decision-makers. Also, it is very hard to determine which indicators are effective since the same indicator may be effective at one company and ineffective at another. Therefore, each indicator has to be considered on an individual basis to reflect specific characteristics of different companies.

It is not expected that indicators of sustainable production alone can change the current production pattern. To achieve sustainable production, stimulation by various parties (national regulators, pressure groups, suppliers, consumers, employees, media, trade associations, etc.) will be needed. However, indicators of sustainable production can provide companies with assessment metrics to determine their actual situation with respect to a sustainable production, to raise their awareness and to set their goals.

References

Afgan NH, Carvalho MG, Hovanov NV (2000) Energy system assessment with sustainability indicators. Energy Policy 28: 604

AIChE (2002), Center for Waste Reduction Technologies (CWRT), focus area: sustainability metrics, http://www.aiche.org/cwrt/project/sustain.htm

Azapagic A, Perdan S (2000) Indicators of sustainable development for industry: A general framework. Trans IChemE 78B: 244–246

Brundtland Commission (1987) Our common future - from one earth to one world. Oxford University Press, Oxford

Callens I, Tyteca D (1999) Towards indicators of sustainable development for firms: a productive efficiency perspective. Ecol Econ 28: 43–44

CEFIC (2002), the European Chemical Industry Council. Chemistry - Europe and the future. http://www.cefic.be/allcheme/index.htm

Changing views on change: participatory approaches to monitoring the environment (1998) SARL Discussion Paper 2 (July)

Cornelissen AMG et al. (2001) Assessment of the contribution of sustainability indicators to sustainable development: a novel approach using fuzzy set theory. Agr Ecosyst Environ 86:183

FEM and FEA (1997) A guide to corporate environmental indicators. Federal Environment Ministry, Bonn, and Federal Environmental Agency, Berlin

Finnish Environment (2000) Finland's indicators for sustainable development. http://www.vyh.fi/eng/environ/sustdev/indicat/inds2000.htm

Gallopín G (1997) Indicators and their use: information for decision making, sustainability indicators. Report on the project on indicators of sustainable development. Wiley, Chichester

Geiz D, Kutzmark T (1998) Developing sustainable communities - the future is now. Public Manage Mag. International City/Country Management Association, Washington, D.C.

Goldblatt ME, Eble KS, Feathers JE (1993) Zero discharge: what, why, and how. Chem Eng Prog 89(4): 22–27

GRI – Global Reporting Initiative 2002. Sustainability Reporting Guidelines. http://www.globalreporting.org/

IChemE (2003) Institution of Chemical Engineers. The sustainability metrics: sustainable development progress metrics (recommended for use in the process industries). http://www.icheme.org/sustainability/metrics.pdf

Krotschak C, Narodoslawsky M (1996) The sustainable process index - a new dimension in ecological evaluation. Ecol Eng 6(4): 241

Lowell Center for Sustainable Production (1998) Sustainable production: a working definition. Informal Meeting of the Committee Members

NRTEE (2001) The National Roundtable on the Environment and the Economy. Eco-efficiency indicators: workbook (pdf). http://www.nrtee-trnee.ca/Publications/Eco-efficiency_Workbook/index.html

O'Brien C (1999) Sustainable production - a new paradigm for a new millennium. Int J Prod Econ 60–61: 1–7

SECRU – Scottish Executive Central Research Unit (2001) Environment group research programme research findings no. 13, sustainability indicators for waste, energy and travel for Scotland. Entec UK

Statistics Canada (1989) Quarterly report on energy supply-demand in Canada, 57-003, Ottawa

UN Sustainable Development (2001) Indicators of sustainable development: guidelines and methodologies. http://www.un.org/esa/sustdev/isd.htm

Veleva V, Ellenbecker M (2001) Indicators of sustainable production: framework and methodology. J Cleaner Prod 9: 519–549

Voinov AA (1997) Paradox of sustainability. Institute for Ecological Economics, Solomons, Mo. Zero Waste Alliance (2001) Zero waste - key to our future (the case for zero waste) http://www.zerowaste.org

Remote Sensing as a Tool for Achieving and Monitoring Progress Toward Sustainability

Gilbert L. Rochon[1], Chris J. Johannsen[2], David A. Landgrebe[3], Bernard A. Engel[4], Jonathan M. Harbor[5], Sarada Majumder[6] and Larry L. Biehl[7]

[1] Associate Vice President for Collaborative Research & Engagement, Information Technology at Purdue (ITaP) & Courtesy Professor, Department of Earth & Atmospheric Sciences, and Department of Agronomy, Purdue University, Young Graduate House, Room 420, 151 South Street, W. Lafayette, IN, USA 47906-3560. Tel: (765) 496-2274; fax: (765) 496-2275; E-mail address: rochon@purdue.edu
[2] Professor of Agronomy and Director, Laboratory for Applications of Remote Sensing (LARS), Purdue University
[3] Professor Emeritus of Electrical and Computer Engineering, Purdue University
[4] Professor of Agricultural and Biological Engineering & Director, Center for Advanced Applications of GIS (CAAGIS) Purdue University
[5] Acting Head and Professor of Earth & Atmospheric Sciences, Purdue University
[6] Senior Research Scientist, Information Technology at Purdue (ITaP), Purdue University
[7] Remote Sensing Specialist, Department of Agronomy, Purdue University

Abstract

Airborne and satellite remote sensing have enormous potential for facilitating and monitoring the dynamics of intergenerational natural resource management and built environment sustainability. Based on reviews of 1) the development and current state of the science for deploying remote sensing, and 2) capabilities and limitations of the array of archival and real-time remotely-sensed resources as tools for attaining and monitoring sustainability, we suggest several "Best Practices" for incorporating remote sensing within sustainability plans for selected rural and urban ecosystems. We present sustainability indices for case studies relating to sustainable forestry, agriculture, watersheds, and urbanizing areas, as well as a conceptual prototype for incorporating remotely sensed data within a general purpose decision support system that highlights the benefits of multi-temporal historical change detection in monitoring past and ongoing change, and in visualization of alternative future scenarios. Finally, we examine the sustainability of remote sensing technology, itself, with respect to economic pressures, public policy perturbations, systems resilience, biogenic factors and both inadvertent and deliberate anthropogenic impact.

1 Introduction

With the advent of higher resolution governmental and commercial satellite imagery and the declassification of certain military satellite legacy data, the potential benefit of satellite remote sensing for spatial and temporal analysis of smaller footprint anthropogenic phenomena, such as individuated built structures and man-made earth impoundments, has increased significantly. Moreover, the inclusion of hyperspectral sensors aboard orbital platforms has greatly facilitated species-specific vegetation identification and further development of spectral libraries, to include synthetic impervious surface materials in urban environments. These capabilities, particularly when combined with a renaissance in environmental modeling, watershed scale monitoring with airborne sensors and multidisciplinary team formation enabled by remote sensing, GIS and GPS technologies, have resulted in an unprecedented opportunity for the scientific community to confront the harbingers of unsustainable development, ecological instability, biodiversity depletion, and global climate change.

Remote Sensing can be viewed as a potentially equal opportunity technology, in the sense that the data are available at the same resolution and frequency of coverage for the underdeveloped countries as for the industrialized countries and for both urban and rural areas within countries. On the other hand, disparities in purchasing power for remotely sensed data and for suitable analytical software and hardware mirror the generic digital divide. Even for those with unfettered access to remote sensing technology, ranging from adequate resources to acquire and process the data, on an as needed basis, to the capability to ingest and analyze real-time remotely sensed data, a crucial limiting factor must be addressed before the true promise of these tools can be fully actualized in the service of attaining and subsequently maintaining sustainability.

Specifically, essential to the sustainable utility of remote sensing for terrestrial environmental monitoring is a widely espoused and accepted cross-national and intergenerational ethical standard, by virtue of which grossly disproportionate over-utilization and/or contamination of global natural resources, irrespective of their permanent or transient location within nation-state boundaries, must be mitigated and, in certain cases, entirely curtailed, so as to assure the potential of prosperity for inhabitants of underdeveloped regions and the potential of prosperity for future generations. The identification and measurement of major environmental problems by proximate and/or remote technologies are necessary but insufficient activities. A linkage between enlightened policy formulation and political will to implement best practices in natural resource management and appropriate environmental remediation is propaedeutic to attaining & safeguarding sustainability.

2 Facilitating and Monitoring

The moral philosophy of natural resource management and sustainable growth entails an evolving imperative to sustain the integrity of ecosystems for future generations, while providing immediate goods and services for an increasingly diverse public. This necessitates the need for broad level integrated assessments and monitoring for maintenance or development of resources, while meeting societal expectations within the limits of the land's ecological potential. In most cases, a first step toward sustainable development is through legal instruments like the National Environmental Protection Act (NEPA), which mandates ecological and environmental impact assessments for all projects with federal jurisdiction. Most common ecological assessments are narrowly focused for specific projects, called "tactical assessments," and they can be limited in spatial scale and address a limited set of issues. They have been criticized for failing the spirit of sustainable development because they do not account for full disclosure of cumulative effects and relation of the proposed action to other resources. This makes it increasingly important to develop in depth or "strategic assessments" at regional scales that would provide the context for localized project level assessments (Jensen and Bourgeron 2001; Treweek 1999; Ortolano and Shepherd 1995).

Strategic assessment of resources at a regional scale is challenging because ecosystems are dynamic and hierarchical systems that may not behave in a predictable manner. In this case, higher levels provide the context, or the environment, within which lower levels evolve. Heterogeneity at one level can be translated into homogeneity at a higher level. Ecosystems are not always in equilibrium. There might be discontinuities and unexpected changes because they are following constantly changing environmental conditions. Environmental conditions are often a result of, as well as the precursor to, societal changes and upheavals, especially in the developing world. Predictability and sensitivity to change in ecosystems varies over spatial and temporal scales (O'Neill and King 1998; Peterson et. al. 1998; Costanza et. al. 1993; Walters and Holling 1990). Dealing with this requires intensive monitoring and adaptive management practices which can adapt to the uncertainties in current and future ecological conditions and changes in societal needs, values, and expectations.

3 Remote sensing applications for sustainable development

Adequate monitoring of the ecosystem is thus the cornerstone of adaptive management for sustainable development. Remote sensing is the tool of choice in this regard because of the capacity of remote sensors to observe the Earth at regional and global scales from various vantage points. With a large number of remote sensors on board the various satellites orbiting the Earth, operated by various public and private agencies, the globe is being monitored at fairly high spatial, spectral and temporal resolution. The data derived from such monitoring are being used in a host of applications such as precision site specific farming, climate change, epidemiological prediction, controlled and uncontrolled urban development, intensity and extent of natural disasters like hurricanes, floods, forest fires, and erupting volcanoes. Often, such applications are interrelated, for example climate change prediction is often a precursor to predicting the growth and spread of epidemics. Monitoring of urban growth is a precursor to the planning and construction of facilities, as well as crafting growth management ordinances. Taken together, these applications form the basis of sustainable development, that of managing our ecosystem resources in concert with their natural potential, while managing and providing for societal expectations.

At the heart of many resource management systems are mathematical or statistical models coupled with a geographic information system (GIS). The models provide the basis for future predictions or scenarios based on current inputs and various assumptions about the rate and type of change, based on an understanding of societal needs and expectations of change in management practices. The GIS is used to develop a spatially indexed database, partial analysis, and as a display tool for the visualization and dissemination of the results of the model (Jensen et al. 2002). Such systems can be used in-house or deployed over the worldwide web for easy user access. An example of this is the Long-Term Hydrologic Impact Analysis (L-THIA) model developed at Purdue University, Department of Agricultural and Biological Engineering with the support of US EPA Region V. This model is linked to GIS and a web server that allows an easy user interface to determine the hydrologic impact of changes in land use management decisions (Pandey et. al. 2002). GIS technology has matured to the point where most government resource management agencies maintain their own GIS, data, hardware, software and staff that leverage the technology, in serious attempts to use it in public decision making. Integrating remotely sensed data with GIS is the key in using it as an input to resource management.

A successful monitoring and assessment program in this regard is the Gap Analysis Program (Scott et. al. 1993), which is being implemented throughout the coterminous United States. It seeks to identify areas of high biodiversity content that have been left out of existing biological reserves, and seeks to fill them by establishing new reserves, or changing land management practices. This program uses a combination of satellite images along with aerial photography, and field verification to establish areas of high biodiversity content. Other federal efforts, like the National Land Cover Dataset (NLCD), develop standardized land cover datasets from satellite images that can be used for a number of applications (Homer et. al. 2002; Vogelmann et. al. 1998). Remotely sensed data also lends itself to other derived datasets such as the normalized difference vegetation index (NDVI), and the modified soil adjusted vegetation index (MSAVI), which indicate areas of healthy vegetation. Stereoscopic images are those collected of the same area at different tilt angles, the relative displacement of features within such stereo pairs enable calculation of terrain elevations and object heights. Such stereoscopic images are readily available from the SPOT sensor which has off-nadir viewing capability. In the realm of high radiometric resolution, hyperspectral data can be used to identify specific features, and develop libraries for characteristic spectra of various minerals, plants, soils, rocks, and cultural features for future data interpretation. Such libraries are maintained by the United States Geological Survey (USGS), NASA's Jet Propulsion Laboratory, the U.S. Air Force, the U.S. Environmental Protection Agency (USEPA), and Johns Hopkins University. While such efforts by the major public agencies are useful in helping users adapt to remote sensing technology, the potential is not yet fulfilled in terms of interpreting and using remotely sensed data in public decision making.

4 Review: Development and current state of the science in remote sensing

Simply described, remote sensing is the process of obtaining information about any object from a remote vantage point. Remote sensing of the Earth implies the collection of information about the Earth from airborne multispectral scanners, either aboard airplanes, or satellites. Very often, this is in the form of "passive" sensing which records the radiant energy reflected and/or emitted from the earth at different spatial, spectral and temporal resolutions. Data from these sensors are widely used for monitoring the Earth, and developing sustainable resource management plans, for natural resource mapping applications, monitoring natural hazards, and urban growth among others. "Active sensors" are those which supply their own source of energy and record the energy reflected back to the sensor. Data from active sensors, such as synthetic aperture radar (SAR) are useful for mapping geologic structures, polar and sea ice types, surface drainage features, and archaeological features among others. Data collected by active sensors like SAR, are somewhat more difficult to interpret than passive sensors, frequently requiring radar interferometry techniques, but the results are rewarding for specific applications, such as those that need to distinguish between varying levels of moisture content, and variability in texture among surface features. A large number of remote sensors are currently on board the array of satellites operated by public and private agencies, orbiting the Earth and collecting data at various spatial, temporal and spectral resolutions. This section is not intended to provide an exhaustive review of every sensor, but rather a sampling of the various data types available. The reader is referred to the excellent text by Lillesand and Kiefer (2000), for additional details. Rochon (2002) and Rochon, Johannsen, et al. (2002) offer historical background and a look into the future for remote sensing technology and applications. Landgrebe (2003) offers practical methods for the optimal analysis of multispectral and hyperspectral image data.

Attempts at aerial photography were made soon after the invention of photographic techniques. The earliest existing aerial photograph was that over the city of Boston, taken from a balloon in 1860. Panchromatic film is the "standard" film type for aerial photography. Spectral sensitivities of panchromatic film extend over the ultraviolet and the visible part of the spectrum. In contrast black and white infra-red photographs are sensitive to the ultraviolet, visible, and near infra-red part of the spectrum. Color aerial photographs can be used to distinguish many more features than panchromatic ones, and color infra-red is widely used to detect vegetation presence and quality, since healthy vegetation reflects highly in the infra-red part of the spectrum (Lillesand and Kiefer 2000).

The USGS has a program called the National Aerial Photography Program (NAPP) to collect panchromatic, and color infrared photographs over the entire United States, cycling over areas about once a decade. The current cycle through 1997-2003, is in the process of completion. The earliest known space imaging programs included the CORONA, LANYARD and ARGON sensors developed by the U.S. military. The program, authorized by President Dwight Eisenhower in 1958, lasted from 1960 through 1972, and provides the earliest public record of land from space. The potential for CORONA images for the study of land use change detection is just now coming into full play, since these images were declassified in the mid nineties by President William Clinton (http://www.usgs.gov).

Some high visibility remotely sensed products are our weather maps from the Geostationary Operational Environmental Satellites (GOES) operated by the National Oceanic and Atmospheric Administration (NOAA). Other widely used passive sensor data have been derived from Landsat Thematic Mapper (TM), Enhanced Thematic Mapper (ETM), and Multispectral Scanner (MSS), the French Space Agency's (CNES) Systéme Probatoire pour L'Observation de la Terre (SPOT), the Advanced Very High Resolution Radiometer (AVHRR), *inter alia*. Since there are tradeoffs between the spatial, spectral, and temporal resolution of satellite data, data from different sensors are suitable for different purposes. For example, GOES data, which is acquired from a geostationary satellite has a low spatial resolution of 1km in the visible band, but high temporal resolution, the repeat frequency being limited only by the time it takes to scan and relay an image, making it ideal for short term weather mapping, and frequent large area analysis. By contrast, SPOT and Landsat are sun synchronous satellites which are overhead at each location on the Earth at approximately the same local time during each cycle. Both SPOT and Landsat have been extensively used for resource mapping and remote sensing applications because they have been in use for a long time, enabling multi temporal analysis, and they provide affordable data in an adequate number of spectral bands at reasonable spatial resolution.

The satellites in the Landsat series have been the flagship U.S. satellites operated by NASA, while SPOT is the French flagship satellite. Landsat images are widely used in the USA and have a number of spectral bands at 30m resolution, along with a thermal band at 60m. The Enhanced Thematic Mapper (ETM+) sensor aboard Landsat 7 collects a total of 8 bands, including one thermal band and one panchromatic band. The resolution of the panchromatic band is 15m. A unique feature of the ETM+ sensor is that images can be acquired in high gain or low gain states. The change in state is associated with calibration of the sensor to compensate for expected brightness conditions in the images. If the expected brightness of a scene is high, the sensor is switched to low gain and vice versa (http://ltpwww.gsfc.nasa.gov/IAS/handbook/handbook_toc.html).

The SPOT program was undertaken by the French Government in 1978, in a series of five satellite launches, SPOT 1 through 5. SPOT 1 through 4 have 10m resolution for the panchromatic band and 20m resolution for multispectral bands. The data from new SPOT 5 has a panchromatic band with 5m spatial resolution, and multispectral bands with 10m spatial resolution. Pointable optics of the SPOT system enable off-nadir imaging capability at an angle of twenty seven degrees providing higher temporal resolution than could otherwise be achieved, along with stereoscopic imaging. Interpretation of SPOT data is facilitated by combining the higher resolution panchromatic band with the multispectral bands, high geometric fidelity and multi-date and stereo imaging (http://www.spotimage.fr/spot5/spot5_eng.html).

Other widely used satellite data include Advanced very High Resolution Radiometer (AVHRR), and Moderate Resolution Imaging Spectro-Radiometer (MODIS). AVHRR data are obtained daily from NOAA satellites at a spatial resolution of 1.1 km, and along with GOES are widely used in large area mapping, calculating indices such as the NDVI on a world wide basis, as well as regional climate change studies. There are also a number of remote sensing systems oriented to ocean remote sensing, such as the Sea-viewing Wide-Field-of-View Sensor (SeaWiFS). This is a joint venture of NASA and the private industry and specifically designed for the study of ocean biogeochemistry. The highest spatial resolution of 1m, and 0.6m, in a number of different spectral bands can be obtained from the IKONOS-2 and QuickBird satellites, respectively, which are operated by private companies. As expected, this data are more expensive than the satellites operated by public agencies.

In addition to satellite images, infra-red photography is obtained from digital and film system cameras aboard aircrafts for detailed large scale small area analysis of specific phenomenon, such as mapping of wetlands. Hyperspectral data, collected in a large number of fine spectral bands (usually >100 bands) allows for detailed identification of materials that cannot be distinguished from coarser spectral resolution of satellite images. It allows for the identification of characteristic spectra of various minerals, specific vegetation, and man-made materials. This makes it ideal for use in a number of specialized applications, such as pollution monitoring.

5 Best Practices for incorporating remote sensing data in sustainable development plans

Given the need for developing sustainable management plans for natural resource utilization, and the cornucopia of available data, we feel that there is a greater need for education and awareness of how to incorporate and integrate remotely sensed data into sustainable development plans. The list here is a start, and is not meant to be exhaustive.

- In large area studies, use a combination of multistage and multi-temporal data. Datasets with low spatial resolution and high temporal resolution like AVHRR can be used for monitoring conditions over large areas, while high resolution data like SPOT or IKONOS can be used to zero in on specific areas. This would provide cost optimization as well. For detailed spectral resolution, NASA's MODIS data with thirty six spectral bands might be sufficient. If not, hyperspectral data needs to be collected, usually by designing flyovers for the area of interest. While analyzing

hyperspectral data, use or add to the libraries of characteristic spectral signatures (or responses) maintained by various agencies.
- For data classification or identification of features, use a combination of field verification and ancillary data. While using either, awareness is needed of the scale at which one is working. If data have high spatial resolution, detailed feature extraction is possible, and field data should include all features that could possibly be resolved within the scene. This is especially true for any interpretation methods using supervised classification or any combination thereof.
- After any classification, or feature identification, a detailed accuracy assessment should be done, and requisite statistics for accuracy estimates should be calculated. It is important that field data used in accuracy assessment be different from field data used in training and feature extraction. Otherwise this will yield a biased accuracy assessment. Or if there just is not enough field data available, use a method like the leave-one-out technique or bootstrapping, to get as close to an unbiased estimate as possible and document the method used.
- Finally, while preparing sustainable natural resource plans, care must be taken to evaluate the land use and resource use history of the region, and any indigenous methods or landscape detail and practices that might provide clues to original use of land and sustainability. If such signs are conducive to long term sustainable development, every effort should be made to incorporate them into management plans.

6 Sustainability indices for various resources

This section provides a brief overview of current and potential sustainability indices that can be developed from remotely sensed data, or used in conjunction with remotely sensed data. It has been mentioned before that remote sensing technology is an "equal opportunity" technology, and this is especially true in case of environmental, social and demographic monitoring in developing countries, in far flung areas where cash strapped governments are unable to keep track of the exploitation of their natural and human resources. Efforts to integrate economic development and environmental planning are still in the early stages for most developing countries while environmental degradation and poverty at a global scale is worsening (National Research Council 2002).

Remote sensing data of high temporal resolution, especially in conjunction with GIS data layers like elevation, slope, demographics, can provide predictions of the occurrence of disease vectors and estimates of the number and characteristics of people likely to be affected. Remotely sensed data in conjunction with climate change models can provide estimates of rainfall, temperature, and weather patterns, factors which affect crop yields, and can be used for foreseeing and planning for potential calamities like floods, droughts, and famines. Natural hazards like volcanic eruptions, escalating forest fires, hurricanes or typhoons, can be predicted by using remotely sensed data and used for leading people out of harm's way.

From a natural resource protection point of view, high temporal resolution remote sensing can be used to monitor legal or illegal clear cutting, create comparisons or ratios between old growth and new growth forests, identify various species of trees and deforestation, monitor legal or illegal mining or strip mining operations, as well as legal or illegal human settlements and movements across borders. Remotely sensed data along with hyperspectral data can track characteristic spectra of specific plants and chemicals which can be used to develop illicit drugs. High resolution imagery can be used to track endangered species, monitor for signs of poaching, and illegal water use. Recent developments in ocean sensing, for example data from SeaWIFs, can be used to track ocean currents and, as such, direct and manage fishing fleets for sustainable harvesting.

Indicators from remotely sensed data can also be used to chart social progress and or dislocations, especially after natural disasters. Growth in sprawl and slums indicating recent migrations from rural areas can be mapped at high temporal resolution. The land use impacts like changes in proportion of impervious area to vegetated area, quality and quantity of riparian zone vegetation, bank erosion and siltation of streams and rivers, can also be monitored.

7 Incorporating remotely sensed data with a general purpose decision support system

The prototype Temporal Analysis, Reconnaissance and Decision Integration System (TARDIS) was developed in response to the observation that many decisions affecting natural resource management were apparently made in ignorance as to what had previously transpired within a region or site of interest as well as with innocence or even insouciance as to future outcomes of any specific proposed disturbance. Accordingly, the TARDIS concept enables the decision maker to have the benefit of virtual hindsight, through visual exploration of the archive of remotely sensed data, infusion of contemporary remotely sensed and surficial data, and virtual foresight, through immersive visualization of alternative future scenarios.

Specifically, the "hindsight" module would allow the decision maker to investigate the satellite archive for the area of interest (AOI) from the vantage point of a virtual time machine traversing a three dimensional terrain with temporally sequenced actual synoptic satellite imagery and on-the-fly change detection for features of interest (FOI). The "contemporary module" would facilitate integration of near-present or real-time data both from *in situ* measurements and satellite sensors, as well as data fusion from other relevant sources (e.g. demographic data from census TIGER files). The "foresight" module generates virtual worlds, enabling visualization of alternative future scenarios and an array of what-if immersive simulations. Finally, with the "decision" module, one has the choice of rule-based or quantitative decision support models to assist in identifying the optimal sustainable path or, at least, eschewing practices, interventions or paths that lead to inherently unsustainable consequences or negative impact on human health and well being, on biodiversity, or on intergenerational resource availability.

8 Sustainability of remote sensing technology

Given that attainment and maintenance of a sustainable biosphere is fundamental to the survivability of plants, animals and the human species, it follows that the adoption of any array of technologies for the purpose of accelerating and monitoring biospheric sustainability must perforce take account of the robustness and reliability of such technological array. Accordingly, the remote sensing tool array (i.e. airborne and orbital sensors) is no exception. Six constraining factors have been identified that can potentially impinge upon the general acceptance by the scientific community of remote sensing technology as a trustworthy facilitative device for accelerating and monitoring sustainability: economic pressures, public policy perturbations, systems resilience, biogenic factors, inadvertent anthropogenic impact and deliberate anthropogenic impact.

8.1 Economic Pressures

Access to certain older data archives (e.g. AVHRR, Landsat MSS) has been rendered more affordable to researchers due to price reductions and the declassification and subsequent distribution of data from formerly secret military sensors such as, CORONA, LANYARD, and ARGON for nominal fees has greatly improved the potential for affordable multi-temporal high resolution change detection. On the other hand, current commercial pricing structures for certain higher resolution data are beyond the reach for many investigators, whether in the industrialized or underdeveloped countries. Current stock market trends have also impacted budgetary allocations from the private industry, government agencies (federal, state and local) and private foundations to academic institutions, limiting resources for remotely sense data acquisition, *inter alia*.

8.2 Public Policy Perturbations

The policy of privatization during the Reagan administration had an immediate and devastating impact upon research utilizing the Landsat MSS and TM archives. Not only did prices soar beyond the reach of many scientists; but also the decision to discontinue the long-standing MSS program was a severe blow to multi-temporal investigations. Similarly, the decision to deploy the hyperspectral sensor, Hyperion, only on a short-term experimental basis, rather then as a sustainable program of hyperspectral monitoring will deprive researchers of a major opportunity to incorporate hyperspectral data as a regular component of analysis. Reliance upon airborne hyperspectral sensors (e.g. Airborne Visible/Infrared Imaging Spectrometer (AVIRIS); Compact Airborne Spectrographic Imager (CASI) for long-term ecological monitoring or for very large areas of interest would be prohibitive due to both scheduling and cost. For purposes of georeferencing, the decision by then President William Clinton to abandon selective availability (SA), the deliberate distortion of GPS signal, was a significant advantage to academic researchers, since it obviated the necessity to invest in both a base station and rover for post correction. The unaltered availability of the Russian GLONAS geopositioning data and anticipated European Space Agency's GALILEO geopositioning data may have accelerated the demise of SA. Yet the kind of security concerns which led to the creation of SA may likewise now imperil critical research, predicated on the continued availability of high resolution satellite data (e.g. for plate tectonics, urban studies, species-specific vegetation studies). Also imperiled would be time-critical civilian research, (e.g. for biogenic disaster intervention, such as emergency response to hurricanes, earthquakes, floods, forest fires and for responses to anthropogenic disasters like oil and chemical spills as well as acts of terror). This neo-SA (neo selective availability) phenomenon is apparent in the twenty four hour delay, financed by the Department of Defense, in distribution of the commercial Quickbird data, which will mitigate the utility of Quickbird data for such time-sensitive civilian disaster response applications.

8.3 Systems Resilience

An ideal remote sensing system would record a uniform energy source without any atmospheric interference with a "supersensor" that is equally sensitive to all wavelengths at high spatial resolution and be accurate and economical to operate. However, the ideal system and the "supersensor" do not exist, and real world systems are limited by inefficient energy emission, atmospheric interactions and limitations on the spatial and spectral sensitivity of the sensor. Not all remote sensing systems are designed resilient and, as such, some fail during launch or in orbit. Such was the case with LandSat 6 in 1993, with the Enhanced Thematic Mapper (ETM) on board. ETM was designed to collect a panchromatic band with 15m spatial resolution in addition to the multispectral bands already present in data from LandSat 4 and 5. However, LandSat 6 had a launch failure in 1993, and the 15m panchromatic band was implemented in LandSat 7 six years later in 1999. The Advanced Earth Observing Satellite (ADEOS) of Japan suffered structural damage to its solar array in 1996, seven months after launch. At that time the JERS-1 also launched by Japan was operational. However, the JERS was discontinued in the late nineties, subsequent to on-board storage capacity failure, and Japan is launching a new satellite ADEOS II in December 2002, to extend the mission of ADEOS. The erstwhile Soviet Union was the first country to operate an earth observing radar system, the Almaz-1. This returned to earth in 1992, and although future Almaz missions had been planned, the Soviet Union was dissolved before these could be executed. On the commercial side, DigitalGlobe, the company which manages the Quickbird satellite offering 0.6m resolution data, saw its fortunes fall and its existence at stake after the launch failure of their EarlyBird satellite, the precursor of Quickbird, in 1998. This shows that strategic planning and design, as well as financial and political stability and commitment, is necessary for long term efficient functioning of satellite systems.

8.4 Biogenic Factors

Cloud cover continues to present problems for region-specific research, for small footprint investigations and for time-critical interventions. While decadal cloud masking has been sufficient for large scale course resolution projects, it may mask small scale anomalies vital to some studies. The cloud penetration capability of radar sensors could theoretically overcome this problem; however, a radar image does not offer the same input to research projects dependant upon multispectral or hyperspectral data. Moreover, long term ecological research (LTER) projects can ill afford dependence upon sporadic shuttle imaging radar (SIR A, B & C) missions, ill fated radar missions (e.g. Russia's Almaz and Japan's JERS) or the expenses associated with RADARSAT I data acquisition. Other biogenic factors include catastrophic volcanic eruptions, such as the Mt. Pinatubo eruption, which created an unrecoverable gap in the high resolution picture transmission (HRPT) satellite data archive.

8.5 Inadvertent and Deliberate Anthropogenic Impact

It is impossible to prevent all forms of human error. That having been said, a greater proliferation of real-time multi-sensor remote sensing receiving stations can offer redundancy and an additional level of comfort, resulting from coverage overlap and data sharing agreements (Rochon et al. 2002). In the era post September 11, 2001, the unthinkable suddenly emerges as that which must be taken into account. The vulnerability of earth observing satellites, their sensors, data transmissions, imagery archives and processing facilities must be assessed and appropriate countermeasures instituted, so as to safeguard against an array of potential willful attacks, whether from mischievous hackers or from cyber terrorists. Countermeasures could include a range of procedures such as archive redundancy, tamper proofing, magnetic pulse shielding, and security protocols designed to protect the remote sensing assets, without obstructing application development, pedagogy and legitimate scientific research.

Taking all these constraints into account, what appears to be needed is the existence of a data source that has adequate spectral and spatial content, is reliable, is predictable, and is of zero or low cost to the (operational) user. The dominate problem is not science but that no one can count on being able to get the data they need from year to year or even from week to week. There is not an acceptable aircraft or satellite system that meets these needs. Without a reliable data source, no operational agency is going to commit to remote sensing technology in any serious way (Miller et al. 2002). At some point, aspects of this technology must begin to pass from being a scientific play-toy to a well-used tool that consistently returns some measurable benefit to the community (Landgrebe 2002).

References

Costanza R, Wainger L, Folke C, Maler K (1993) *Modeling complex ecological economic systems*. Bioscience 43: 545 – 555

Homer CG, Huang C, Yang L, Wylie B (2002) Development of a circa 2000 landcover database for the United States. *Proceedings of the American Society of Photogrammetry and Remote Sensing Annual Conference*, April 2002. Washington, DC.

Jensen JR, Botchwey K, Braennan-Galvin E, Johannsen CJ, Juma C, Mabogunje AL, Balstad-Miller R, Price KP, Rening PAC, Skole DL, Stancioff A, Taylor DRF (2002) Down to Earth: Geographic Information for Sustainable Development in Africa. National Research Council, Washington, DC. 155 pp.

Jensen ME, Bourgeron PS (eds.) (2001) *A guidebook for integrated ecological assessments*. Springer Verlag: New York, NY

Landgrebe DA (2002) Personal Communication. Purdue University

Landgrebe DA (2003) (forthcoming) *Signal Theory in Multispectral Remote Sensing*. Wiley-Interscience: New York, NY

Lillesand TM, Kiefer RW (2000) *Remote sensing and Image Interpretation*. John Wiley and Sons: New York, NY

Miller RB, Abbott MR, Harding LW, Jensen JR, Johannsen CJ, MaCauley M, MacDonald JS, Pearlman JS (2002) "Toward New Partnerships: Government, the Private Sector and Earth Science Research." Space Studies Board, National Research Council, 81 pp.

National Research Council (2002) Down to Earth: Geographic Information for Sustainable Development in Africa. The National Academies Press, Washington, D.C.

O'Neill RV, King AW (1998) Homage to St. Michael; or why are there so many books on scale? In Peterson DL, Parker VT (eds.) *Ecological scale: theory and applications*. Columbia University Press: New York, NY

Ortolano L, Shepherd A (1995) Environmental impact assessment. In Vanclay F, Bronstein DA (eds.), *Environmental and Social Impact Assessment*. John Wiley and Sons: Chichester, UK

Pandey S, Harbor J, Choi JY, Engel B (2002) Internet based planning decision support system. Presented at the 40th *Annual National Conference and Exposition of the Urban and Regional Information Systems Association (URISA)*, Chicago, Illinois.

Peterson G, Allen CR, Holling CS (1998) Ecological resilience, biodiversity and scale. *Ecosystems* 1: 6-18

Rochon GL (2002) Next generation remote sensing. Presented at the 7th International Seminar and Workshop on GIS, Seoul, Korea. Nov 8th 2002

Rochon GL, Johannsen C, Landgrebe D, Engel B, Harbor J, Majumder S, Biehl L, (2002) The Evolution of remote sensing and the Purdue Terrestrial Observatory. Presented at Research in Indiana, Supercomputing 2002, Baltimore, MD: Nov $18^{th} – 24^{th}$

Scott JM, Davis F, Csuti B, et. al. (1993) Gap Analysis: A Geographic Approach to Protection of Biological Diversity. Wildlife Monographs No. 123. Supplement to the Journal of Wildlife Management, 57(1)

Treweek J (1999) *Ecological impact assessment*. Blackwell: Oxford, UK

Vogelmann JE, Sohl T, Howard SM, Shaw DM. (1998) Regional land cover characterization using Landsat Thematic Mapper data and ancillary data sources. *Environmental Monitoring and Assessment*, 51: 415-428

Walters CJ, Holling CS (1990) Large scale management experiments and learning by doing. *Ecology* 71(6): 2060-2068

Summary of Panel Discussions

Kristan Cockerill

Secretary, International Society for Industrial Ecology, P.O. Box 7731, Albuquerque, NM 87194, E-mail address: kristan5@unm.edu

The 23 formal presentations at the NATO workshop, Technological Choices for Sustainability, ranged in topic from thermodynamics to pedagogy to alternative food sources and provided an incredibly rich experience for all attendees. The wealth of information and diversity of perspective were wonderfully expansive. Attempts to summarize the information presented or to describe the milieu of the meeting face a challenge similar to attempts to define "sustainability." There was simply too much to encapsulate the concepts and interactions with a pithy statement. Yet, on the final day, as attendees gathered to provide last comments and suggestions, the discussion had a single recurring theme: communication. While there were certainly specific suggestions related to methods and tools, the comments did not dwell on technology, although this was the primary concept in the meeting's title. The discussion rather, seemed more in touch with the second word in the title, choices.

Meeting attendees fully recognized that we are making and we must make choices, often difficult and painful choices. To ensure that the choices we make are most suitable and worth any pain suffered, we need to improve communication. There was a strong sense that we need better communication in multiple venues – among scientific disciplines, among datasets, between science and technology emphases, and with broader society. As someone who has dedicated her career to studying and practicing communication in diverse manifestations, I left Maribor extremely encouraged to hear so many scientists and engineers raise communication as a crucial element within our efforts to make choices about technology and to provide ideas for sustainability.

I, however, have left numerous meetings encouraged by what I heard. The proof of having learned something will be in applying ideas discussed in Maribor and beginning to communicate by maintaining contact with our new friends and colleagues. Putting communication principles into practice is difficult and the reality is that we return to busy lives and it is much easier to remain focused on what is familiar than to venture outside our comfort zone. To truly begin to communicate involves taking risks. It can be a frustrating process and it may not provide the resolution that many seek. In fact, initial efforts to improve communication often increase the level of controversy. To successfully communicate across disciplines, as well as with the general public and decision-makers, it takes time to develop trust, to develop respect for another way of doing research or interpreting and applying new information.

Additionally, there is one form of communication that is rarely considered and is perhaps most crucial to the topic of sustainability and perhaps the most difficult to pursue. Communicating with others is often less taxing than efforts to truly communicate within our own disciplines, to communicate with ourselves. Making choices with sustainability in mind may require serious self-examination as to what our own disciplines emphasize and why that is so. It is entirely too easy to accept our own disciplines as "right" and "necessary" and to view other disciplines as "wrong" or "unnecessary" and to never question the assumptions within our field. With such attitudes, communication across disciplines becomes untenable and hence, communication throughout society about issues as complex as sustainability remains an elusive goal. Expanding the idea of self-communication to include examining our communities (note the same root word – common – in communication and community), and to examining our cultures, increases the difficulty, but also increases the opportunities.

Communication is clearly perceived to be crucial to our efforts to make technological choices and to pursue the idea of sustainability. Advancing communication at all levels is a worthy goal and will require work, but the dividends can be quite substantial. Perhaps we need to begin the communication process from within our own disciplinary paradigm (yes, all disciplines have paradigms!) and from within our broader social/cultural paradigm. Perhaps we need to be asking why we have particular technologies and why we have applied scientific information in particular ways and whether these are appropriate choices for sustainability. Perhaps a more critical examination from within our disciplines and our cultures will ease efforts to begin communicating across disciplines and across cultures (both C.P. Snow's ideas of "cultures" as well as cultures more typically defined with geographic boundaries). Perhaps we can begin to focus less on our differences and more on what we have in common. Perhaps we will then more readily identify new connections and hence new alternatives, new possibilities, new opportunities. Perhaps then we will make the best choices about technology and find sustainable paths for an increasingly global society.

In an attempt to get us started on communicating our way to sustainability, the final day featured presentations to remind us of all we'd heard. Drs. David Shonnard, Gilbert Rochon, and Roland Clift presented summary thoughts on the workshop's primary topics: Issues in Sustainability, Sustainable Pathways, and Sustainable Metrics. Consistent with my struggle to condense the meeting in a meaningful way, they noted the difficulty in truly capturing the breadth and depth of the material covered and the issues raised in the various sessions. The presentations and discussions had ranged from raising broad philosophical questions to identifying very specific research results and/or research needs. These final summaries did, however, provide numerous key ideas that participants agreed warranted further discussion and/or research. Following each overview, Dr. Subhas Sikdar of the US Environmental Protection Agency moderated the flow of comments from participants that raised more relevant points. Following are the further summarized **key ideas and issues**:

- We need to better understand how and why people change and how they make decisions. What are the connections between scientific knowledge and adaptation in the social realm?
- The body of knowledge on public perception has developed tremendously in recent decades and what we often see is that there are very rational reasons why people behave as they do, such as not moving in directions that scientists keep pointing. From the public side, they are saying, why do these scientists insist on remaining in their ivory tower and not work on what really concerns us?
- We need to consider the Churchill quote, "Scientists should be on tap, but not on top." Scientific inputs are essential to an adequate understanding of sustainability. The decisions societies make on their journey to sustainability, however, will arrived at by compromising among societal values, economic issues governing the choices, political factors and scientific facts.
- A fundamental assumption in economics is that people follow a certain goal, for instance maximum personal material benefit. Research has shown that this is not usually the case because people cannot completely analyze complex situations. Instead, people follow the rules of their experience. The sustainability problem is a complex situation. Therefore, communication to the public, especially explaining scientific results and problems with sustainability, should start from "our" scientific experience and should try to rely on the public's experience.
- We must 'institutionalize' sustainability, meaning that the goal of sustainability must be encoded in the institutions of our society, not simply in the way in which business is done, or the platform of a particular politician.
- One issue related to public reaction and sustainability is that we are talking about decisions that do not have immediate impacts upon the people making the decisions, the impacts will be felt in the future. People want proof before acting, but there is tremendous difficulty in making predictions, even with models. So, we have a communication problem. We therefore end up in crisis-driven or competition-driven change. We react when we begin to see the effects of something.
- Perhaps in communicating we need to take a positive approach, not a disaster-based approach. Telling "disaster stories" is no longer effective because there has not been a "disaster" to prove the stories.
- A possible positive approach is to begin saying that sustainability is the path we must take to enhance our standard of living. This is language being used in the ex-communistic countries. If we could show that sustainability is the choice for society and there is actually no other way, perhaps both the politicians and the public will pay special attention to this and open the door to introducing sustainability issues in a broader way.
- Catastrophes can be viewed as opportunities. Our politicians and institutions must examine the issues at the root of major setbacks and warnings as they occur, whether from the environment or politics or society.

Moreover, we have a tendency to offer simple analyses and solutions to complicated problems out of expediency or politics. We need to try and overcome this tendency.
- Sustainability is ultimately about human values. Therefore, we must involve more people in developing the science of sustainability. This includes participation from vested interests in developing sustainability goals, as well as the questions that scientists must address in order to contribute to a sustainable society.
- There seems to be confusion between indicators and metrics. A useful distinction is that indicators indicate direction and metrics indicate movement. Metrics are things which can be used locally, subjectively. Metrics can only support qualitative reasoning and subjective judgement. Therefore, if we pay too close an attention to metrics as a basis, we can easily fool ourselves. Additionally, the way in which we might use metrics will differ according to where we are.
- Several speakers emphasized the importance of security. Security provides an example of metrics being different according to where you are. An example is in the debate over genetically modified organisms being used to abate famine. If the indicator is being properly fed, the metric depends on where you are. If you are starving, any food is good. If you are reasonably fed, but a bit undernourished you start to worry about the nutritional value of food. If you are well fed, then quality and risk issues surface and if you are getting too much food, this is bad for you.
- We do have completely different metrics in various places. The concept of sustainability in this meeting is not relevant to some poor Bangladesh family. One important message from this workshop is to recommend the development of institutional and technical frameworks for international discussions and to prepare international agreements.
- The economic and environmental dimensions of sustainability are in reasonably good shape and there is a considerable measure of convergence of the two. Where we have more difficulty is in the other dimension, the social one.
- Technology has a central role in improving resource efficiency, but as engineers, scientists, we also have roles as citizens. We can contribute technological improvements, but we also have to inject a measure of physical reality into discussions. However, all the technological developments in the world cannot compensate for human greed. You can introduce a technological development but people have ways of getting around it. For example, in the UK the fuel efficiency of cars on the road has barely changed for 50 years. The reason is that every time there's an improvement in engine design people compensate by buying bigger cars.
- As scientists, it is a challenge to communicate the concepts and we need to use indicators and metrics. We also see that even among scientists, we have different ways of understanding sustainability. In this group we have agreed that sustainability is about three areas: economics, social and environmental analysis. But if we talk to an economist, then sustainabil-

ity may be something else. So, it is important to have a clear way of communicating. When we communicate to the public, we need to have simplified indicators and this is an important challenge.
- An old joke about three laws of thermodynamics is relevant to the sustainability discussion. The first law being you cannot win, the second, you cannot get even and third, you cannot quit the game. So, we seem to be in a sort of no win situation, but we are in the middle of the game and cannot quit.
- Citing a Hungarian writer, the common fallacy of all prophets is that people are listening when you talk to them. There is a basic fallacy in the ways we try to communicate. Explaining, interpreting - it somehow doesn't stick. It is imperative that we find ways of communicating which will be somehow, someway absorbed by those to whom it is directed.
- Relying on biogenic process for carbon sequestration may be insufficient in the face of tepid response from industrialized countries and the perceived need for rapid industrialization, urbanization and deforestation within Africa, Asia and Latin American and therefore continued research on more proactive creation of carbon sinks may be necessary.
- New and dynamic modeling applications for sustainability as a research tool and as mechanisms for decision support may be extremely critical.
- Vulnerability risk assessment is vital, especially given the disproportionate impact of biogenic and anthropogenic disasters on the poor.
- Sustainability indices must transcend narrow disciplined bound perspectives to include a more holistic inter-disciplinary view is key to collaborative research.
- The next generation of environmental investigators must be trained differently from our own generation. Curricula will need to include emerging fields, such as green engineering and industrial ecology and will also need to include a sense of ethics that is both intergenerational and intercontinental. It will also need to be intra-regional to avoid creating disparities within countries that can give rise to ideological isolation of particular groups.
- We need to add architecture to the list of educational programs that have a large role in sustainability. Urban design warrants the same level of attention granted to agricultural lands, rangelands and forests.
- We must address the typical university system in which each discipline is tightly bound within itself and fluidity on a horizontal scale is virtually non-existent. Without interdisciplinary approaches, it is too easy to abrogate responsibility and to put the burden for collective action and moving from theory to practices on others. All disciplines have a responsibility to contribute to sustainability. One thought is that GIS helps promote interdisciplinary efforts and could become the "lingua franca" of the scientific community.

Closing Discussion

Continuing the inclusive approach used throughout the meeting, the final session was an open-ended discussion. Dr. Ravi Jain from the University of the Pacific in California moderated the session and opened by noting that there are two ethos of the scientific community: 1) sharing ideas and 2) participating in organized skepticism that art relevant to this session. With that in mind, he and the organizers requested that attendees provide comments, **recommendations and suggestions for future research** and conference themes. Following are the specific ideas generated:

- This meeting did not have a strong focus on uncertainty and risk issues. Human judgment and decision-making under uncertainty should be considered. Scientists are the first to know that perfect prediction is not possible. For future efforts, we need to link sustainability metrics with risk metrics and study uncertainty relative to sustainability.
- Uncertainty is an important topic and is one of the key links among science, public perception, politics and policy.
- We need to add indeterminacy (meaning we cannot predict) to ideas about uncertainty and then add incommensurate to identify where we are not able to make connections in a quantitative way because there are no sensible measures (or even surrogates) to use.
- Politicians and decision makers have to listen to the scientists to understand the limits of scientific knowledge. Where we know we cannot generate certainty or reduce uncertainty, the absence of certainty cannot be used by decision-makers as an excuse for inaction.
- Historical data archives exist but they are not connected. Being able to bring long-term datasets together to make connections is a key question/project to raise for future efforts.
- Sustainable development is a big frame that includes diversity and how to manage diversity. This is tied to knowledge, education and disseminating knowledge. We need to know more about local knowledge, especially as it relates to food sources.
- The idea of sustainability is multi-dimensional and cannot be treated as uni-dimensional for convenience. We need all three legs (environmental, economic, social) to come together and communicate.
- We should focus on methods and tools for communication. We need strong educational systems (K-16) to put ideas from the meeting into the system. The MS program being developed in Iceland is a good example – requires 10 core courses across disciplines, then students specialize, and all students are encouraged to study abroad.
- We need to develop case studies to take past data and tools and methods and assess why some cases are sustainable and others are not sustainable.
- Case studies are a good idea. We need to start with all parameters and all "enterprises" to identify all data metrics. Then, we can see if there are

similarities among enterprises. If they are organized, the metrics could benefit all and we can do these types of comparisons for all nations.
- The modeling efforts presented are impressive and we can do social models and then begin to link ecosystem models with industrial system models and social models.
- We need to focus on how to communicate ideas of sustainability to society, how to communicate this with indicators and how to create indicators with a social side.
- We have a lack of information about soil sustainability and ecosystem models need to integrate all activities.
- Future meetings and research should look at defining non-sustainability and defining points where non-sustainability becomes irreversible.
- We need environmental communication research. It is difficult to convey messages, but there have been successful media efforts related to safety issues (e.g. wearing helmets, seat-belts, not smoking) and we need to find a specific example to give a specific message to the media.
- We need to assess beyond material need to mental issues. We need a paradigm shift.
- Framing the issue influences/dictates framing the solution. We need to widen the frame to include cultural aspects. We need to communicate with the public and remember that science is often perceived as ideology. Perhaps probability models could help address this.
- There must be efforts toward informing the public. Perhaps the best way to influence corporate action is to influence public opinion.
- There have been meanderings between questions of scientific nature and perhaps we need to return to the point about what questions to ask. The discussions about communication have focused on one-way communication – scientists to public – maybe scientists need to listen to the public as well.
- There is confusion about what energy is and how it influences sustainability issues. Researchers in thermodynamics need to work with other disciplines.
- The MS program in Iceland is similar to programs in the UK, which also include education about environmentalism and about working with the media. On a related point, using negative messages is failing to communicate. Maybe we should access tendencies toward altruism.
- Altruism may be linked to research in metrics and indicators. Perhaps we need to look at the interconnections in metrics and indicators rather than at the specific properties.
- Simplifications are problematic. The talks at the meeting were good, but still simplifications. Civilizations must be how they are, but they must develop recognizing sustainability issues. We must learn to ask, "am I happy?" not, "how much do I have?"
- Important research areas include:
    Consider system boundaries - because achieving eco-efficient industrial ecosystem may require inefficient individual pro-

- duction units; recycling between nations may require no recycling within a nation.
- Consider categorically non-sustainable activities
- Consider security as a window of opportunity for sustainability (security often overrides economics)
- Consider crises as opportunities for sustainability
- Explore new ideas for process synthesis and design tools in order to accomplish more important goals towards more sustainable industrial processes.
- Improve eco-efficiency of industry and overcome inertia and barriers to change. There is a need to go beyond current measures, to do something more cutting-edge, more innovative and socially useful.
- Investigate ways of helping industry to improve their practices, directing research studies to address the real needs of industry, promoting partnerships between industry and research centers, disseminating current research studies in the field of sustainability or find ways to communicate important research results.
- Develop more sustainable waste management techniques or to promote the use of the existing ones in a more sustainable way, by ensuring that a number of factors reinforce one another in a positive manner rather than working against each other.

As time was quickly waning and all attendees had an opportunity to speak, Dr. Jain closed the session with a comment about our "human predicament" and the need for both technology and education to contribute to our quest for sustainability.

Workshop organizer, Dr. Peter Glavič, added a summary remark noting that we need to refine our efforts toward framing the idea of sustainability; to increase cooperation between disciplines; to identify threads and explain phenomenon; and to develop education designed to transform. Many of these ideas are tied to communication and to metrics, which help make comparisons and assist communication. We need networking efforts with a multidisciplinary theme and an information system to transfer ideas. Related to that point, recommendations for using the Internet to facilitate ongoing interaction, to refine proposals and reach conclusions, were well received.

Subject Index

accumulation factor, 189
aerobic composting, 129
aerosol ambient data, 156
aerosol monitoring, 153
aggregate pollution projection, 292
anthropogenic impact, 426
assessment of competitiveness, 289
assumption of universality, 322

Baltic Sea, 319
behaviour, 14
biochemical processes
    sustainable features, 257
biodegradation, 134
biodiversity, 39
biogenic factors, 426
biological system, 40
Black Sea coastal sediments, 152
brain-drain, 313
    phenomenon, 313
Brundtland statement, 239
business impact assessment, 242

canning industry, 106
carbon capture, 171
carbon storage, 165
carrying capacity, 321, 323
chemical mass balance method, 152
clean wastewater technologies, 105
cleaner production, 92, 94, 141, 142, 216
    centre, 140
    projects, 141
cluster analysis, 149
CO_2 storage
    in leaky reservoirs, 166
COD removal, 111
communication, 430
composite indicator, 340
composting process, 130
computer-aided toolbox, 197
consumption surplus, 273
    index, 273
contemporary literature, 78
corporate social responsibility, 385

data, 339

collecting, 341
    set structure, 160
decision-making, 24
decomposition of flows, 187
degree of substitutability, 42
demand cost, 190
depletion time, 260
dirty industries, 297
disturbances, 40
domestication, 40

eco-efficiency, 216, 371, 374
ecological sustainability, 39
ecological sustainability index, 273
ecology, 38
economic aspects, 168
economic development competitiveness, 286
economic exchange index, 275
economic functions, 279
    consumption surplus, 279
    production surplus, 279
    survivability, 279
economic performance indicators, 373
economic pressures, 424
economic sustainability, 268
economics of leakage, 166
economy, 38
ecosphere, 258
ecosystem components, 38
ecosystems, 39
ecosystems, 39
educational programs, 81
efficiency, 277
    indicator, 263
    indices, 278
effluent, 111
electronic goods, 249
emissions
    atmospheric, 107
end-of-life products, 245
energy accumulation factor, 189
energy cost, 188
energy penalty, 171
energy use, 408
engineering education, 66
environmental accounts, 278

environmental assessment
  methodologies, 69
environmental impact, 362
  factors, 194
environmental management systems,
  141, 218
environmental monitoring, 416
environmental problems, 107
environmental science programs, 86
environmental services, 287
environmental sustainability, 84
environmentally-conscious design, 66
environmetric approaches, 148
environmetric strategies, 148
environmetrics, 148
ethical norms, 313
evaluating environmental hazards of
  chemicals, 68
evaporation assays, 121
exergy, 256, 258
  analysis, 256
  and sustainability, 258
  calculations, 256
exergy
  and sustainability, 258
exergy analysis, 257
exergy efficiency, 261
exports, 291
  factor content, 291, 292
  polution-intensive, 297

factor content, 293
fast-moving consumer goods, 250
financing of investments, 97
fish canning industries, 103
Fisher Information, 45, 46, 48
food production, 41
future research, 434
fuzzy set theory, 399

gap analysis program, 418
generation of alternatives, 191
geologic CO_2 storage, 166
geophysical aspects, 168
global reporting initiative, 372, 399
  aspects, 240
  categories, 240
globalisation, 284
green engineering
  poster, 73
  program, 73

textbook, 67, 72
GRI indicator framework, 372
  aspect, 372
  category, 372
  indicator, 372

HDA process, 357
health and safety, 219
heat integration, 356
HEN design, 356
higher education, 80, 82
holistic management, 84
human activity, 38
human component, 38
human needs, 245
humans, 40
hydrogeological model, 387

impacts
  of production, 323
import efficiency index, 278
index
  of balanced sustainable development,
    341
indicator
  design, 323
  improving the values, 263
  of environmental compatibility, 262
  of thermodynamic efficiency, 261
indicator-based methodology, 197
indicators, 240, 339
  additional, 373
  aggregation, 340
  air emissions, 411
  combining, 345
  composite, 338
  core, 373
  cross-cutting, 373
  economic, 411
  energy, 187
  for sustainable development, 308
  for sustainable energy development,
    308
  framework, 401
  function of, 400
  headline, 339
  liquid waste, 411
  mass, 187
  material use, 409
  new role, 329
  of sustainability, 259

Subject Index 439

of sustainable production, 398
products, 410
social, 246
solid waste, 410
standardization, 342
systemic, 373
values of, 242
water use, 409
indices, 339
   quality criteria, 340
   of economic sustainability, 273
   of efficiency, 277
   of sustainable economy, 271
industrial ecology, 78, 79, 80, 81, 92
industrialised economies, 66
information overflow, 310
information theory, 45
inherent safety index, 196
inherent safety sub-indices, 196
inhibiting satisfiers, 247
injecting CO_2, 168
insurance companies, 219
integrated computer aided system, 187
integrated emissions, 289
integrated management systems, 92
integrated performance, 373
integrated pollution levels, 292
intergovernmental panel on climate change, 15
ISO 14031, 399
ISO 14040, 376

judgement-making, 14

Kohonen network, 151

Landsat, 420
leakage
   describing, 167
   microeconomics, 169
living cell, 263

Maslow hierarchy, 246
material - value added, 188
mental resources, 309
metrics, 14, 15
model system, 48
modeling of the composting process, 133
monitoring system, 98
morality, 312

multiple regression on principal components, 151

national environmental protection act, 417
natural resource management, 416
natural resources, 41, 287
natural sources and sinks, 271
neural gas network, 151
new materials, 144
new subproducts, 118
new trade theories, 288
normalisation, 242
nutritional value assays, 121

occupational psychosomatic disorders, 310
oil, 119
organizational attention deficit, 310

path dependence, 322, 323
performance indicators, 372
petrochemicals, 248
petroleum, 248
polluting substances, 104
pollution control, 66
pollution intensity, 288, 289
pollution prevention, 186, 216
principal components analysis, 150
process mapping tools, 220
process sector, 248
process sustainability
   aspects, 264
product related measures, 93
projection pursuit, 151
protein concentrates, 120
pseudo-satisfiers, 247
public policy perturbations, 424

quality initiatives, 218

rate processes, 231
raw materials, 109
reaction quality, 189
reactor system, 129
re-capturing leaks, 169
reductionism, 17
remote sensing, 419
research, 15
   activities, 98
   style, 15

researchers, 81
resource availability indicator, 259
resource management systems, 418
responsible care, 248
reuse alternatives, 118
reuse of waste, 143
risk-assessment, 167
role of stakeholders, 92
root cause analysis, 220

safe chemical process, 196
safety
  indicators, 195
  inherent, 196
sampling site similarity, 158
satellite remote sensing, 416
science, 8, 9, 10, 12
  and decision making, 24
  and public, 18
  and sustainability, 13
scientific attitude, 9, 20
scientific knowledge, 9, 20
scientist, 11, 12, 16
screening methods, 68
sea food processing industry, 107
seasonality in lead concentration in rainwater, 158
second law of thermodynamics, 19
segregation of effluents, 112
sensitivity analysis, 191
single trajectory of development, 321
singular satisfiers, 247
social benefits, 245
social factors, 82
social systems theory, 269
societies, 41
solid wastes, 128
solvency, 275
source apportioning, 161
species diversity, 39
steady state, 40
strategic assessment
  of resources, 417
supply chains, 243
sustainability, 10, 13, 17, 38, 44, 88
  aggregated information, 338
  concept, 38
  dimensions, 240
  economic, 42
  related disciplines, 306
  social, 407

society, 40
systems, 43
technology, 42
sustainability, 269
  equity and constraints, 239
  key ideas and issues, 430
sustainability assessment, 401
  system, 327
sustainability indicators, 38, 231, 319, 398
sustainability metrics, 193
  energy, 194
  material, 194
  water, 194
sustainability reporting, 93
sustainability science, 18
sustainable development, 13, 14, 78, 91, 139, 239, 306, 338, 396
  monitoring and reporting, 339
  tools for monitoring, 339
  negentropic perspective, 306
sustainable economic development, 269
sustainable management, 421
sustainable process index, 271, 400
sustainable production, 397
  achievement, 398
synergic satisfiers, 247
systems resilience, 425

technical assistance to enterprises, 97
technical education, 87
technology, 38, 43
textbook project, 71
thermodynamic, 261
thermodynamic analysis, 256
thermodynamic efficiencies, 261
thermodynamic efficiency, 261
time-series analysis, 152
tobacco industry, 127
total value added, 189
trade, 287
transparency, 313
transparency, 312
transportation of CO_2, 166
treatment of CW wastewaters, 117
treatment of HOC wastewaters, 113
treatment of HSC wastewaters, 117
treatment of LOC wastewaters, 115
trend study, 157
truth, 11

underground reservoirs, 166

voluntary activities, 139

WAR algorithm, 194
waste
   generation, 143
   reduction, 144
   solid, 107
waste cost, 188

waste reduction
   algorithm, 194, 356
wastewater, 108
   sea food-processing, 109
   treatment plant, 117
water reuse, 118
wet precipitation monitoring, 153, 157
World Business Council, 243

zero-emission, 104

Printing: Mercedes-Druck, Berlin
Binding: Stein+Lehmann, Berlin